Design of Mechanical Elements

Design of Mechanical Elements

A Concise Introduction to Mechanical Design Considerations and Calculations

Prof. Bart Raeymaekers
University of Utah
Department of Mechanical Engineering
Salt Lake City, UT

This edition first published 2022

Registered Office
John Wiley & Sons, Inc., 111 River Street, Hoboken, NJ 07030, USA

Editorial Office
111 River Street, Hoboken, NJ 07030, USA

For details of our global editorial offices, customer services, and more information about Wiley products visit us at www.wiley.com.

Wiley also publishes its books in a variety of electronic formats and by print-on-demand. Some content that appears in standard print versions of this book may not be available in other formats.

Library of Congress Cataloging-in-Publication Data Applied for:

ISBN: 9781119849919

Cover Design: Wiley
Cover Image: © Mindscape studio/Shutterstock

Set in 9.5/12.5pt STIXTwoText by Straive, Chennai, India

SKY10032621_012022

Contents

About the Author

Bart Raeymaekers is a Professor of Mechanical Engineering at the University of Utah. He received a BS (ing.) in electromechanical engineering from KaHo St. Lieven in Ghent, Belgium (2002), an MSc (ir.) in mechanical engineering from the Vrije Universiteit Brussel in Brussels, Belgium (2004), and an MS and PhD in mechanical engineering from the University of California San Diego (2005 and 2007). He also received an MBA from the Massachusetts Institute of Technology (2009). His research relates to tribology and materials manufacturing and he teaches courses in mechanical design. Dr. Raeymaekers is a fellow of the American Society of Mechanical Engineers (ASME), a member of the Society of Tribologists and Lubrication Engineers (STLE), and a fellow of the Belgian American Educational Foundation (BAEF).

Preface

This textbook provides a concise introduction to design considerations and calculations of mechanical elements. A design of mechanical elements, machine design, or mechanical design course is part of the mechanical engineering curriculum at most universities, typically in the junior year. Hence, this textbook was written with the expectation that the reader understands important engineering concepts such as *statics* and *strength of materials*, which are the foundation of any mechanical design calculation.

Many university courses on design of mechanical elements rely on classic reference works that provide a comprehensive overview of a broad variety of common mechanical elements such as shafts, bolted joints, and welded joints and less common elements, including springs, clutches, and couplings. As such, these reference works are an invaluable resource for students and practicing design engineers. However, undergraduate student feedback gathered over the past decade of teaching this course indicated that these comprehensive reference works lack the entry-level explanations of how to think about the different considerations and assumptions that go into mechanical design calculations, and how to perform these calculations. This textbook attempts to fill this gap and provides an introduction to design of mechanical elements, focused on basic concepts and methodically performing mechanical design calculations. It teaches undergraduate students the knowledge and skills required to perform mechanical design calculations, and it serves as a stepping stone from which they can build advanced knowledge through studying other comprehensive textbooks or specialized reference works.

The chapters in this textbook are organized into three distinct sections. Chapters 1, 2, 3, and 4 cover *basic design techniques*. This includes an overview of the mechanical design process to highlight where the content of this textbook fits into the "big picture" of mechanical design, the definition of important concepts such as the design and safety factor (Chapter 1), in addition to material selection (Chapter 2), statistical considerations (Chapter 3), and tolerances (Chapter 4). The second section covers *strength of materials in the context of design of mechanical elements*. Specifically, Chapter 5 describes static loading problems and provides an overview of the different types of simple loading and common failure criteria to solve combined loading problems of both ductile and brittle materials. Chapter 6 describes dynamic loading problems, and designing mechanical elements for finite and infinite fatigue life. Finally, in the third section of this textbook, we use the *basic design techniques* and *strength of materials* knowledge to design and verify specific, commonly used mechanical elements, including shafts (Chapter 7), bolted joints (Chapter 8), welded joints (Chapter 9), rolling element bearings (Chapter 10), and gears (Chapter 11).

Each chapter starts with an overview of the terminology required to understand the topic, followed by an explanation of the physics underlying the mechanical design problems that will be solved. Then, methodical solution procedures to typical mechanical design problems are discussed. Finally, ample solved example problems are presented to illustrate the theoretical concepts.

I developed the "Design of mechanical elements" course in the fall of 2010 when starting at the University of Utah, and I first taught it during the spring 2011 semester. Since then, the course has evolved substantially, as I have taught it more than a dozen times, each time implementing valuable feedback from students, colleagues, and industry connections.

Since I was an undergraduate student, I have been fascinated with mechanical elements, mechanisms, machinery, and mechanical design calculations. I hope that my passion for this topic shines through in the various chapters of this textbook, and also spurs interest and excitement for this material with the reader.

Salt Lake City, Utah
May 2021

Prof. Bart Raeymaekers

About the Companion Site

This book is accompanied by a companion website:

www.wiley.com/go/raeymaekers/designofmechanicalelements

The website includes the Instructor Solutions Manual.

1

Mechanical Design

1.1 Introduction

In this chapter, we describe the basic concepts of mechanical design. Mechanical design is the branch of mechanical engineering that is concerned with the design of things and systems of a mechanical nature. It covers many different aspects of engineering and is highly interdisciplinary. Mechanical design is the process by which we attempt to satisfy a specific set of requirements to achieve an optimum outcome and functionality. Figure 1.1 schematically illustrates this concept. The mechanical design process starts from an initial conceptual idea, after which the idea is iteratively refined until all requirements have been satisfied.

These specific requirements or *design specifications* can include, e.g. cost, weight, strength, reliability, appearance, among many other requirements that we may need to consider. Sometimes, the requirements can be ambiguous, or even conflicting. For instance, "appearance" might be an ambiguous requirement because it is difficult to quantify. On the other hand, reducing weight while simultaneously increasing strength, or reducing cost while increasing reliability, are examples of conflicting requirements. Thus, the design process typically generates a multitude of solutions that satisfy the design specifications and no unique outcome exists.

1.2 Mechanical Design Process

Figure 1.2 schematically depicts the different phases of the mechanical design process.[1] The feedback loops emphasize the iterative nature of mechanical design, required to achieve optimal outcome and functionality. Indeed, sometimes we find that a concept does not satisfy all requirements, and we must make adjustments prior to moving forward in the design process.

We illustrate the different phases of the mechanical design process (see Figure 1.2), by conceptually designing a high-performance bicycle wheel.

Phase 1: Recognition of a need. Cyclists are very competitive. To gain an edge over other riders, they use lightweight wheels with improved aerodynamic performance compared to regular wheels.

1 For an in-depth discussion of the mechanical design process see, e.g. Ulrich, K.T. and Eppinger, S.D. (2011). *Product Design and Development*, 5e. McGraw-Hill.

Design of Mechanical Elements: A Concise Introduction to Mechanical Design Considerations and Calculations, First Edition. Bart Raeymaekers.

Companion website: www.wiley.com/go/raeymaekers/designofmechanicalelements

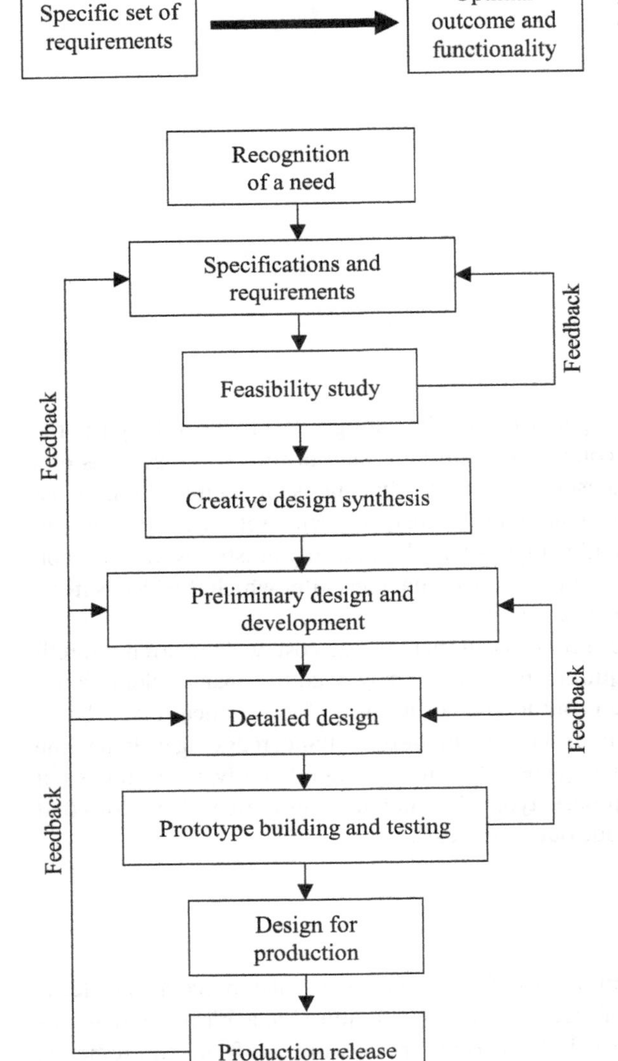

Figure 1.1 The design process relates a specific set of requirements to an optimal outcome and functionality.

Figure 1.2 Overview of the nine phases of the mechanical design process, with feedback loops that emphasize the iterative nature of the mechanical design process.

Phase 2: Specifications and requirements. High-performance bicycle wheels must be lightweight to allow fast acceleration and gain an advantage when riding uphill. In addition, the wheels must be stiff to ensure minimal energy loss and strong to withstand mechanical insults while riding. The wheels should also provide excellent aerodynamic performance. The mechanical design engineer then translates these qualitative requirements into quantitative design specifications. For instance, the diameter of the wheel is 700 mm to fit standard tires, the weight of the wheel set must be below 1200 g, in addition to other quantitative design specifications that define the required strength, reliability, and durability of the bicycle wheel. Aesthetic requirements are sometimes also imposed for marketing purposes. It is important that design specifications are quantitative because the design engineer must evaluate the outcome of the mechanical design process with respect to the quantitative design specifications to determine whether they achieved an optimal outcome and functionality.

Phase 3: Feasibility study. During this phase of the mechanical design process, the design engineer attempts to answer the question "Can it be done?" Typically, they will consider multiple concepts to satisfy the different design specifications. For instance, design concepts may explore and evaluate the use of different materials. Carbon composite materials are lightweight and enable increasing the rim height to benefit aerodynamics, without increasing the rim weight compared to metal alloys. Furthermore, the shape and material of the spokes must be chosen to fit weight and aerodynamic requirements. Hubs can either be custom-built or procured from a vendor. The design engineer will perform "back-of-the-envelope" calculations to verify the feasibility of the different concepts and evaluate whether a concept could satisfy the quantitative design specifications.

Phase 4: Creative design synthesis. Based on evaluating different design concepts during Phase 3, the design engineer defines a final concept that moves forward in the mechanical design process. Often, the creative design synthesis selects the best ideas from different possible concepts to create the final concept. For instance, the design engineer determines that to meet the quantitative design specifications, they must use a 40 mm tall carbon composite rim with 16 aluminum flat spokes and commercially sourced hub.

Phase 5: Preliminary design and development. The design engineer determines the critical aspects of the final concept, i.e. which aspect of the concept is most likely to fail? For instance, they might determine that the carbon composite material might fracture when hitting a 15 cm deep pothole. To evaluate whether the concept will work, the design engineer will implement a "critical function prototype." This could for instance be accomplished by implementing a straightforward experiment, where a carbon composite specimen is impacted by a force similar to that of the rim hitting a pothole. If the critical function prototype is successful, i.e. the design concept does not fail where it is most likely to do so, then the design engineer can confidently advance the final concept to the next phase of the mechanical design process.

Phase 6: Detailed design. After the preliminary design demonstrates feasibility of the concept, the design engineer performs an in-depth analysis of the final concept. They will calculate the mechanical stresses in all components of the design (rim/hub/spokes) and consider how these stresses change with the environment, such as riding the bicycle wheel on a flat road, on cobble stones, or on a hill. Carefully determining the boundary conditions and external loading of the wheel is paramount to performing an accurate analysis. Furthermore, the design engineer can choose to use numerical simulations to assist with the calculations and/or to optimize the mechanical design in terms of the quantitative design specifications. The end of the detailed design phase results in a set of detailed drawings of all individual parts or components and their dimensions, to allow prototype building.

Phase 7: Prototype building. The design engineer implements a prototype of the bicycle wheel for lab and field testing, based on the results of the detailed design. Importantly, they should instrument the prototype to measure key parameters of the design during testing to validate the calculations of Phase 6. Furthermore, evaluating the prototype verifies whether the product performs as expected based on the calculations and whether it satisfies the quantitative design specifications.

Phase 8: Design for production. After extensive product testing and iteratively tweaking the design based on quantitative test results, the design engineer will evaluate the design in light of mass production. Tolerances and assembly/disassembly are important parameters to consider at this stage of the design process. This phase also includes designing all tools and equipment necessary for the production process of the new bicycle wheels.

Phase 9: Production release. The design engineer releases the wheels for mass production and consumer use. Consumer feedback informs future designs or important updates to the product.

This example illustrates that mechanical design is more than just drafting, numerical analysis, programming, or optimization. Instead, mechanical design is the synthesis of all these areas. Most importantly, mechanical design still requires human experience, which cannot be provided by a computer, despite several attempts in the artificial intelligence field. Design engineers can generate ideas and concepts, interpret computer simulations and results, and use their engineering judgment to validate the results.

In this textbook, we focus primarily on *Phase 6: Detailed design* of the design process, i.e. performing detailed calculations of the mechanical elements that make up a mechanical design.

1.3 Mechanical Elements

Mechanical elements or machine elements comprise both individual components such as a shaft, a bolt, a gear, or involve assemblies of different individual components that fit together to perform a specific function or obtain a particular outcome. Table 1.1 shows five categories of mechanical elements.

1.4 Standards and Codes

To assist with the multitude of possibilities and outcomes resulting from the mechanical design process, we use *standards* and *codes*.

A *standard* is a set of technical definitions and guidelines, written by experts, which establishes uniform engineering or technical criteria, methods, processes, and practices. The purpose of a standard is to limit the number of possible outcomes resulting from the design process. For instance, the ISO 68-1:1998 standard[2] describes the geometry of metric screw thread. Hence, by adopting this

Table 1.1 Five categories of mechanical elements.

Category	Example mechanical elements
Joint elements	Bolt, spring, pin, welding, soldering
Bearing elements	Rolling element bearing
Power transfer elements	Shaft, gear, pulley, clutch, chain
Gasket elements	Static and dynamic gaskets
Lubrication	Oil, lubrication fat, solid lubricant

2 ISO 68-1:1998. *General Purpose Screw Threads – Basic Profile – Part 1: Metric Screw Threads*. International Organization for Standardization.

standard, manufacturers of metric fasteners reduce the number of possible screw thread geometries they may design, to a standard screw thread geometry as defined in this standard. Adopting this standard also ensures that fastener products from one manufacturer are compatible with those of other manufacturers, e.g. a bolt from manufacturer A is compatible with a nut from manufacturer B, if both manufacturers adhere to the standard.

Several professional organizations oversee committees of industry and academic experts that develop these voluntary consensus standards, such as the International Organization for Standardization (ISO), American National Standards Institute (ANSI), American Society of Mechanical Engineers (ASME), American Society of Testing and Materials (ASTM), American Iron and Steel Institute (AISI), and the National Institute of Standards and Technology (NIST), to only name a few. They also manage websites that have an index and repository of all available standards, which is helpful when searching for information about, e.g. how to design a specific mechanical element or assembly.

A *code* is a standard with the force of law behind it. The purpose of a code is to achieve a specific degree of safety. For that reason, and in contrast to a standard that is voluntary, a code must be followed.

1.5 Uncertainty in Mechanical Design

We must consider uncertainty in any mechanical design because not all information about the design of the mechanical element, loading, material properties, and the environment in which it will operate, is known *a priori*. In the context of designing a mechanical element for strength, we categorize sources of uncertainty as follows:

1. Uncertainty about the loading of the element. This may include the effect of intensity and distribution or direction of loading, intensity of local stress concentrations, and effect of changes of loading with time.
2. Uncertainty about the strength of the element. This may include the effect of nonuniform properties of the material, manufacturing defects that have an effect on the strength of the material, effect of environmental parameters, and effect of corrosion and wear.

For instance, consider a guardrail alongside a hairpin turn on a mountain road. During the design process, we consider the types of vehicles that drive on the mountain road and the speed at which they travel. Furthermore, we consider the strength of the material of the guardrail from either a standard table with mechanical properties or from a data sheet provided by the material supplier. This information enables designing the dimensions of the guardrail such that the guardrail can adequately perform its function. However, we are never certain about the loading of the guardrail, as it is impossible to predict how fast someone may crash their vehicle, or even the types of vehicles that may travel on the mountain road in the future. Second, we are never certain about the material properties of the guardrail, as manufacturing and material defects are always possible.

To account for uncertainty in mechanical design, we use a *design factor*. The magnitude of the design factor reflects the level of uncertainty about both the loading and the strength of the mechanical element. Thus, we define the design factor n_D as

$$n_D = \frac{\text{loss-of-function parameter}}{\text{maximum value of parameter in the design}}. \tag{1.1}$$

The *parameter* term in Eq. (1.1) can represent any measurable property. For instance, when designing a mechanical element for strength (e.g. a steel guardrail), we would select the yield stress σ_y as the loss-of-function parameter and compare it to the maximum stress occurring in the design σ_{max}, resulting in a design factor $n_D = \sigma_y/\sigma_{max}$. Indeed, when the maximum stress in the element is equal to the yield stress of the material, the element yields and can no longer perform its function. Thus, the design factor physically represents how much we *overdesign* a mechanical element, i.e. how much thicker than needed we design the element to account for uncertainty. Evidently, we must choose the design factor carefully because while overdesigning the dimensions of an element generally reduces the likelihood of failure, it also adds weight and cost, which is undesirable. As such, the magnitude of the design factor reflects the level of uncertainty about the loading and the mechanical properties of the element. Detailed engineering analysis can reduce uncertainty and, thus, reduce the magnitude of the design factor.

It is important to emphasize that we should only use the design factor to account for uncertainty or unknowable aspects of the design. We must include all knowable aspects of the design, such as stress concentrations, among others, in the design calculations. It is incorrect to account for knowable quantities by means of the design factor.

We consider a multitude of parameters when selecting a design factor, such as understanding how much uncertainty exists about loading and mechanical properties (and also consider that environmental effects could alter the mechanical properties). Additionally, we consider the consequence of failure of the mechanical element. For instance, if failure of the mechanical element results in substantial injury or economic loss, we often choose a design factor of 4 or higher. In contrast, we may use a design factor of 2 for a less-critical mechanical element. Risk analysis and *failure mode and effects analysis* (FMEA) are helpful tools to determine the risks and consequences of failure. Also note that mechanical elements used in specific applications might be subject to a *code*, in which case we could be required to use a specific, prescribed design factor.

The following are a few examples of design factors.[3] Structural elements in buildings ($n_D \cong 2$) use a relatively low design factor because the properties of the materials are well documented and the loading is well understood. Pressure vessels ($n_D \cong 4$), automobiles ($n_D \cong 3$), and aircraft and spacecraft ($n_D \cong 1.2$–3) use different design factors depending on the specific element and material choice. Aerospace elements generally require a low design factor because the weight penalty is high. However, this low design factor is mitigated with detailed engineering analysis, rigorous quality control, and aggressive preventative maintenance schedules to detect defects before failure occurs.

After we complete a mechanical design, we calculate an *actual* design factor, which we refer to as the *safety factor n*. When calculating the dimensions of a mechanical element considering a specific design factor, we might round up the dimensions resulting from the design calculations to the nearest standard size, to minimize manufacturing cost. As such, the dimension of the actual element may be slightly larger than it would be based solely on the design factor we specified *a priori*. The definition of the safety factor n is identical to that of the design factor n_D (see Eq. (1.1)), i.e.

$$n = \frac{\text{loss-of-function parameter}}{\text{maximum value of parameter in the design}}. \tag{1.2}$$

However, in the case of the safety factor, the "maximum value of the parameter in the design" is based on the actual dimensions of the mechanical element, after its design. Thus, the safety factor represents a quantitative measure of the implicit safety of the design. Alternatively, some textbooks also define the *margin of safety* $m = n - 1$, which contains the same information as the safety factor, but presents the "excess capacity" of the mechanical element before failure occurs.

3 For a discussion on how to select an appropriate design factor for a mechanical design see, e.g. Ullman, D.G. (2017). *The Mechanical Design Process*. 6e. David Ullman LLC.

It is important to emphasize that a safety factor $n > 1$ does not guarantee that the mechanical element will not fail. However, the likelihood of failure decreases with increasing safety factor. Absolute safety does not exist in mechanical design because uncertainty cannot be fully eliminated. Ultimately, both the *loss-of-function parameter* and the *maximum value of the parameter in the design* are statistically varying quantities.

Example problem 1.1

Figure 1.3 shows a cantilever beam with square cross-section $b \times b$. The cantilever beam is made from steel with yield stress $\sigma_y = 200$ MPa, and is subject to an axial load $F = 25$ kN. Calculate b to avoid yielding, and consider a design factor $n_D = 1.5$. Select a standard size of square bar stock from the supplier's catalog shown in Table 1.2. Calculate the resulting safety factor.

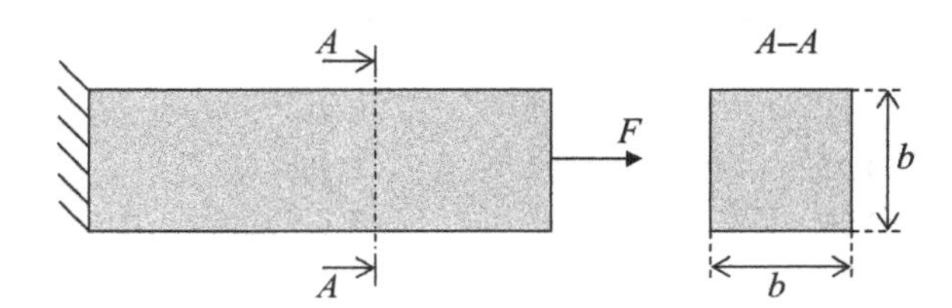

Figure 1.3 Cantilever beam with square cross-section.

Table 1.2 Excerpt from supplier catalog.

Square bar stock	
***b* (mm)**	***b* (mm)**
2	10
3	15
4	20
5	25

Solution

Starting from the definition of the design factor (Eq. (1.1)), we write that $n_D = \sigma_y/\sigma_{max}$ and, thus, $n_D = \sigma_y\, b^2/F$.

Hence, $b = \sqrt{\frac{n_D F}{\sigma_y}}$, or $b = \sqrt{\frac{1.5 \times 25 \cdot 10^3}{200 \cdot 10^6}} = 13.69$ mm.

We select the nearest larger standard size of square bar stock from the supplier catalog (see Table 1.2), i.e. $b = 15$ mm.

The resulting safety factor $n = \sigma_y b^2/F = 200 \cdot 10^6 \times 0.015^2/25 \cdot 10^3 = 1.8$.

The margin of safety $m = n - 1 = 0.8$.

Note that we always round up to the nearest standard size to ensure we reach at least the required design factor. Hence, the safety factor n is always equal to or larger than the design factor n_D.

Example problem 1.2

Figure 1.4 shows a circular tube with outer diameter D and inner diameter d. The tube is made from titanium (Ti6Al4V) with yield stress $\sigma_y = 724$ MPa, and is subject to an axial load $F = 10$ kN. Design the tube to avoid yielding and consider a design factor $n_D = 2.0$. Select a tube from the supplier's catalog shown in Table 1.3. Calculate the resulting safety factor.

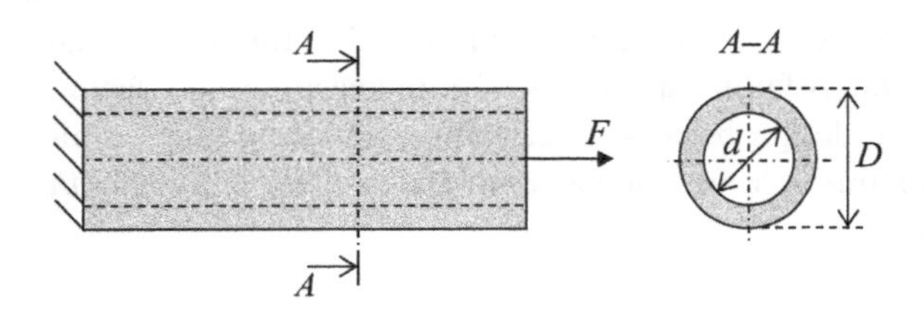

Figure 1.4 Circular tube.

Table 1.3 Excerpt from supplier catalog.

Circular tube	
D (mm)	**d (mm)**
6.35	5.54
6.35	5.33
9.53	8.56
9.53	8.10
12.70	11.40
15.88	14.25
19.05	17.07
25.40	22.81

Solution

Starting from the definition of the design factor (Eq. (1.1)), we write that $n_D = \sigma_y/\sigma_{max}$ and, thus, $n_D = \sigma_y A/F$. Here, $A = \pi(D^2 - d^2)/4$.
Thus, $A = n_D F/\sigma_y = 2 \times 10 \cdot 10^3/724 \cdot 10^6 = 2.76 \cdot 10^{-5}$ m^2.
The cross-sectional area of the circular tube must exceed 27.6 mm^2 to satisfy the design factor requirement. Table 1.4 shows the cross-sectional area for each circular tube listed in the excerpt of the supplier catalog of Table 1.3. We select the nearest standard size with cross-sectional area larger than $A = 27.6$ mm^2 from the supplier catalog, i.e. $D = 15.88$ mm, and $d = 14.25$ mm.
The resulting safety factor $n = \sigma_y A/F = 724 \cdot 10^6 \times 38.46 \cdot 10^{-6}/10\,000 = 2.78$.
The margin of safety $m = n - 1 = 1.78$.

Table 1.4 Cross-sectional area of the circular tubes.

D (mm)	d (mm)	A (mm^2)
6.35	5.54	7.59
6.35	5.33	9.32
9.53	8.56	13.70
9.53	8.10	19.69
12.70	11.40	24.98
15.88	14.25	38.46
19.05	17.07	56.20
25.40	22.81	98.08

Note again that we always round up to the nearest standard size to ensure we reach at least the required design factor.

1.6 Design for Safety

During the mechanical design of an element, product, or assembly, we consider multiple factors, including functionality, manufacturability, assembly, and cost, among many others. For instance, a design that does not perform its function, or cannot be manufactured, or is not economical, ultimately cannot be a viable design. However, it is paramount that engineers also consider safety during any mechanical design.[4] A design for safety approach requires identifying potential hazards of a design and quantifying the risk associated with each hazard. While absolute safety may not be achievable due to unforeseeable misuse of the element or product, adopting design for safety practices can prevent many accidents by considering safety throughout the design process.

Several standards exist to assist with incorporating safety into mechanical design, including ISO 10377: "Consumer product safety – Guidelines for suppliers," and ISO 12100: "Safety of machinery – General principles for design – Risk assessment and risk reduction," among other standards.

The National Safety Council publishes the so-called *safety hierarchy*, which provides a framework for how to design products with safety in mind:

1. **Design to eliminate hazards and minimize risk**: The most effective way to improve safety is to change the mechanical design to eliminate any hazard or reduce the risk associated with its occurrence.
2. **Incorporate safety devices**: If we cannot eliminate the hazard by changing the mechanical design, we attempt to reduce the risk to an acceptable level by using safety devices such as guards and shields.
3. **Provide warning devices**: If we cannot eliminate the hazard by changing the design, and we cannot reduce the risk to an acceptable level by implementing safety devices (for instance because the safety devices would interfere with the functionality of the design), then we must warn the operator about any hazards that exist by selecting proper warning devices, including light, sound, and labels.
4. **Develop and implement safe operating procedures and operator training programs**: Safe operating procedures and training must be part of any safety culture and inform the user about hazards and risk of operating a machine, and how to prevent injury.
5. **Use personal protective equipment**: Personal protective equipment (PPE) is a last resort to protect a user from hazards involved with operating a product, machine, or mechanical design.

1.7 Key Takeaways

1. The mechanical design process starts from an initial conceptual idea. Then, the idea is iteratively refined until all design specifications have been satisfied. Design specifications must be quantitative, so that we can evaluate the outcome of the design process against those specifications.
2. Standards (voluntary) contribute to reducing the infinite number of possible outcomes of the mechanical design process to a finite number. Codes (mandatory) attempt to ensure a minimum level of safety.

---

4 See, e.g. d'Entremont, K.L. (2020). *Engineering Ethics and Design for Product Safety* 1e. McGraw-Hill.

3. The design factor accounts for uncertainty in the mechanical design process. The magnitude of the design factor reflects the level of uncertainty about the loading of the element and the strength of the material.
4. The safety factor quantifies the implicit safety of a design by comparing the loss-of-function parameter of a design to the maximum value of that parameter occurring in the design.
5. Safety must be a primary consideration of any mechanical design, and *design for safety* principles must be implemented throughout the mechanical design process.

1.8 Problems

1.1 Figure 1.5 shows a circular tube with length L, attached to a rigid structure and subject to an axial load F. The tube is made from low-carbon steel with $\sigma_y = 165$ MPa. $L = 0.2$ m, $D = 50$ mm, $d = 44$ mm, and $F = 37.5$ kN. Calculate the safety factor against yielding.

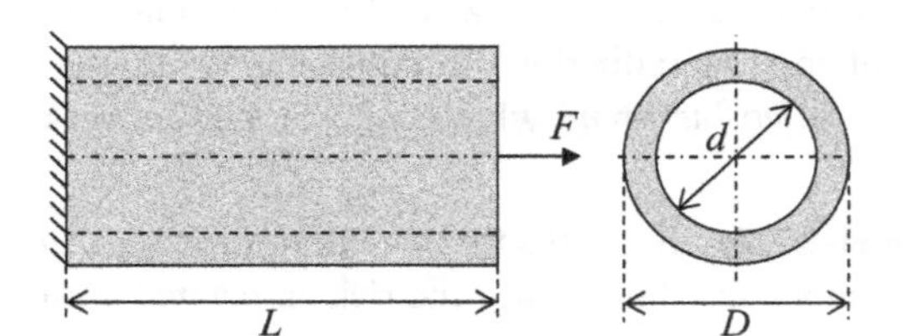

Figure 1.5 Circular tube made from low-carbon steel.

1.2 Figure 1.6 shows a circular rod with length L, attached to a rigid structure, and subject to an axial load F. The rod is made from low-carbon steel with $\sigma_y = 165$ MPa. $L = 0.2$ m, $D =$ 20 mm. Calculate the maximum load F that can be applied to this design to maintain a safety factor against yielding $n = 3$.

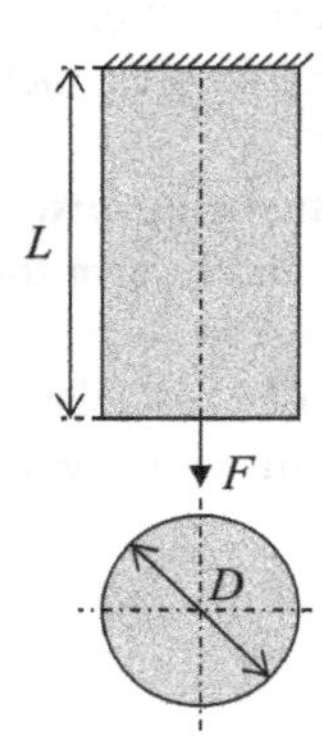

Figure 1.6 Circular rod made from low-carbon steel.

1.3 Figure 1.7 shows a bar with rectangular cross-section and with length L, attached to a rigid structure, and subject to an axial load F. The bar is made from aluminum with $\sigma_y = 230$ MPa. $L = 0.2$ m, $w = 25$ mm, $t = 5$ mm. Calculate the maximum load F that can be applied to this design to maintain a safety factor against yielding $n = 3.5$.

Figure 1.7 Rectangular bar made from aluminum.

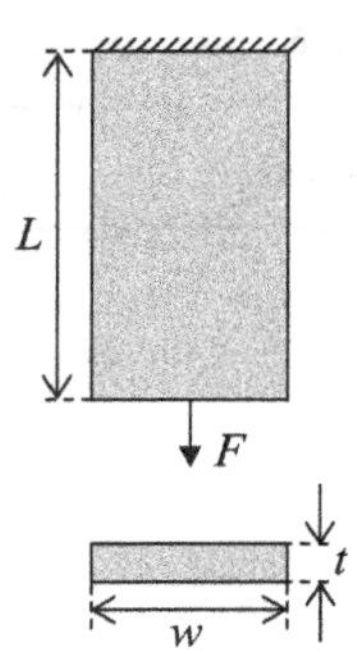

1.4 Figure 1.8 shows a "T-bar" with length L, attached to a rigid structure and subject to an axial load F. The bar is made from 304 stainless steel with $\sigma_y = 207$ MPa. $L = 0.5$ m and $F = 20$ kN. Calculate which T-bar to select from the supplier catalog, considering a design factor of $n_D = 2$. Table 1.5 shows an excerpt from the supplier catalog. Calculate the safety factor against yielding.

Figure 1.8 T-bar made from 304 stainless steel.

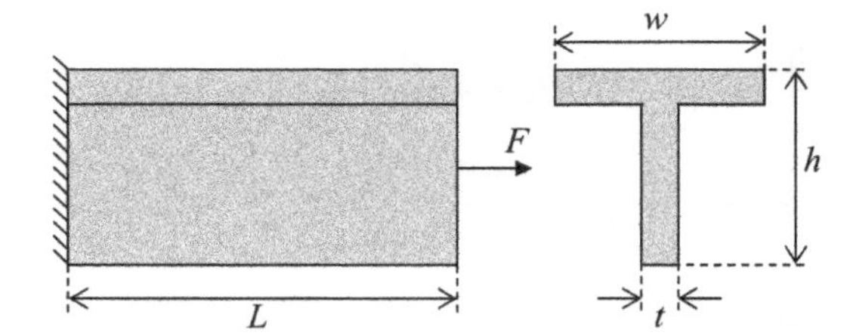

Table 1.5 Excerpt from supplier catalog.

h (inch)	*w* (inch)	*t* (inch)
1	1	1/8
1 1/2	1 1/2	3/16
2	2	1/4
3	3	1/4

1.5 Figure 1.9 shows a square tube with length L attached to a rigid structure and subject to an axial load F. The tube is made from carbon steel with $\sigma_y = 220$ MPa. $L = 0.5$ m, and $F = 20$ kN. Calculate which square tube to select from the supplier catalog, considering a design factor of $n_D = 2$. Table 1.6 shows an excerpt from the supplier catalog. Calculate the safety factor against yielding.

Figure 1.9 Square tube made from carbon steel.

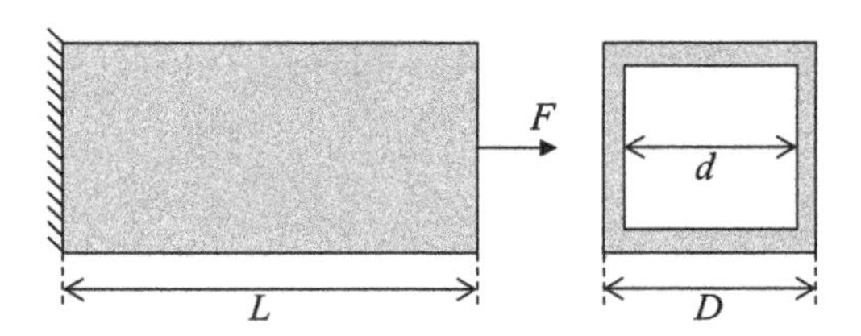

Table 1.6 Excerpt from supplier catalog.

D (mm)	*d* (mm)
12	10
25	21
55	51
100	94

1.6 Figure 1.10 shows a hexagonal bar with length L, attached to a rigid structure, and subject to an axial load F. The hex bar is made from carbon steel with $\sigma_y = 200$ MPa. $L = 0.5$ m and $F = 20$ kN. Calculate which hexagonal bar to select from the supplier catalog, considering a design factor of $n_D = 3$. Table 1.7 shows an excerpt from the supplier catalog. Calculate the safety factor against yielding.

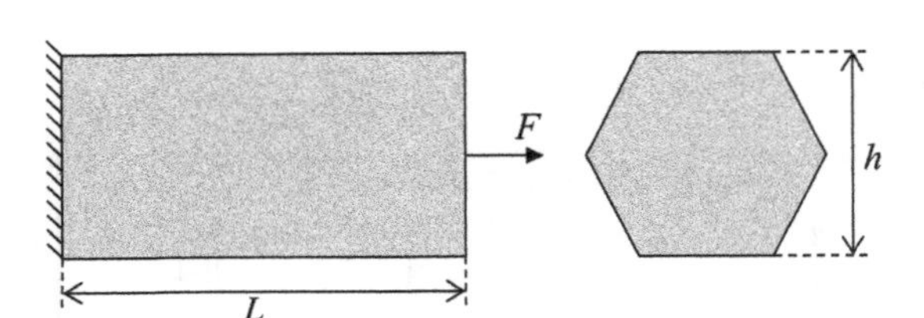

Figure 1.10 Hexagonal bar made from carbon steel.

Table 1.7 Excerpt from supplier catalog.

h (mm)
5
6
8
10
12
14
22
24
30

2

Material Selection

2.1 Introduction

In this chapter, we describe the considerations related to selecting a material for the design of a mechanical element. Selecting a material is often one of the most important decisions during the design process because it immediately determines and limits some of the functionality of the mechanical element. For instance, if the mechanical element will operate at high temperature in a corrosive environment, then it is important to consider the melting point of the material as well as its resistance to corrosion. If the mechanical element must conduct electricity or heat, then we have to consider electrical and thermal properties of the material. Many other material properties may be important and require consideration depending on the mechanical design, the intended function of the mechanical element, and the environment in which it will operate. Table 2.1 illustrates a few examples of different categories of material properties we might consider during the design of a mechanical element.

2.2 Material Classification

We can classify or categorize materials in different ways, according to, e.g. physical, chemical, or mechanical properties. We refer to materials science textbooks for an overview of material classification systems using different metrics. However, a common way to classify materials is according to their microstructure.[1] Figure 2.1 illustrates this classification and identifies five independent material categories, and a hybrid category, which constitutes a combination of any of the five independent categories. These categories include:

- **Ceramics**: Nonmetallic, inorganic solids
- **Glasses**: Amorphous solids
- **Metals**: Pure or combined chemical elements with specific chemical bonds
- **Polymers**: Materials based on long carbon or silicon chains
- **Elastomers**: Viscoelastic polymers, i.e. polymers with high yield strain
- **Hybrids**: Combinations of multiple materials, for instance, composite materials.

Here, we are particularly interested in the *mechanical properties* of materials. We also limit the discussion to metals and primarily steels because a large number of mechanical elements are made

1 See, e.g. Ashby, M., Shercliff, H., and Cebon, D. (2014). *Materials*. 3e. Elsevier Ltd.

Design of Mechanical Elements: A Concise Introduction to Mechanical Design Considerations and Calculations, First Edition. Bart Raeymaekers.

Companion website: www.wiley.com/go/raeymaekers/designofmechanicalelements

Table 2.1 Example categories of material properties.

Mechanical	Tribology	Thermal	Economical
Strength	Friction coefficient	Conductivity	Cost
Stiffness	Surface topography	Melting temperature	Manufacturability
Toughness	Wear	Thermal expansion	Aesthetics
Resilience	Corrosion	Combustibility	Assembly

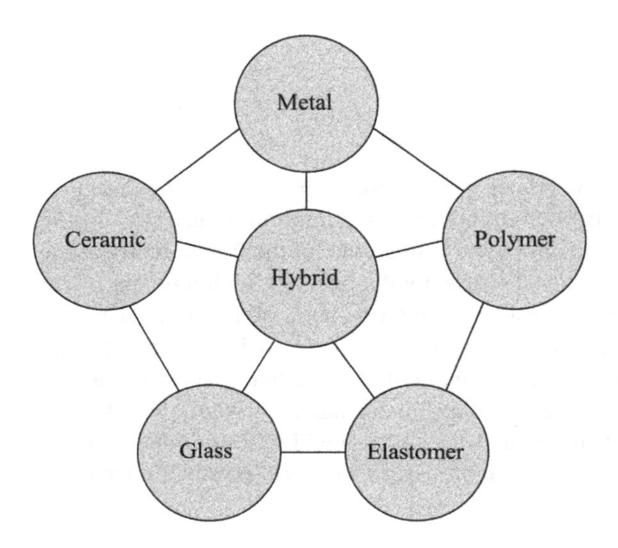

Figure 2.1 Material classification by microstructure.

from steel. Information about the mechanical properties of other materials is available in the literature and reference works.

The material choice determines the mechanical properties we will use during the design calculations and, thus, immediately defines limitations to the design of the mechanical element. For that reason, the material selection is typically one of the first decisions that we make during the design of a mechanical element. Indeed, the geometry and dimensions of the mechanical element that result from the design calculations depend on the mechanical properties of the material and, thus, we must choose the material prior to performing these calculations.

Economics may also play a role in material selection. In the context of steels, the material cost usually increases with increasingly exotic alloying elements and with increasing quantity of alloying elements. Thus, we first evaluate whether using an inexpensive carbon steel allows meeting all design requirements. Alternatively, we use material processing methods, such as cold work, to improve the mechanical properties of the material to meet the design requirements. Finally, we consider alloy steels to obtain, e.g. high ultimate tensile stress, corrosion resistance, a combination of strength and toughness, or high strength at elevated temperature, to only name a few.

2.3 Mechanical Properties

2.3.1 Strength and Stiffness

The stress–strain diagram relates the stress in a material to the corresponding strain during a uniaxial tensile test. A standard tensile test, as for instance described in the *ASTM E8/E8M – 16ae1*

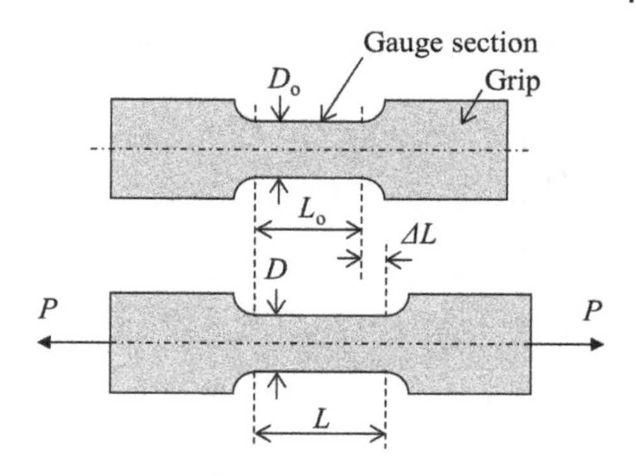

Figure 2.2 A dog-bone specimen for a uniaxial tensile test.

Standard Test Methods for Tension Testing of Metallic Materials standard, uses so-called "dog-bone specimens." Figure 2.2 schematically shows a typical dog-bone specimen, indicating the gauge length L_0 and diameter D_0. This specific specimen shape ensures that fracture occurs in the thin gauge section of the dog-bone specimen, away from the grip section, and the rounded edges minimize stress concentrations. When we load the dog-bone specimen axially with force P, the gauge length extends from L_0 to L, whereas the gauge diameter reduces from D_0 to D (assuming a positive Poisson coefficient ν, e.g. for steel $\nu_{steel} \cong 0.3$). The difference between the length of the dog-bone specimen before and after we apply the axial force P is the elongation $\Delta L = L - L_0$.

Thus, we define the engineering stress as

$$\sigma = \frac{P}{A_0}, \tag{2.1}$$

where A_0 is the cross-sectional area of the gauge section of the dog-bone specimen, before loading. For instance, if the cross-section is circular with diameter D_0, then $A_0 = \pi D_0^2/4$. Correspondingly, we define the engineering strain as

$$\varepsilon = \frac{\Delta L}{L_0}. \tag{2.2}$$

Practically, we measure the axial force P using a load cell during the tensile test, whereas we determine the strain using, e.g. strain gauges, an extensometer, or digital image correlation (DIC).

Figure 2.3 shows a typical engineering stress–strain diagram for (a) a ductile material and (b) a brittle material. Figure 2.3 (a) shows that stress and strain are linearly related until they reach the proportionality limit (*PL*). The proportionality constant is the elastic modulus or Young's modulus E. The elastic limit (*EL*) reaches just beyond the proportionality limit and marks the inception of plastic deformation, i.e. the material yields. Hooke's law ($\sigma = E\varepsilon$) is valid within the elastic region of the stress–strain diagram.

Practically, it is difficult to experimentally determine the elastic limit *EL* because the inception of plastic deformation may occur anywhere in the gauge section of the dog-bone specimen and it is difficult to observe or measure. Therefore, the strain that corresponds to 0.2% plastic deformation, which is straightforward to measure, and the corresponding 0.2% yield stress $\sigma_{y,0.2\%}$ is commonly used to determine yielding. The 0.2% yield stress and the *EL* are very close to each other on the stress–strain diagram and, thus, we consider $\sigma_y \cong \sigma_{y,0.2\%}$. Indeed, the values of σ_y in material property tables are in fact $\sigma_{y,0.2\%}$ values. The strain that corresponds to the yield stress σ_y is the yield strain ε_y.

When we load the dog-bone specimen to the *EL* and subsequently remove the load, the specimen regains its original shape and dimensions because the deformation caused by the load is purely elastic. Increasing the stress in the dog-bone specimen beyond the yield stress causes the material to

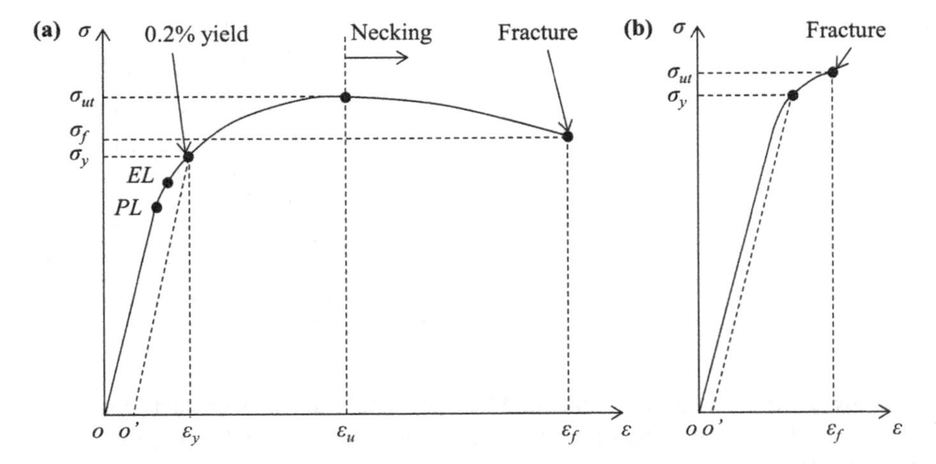

Figure 2.3 Engineering stress–strain diagram of (a) a ductile and (b) a brittle material.

deform plastically. Hence, when we load the dog-bone specimen beyond the *EL* and subsequently remove the load, the material does not regain its original shape and dimensions because plastic deformation has occurred. We observe this in the stress–strain diagram as follows. When loading the specimen in the elastic region and removing the load, we start from and end back in the origin o of the diagram. However, when loading the specimen into the plastic region and removing the load, we follow a line parallel to the proportional part of the stress–strain diagram, from the plastic stress/strain combination that corresponds to the load, down to $\sigma = 0$, i.e. the intersection with the strain axis at o', as shown in Figure 2.3a.

The ultimate tensile stress σ_{ut} and the corresponding uniform strain ε_u mark the inception of necking. When the strain ε is smaller than ε_u, it is uniform in the entire gauge section of the dog-bone specimen. However, if the strain ε exceeds ε_u it is no longer uniform in the specimen, and "necking" occurs. Locally, the dog-bone specimen experiences more strain than elsewhere, which results in a locally reduced cross-section and, thus, "necking." Ultimately, the dog-bone specimen fractures where the local stress exceeds the fracture stress σ_f. We refer to the strain that corresponds to the fracture stress σ_f as the fracture strain ε_f.

Figure 2.3b shows a typical engineering stress–strain diagram of a brittle material, such as a ceramic material. The difference with the stress–strain diagram of the ductile material in Figure 2.3a is apparent. Specifically, a brittle material does not experience necking or any substantial amount of plastic deformation. As opposed to a ductile material, fracture occurs at the ultimate tensile stress and, thus, $\sigma_f = \sigma_{ut}$. We also note that a ductile material typically displays identical mechanical properties in tension and in compression, i.e. the ultimate tensile stress σ_{ut} is identical to the ultimate compressive stress σ_{uc}. In contrast, a brittle material is typically stronger in compression than in tension, i.e. $\sigma_{uc} \gg \sigma_{ut}$.

2.3.2 Elastic Versus Plastic Strain

Figure 2.4 shows a stress–strain diagram of a ductile material, where we indicate a point i located in the plastic region, i.e. $\sigma_i > \sigma_y$. When applying a tensile force to the dog-bone specimen, the stress–strain relationship follows the stress–strain diagram starting in the origin o until reaching point i. When removing the tensile force, part of the deformation restores (elastic deformation), whereas the other part of the deformation does not restore (plastic deformation). We can quantify

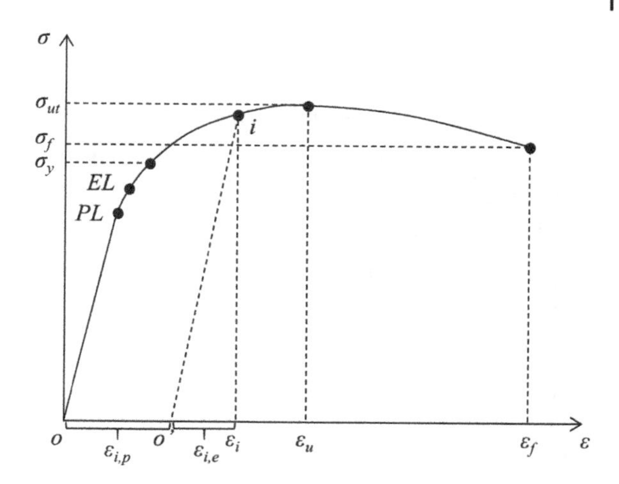

Figure 2.4 Engineering stress–strain diagram of a ductile material, graphically indicating elastic strain $\varepsilon_{i,e}$ and plastic strain $\varepsilon_{i,p}$.

this information from the stress–strain diagram as follows. The orthogonal projection of point i onto the strain axis gives the total strain ε_i, i.e. the sum of the plastic strain $\varepsilon_{i,p}$ and the elastic strain $\varepsilon_{i,e}$, when loading the specimen to point i. When we remove the axial force from the dog-bone specimen, the elastic strain vanishes, i.e. only the plastic strain remains. Thus, the intercept o' between a line from i parallel to the proportional region of the stress–strain diagram and the strain axis indicates the plastic strain $\varepsilon_{i,p}$. The difference between the total strain ε_i and the plastic strain $\varepsilon_{i,p}$ denotes the elastic strain $\varepsilon_{i,e}$.

2.3.3 Resilience

The resilience is the energy that a material can absorb per unit volume without incurring plastic deformation. It is the area under the stress–strain diagram between the origin o and the elastic limit EL, which we indicate as the shaded area in Figure 2.5. We quantify the resilience of a material by means of its *modulus of resilience* U_R. Thus, we calculate the area under the stress–strain diagram as

$$U_R = \int_0^{\varepsilon_{EL}} \sigma \, d\varepsilon. \tag{2.3}$$

This integral is difficult to calculate because we require a mathematical relationship between σ and ε in the elastic region of the stress–strain diagram. While this relationship is linear in the proportional region of the stress–strain diagram, it is nonlinear between PL and EL. However, we approximate the integral by recognizing that the area under the stress–strain diagram between the origin o and the elastic limit EL approximates a triangle, as shown in Figure 2.5 by means of the dashed triangular contour. Hence,

$$U_R = \int_0^{\varepsilon_{EL}} \sigma \, d\varepsilon \cong \frac{\sigma_y \varepsilon_y}{2} = \frac{\sigma_y^2}{2E}. \tag{2.4}$$

Equation (2.4) implies two approximations. First, as discussed previously, a small difference exists between the elastic limit EL and the yield stress σ_y. Second, the area under the stress–strain diagram between the origin and the EL is not a perfect triangle, since there is a small nonlinear section between the PL and the EL (or σ_y in the approximation). Nevertheless, Eq. (2.4) provides a good approximation of U_R for most steels.

The modulus of resilience is an important metric in the design of mechanical elements. For instance, a bicycle frame and an airplane landing gear are examples of engineering applications in which resilience plays an important role. Upon impact with a bump in the road, a bicycle frame

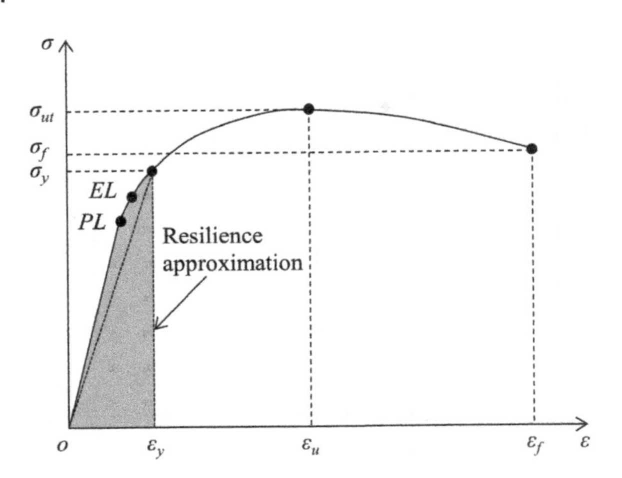

Figure 2.5 Engineering stress–strain diagram graphically indicating the resilience as the shaded area, and its approximation as the dashed triangular contour.

must not plastically deform. Similarly, upon landing and impact with the runway, the landing gear of an airplane must only deform elastically.

2.3.4 Toughness

The toughness is the energy that a material can absorb per unit volume without fracture. It is the area under the stress–strain diagram between the origin o and the fracture strain ε_f, which we indicate as the shaded area in Figure 2.6. We quantify the toughness of a material by means of its *modulus of toughness* U_T.

$$U_T = \int_0^{\varepsilon_f} \sigma \, d\varepsilon. \tag{2.5}$$

Similar to the modulus of resilience, this integral is difficult to calculate because we need a mathematical relationship between σ and ε in both the elastic and plastic region of the stress–strain diagram. This relationship is nonlinear in the plastic region (and in a small section of the elastic region). However, we can approximate the integral by recognizing that the area under the stress–strain diagram between the origin o and the fracture strain ε_f approximates a rectangle, shown in Figure 2.6. Hence,

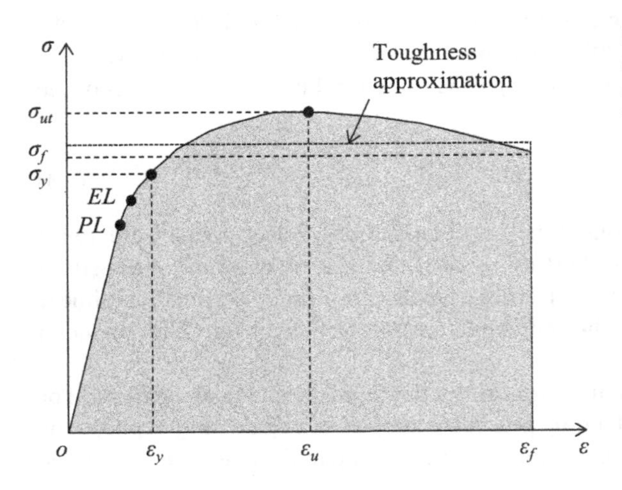

Figure 2.6 Engineering stress–strain diagram graphically indicating the toughness as the shaded area and its approximation as the dashed rectangular contour.

$$U_T = \int_0^{\varepsilon_f} \sigma \, d\varepsilon \cong \frac{\sigma_y + \sigma_{ut}}{2} \varepsilon_f. \tag{2.6}$$

The modulus of toughness is an important metric in the design of mechanical elements. For instance, a helmet is an engineering application in which toughness plays an important role. Upon impact, the helmet must absorb as much energy from the impact as possible, minimizing the energy that transmits into the head of the person who wears the helmet. Indeed, one is not concerned with plastic deformation of the helmet as it needs to save the wearer's life and can be discarded after impact.

2.3.5 Engineering Stress–Strain Diagram Summary

Figure 2.7 summarizes how different mechanical properties of a ductile material appear in the stress–strain diagram, specifically indicating how the shape of the stress–strain diagram affects those mechanical properties. Hence, by comparing the stress–strain diagram of two different ductile materials relative to each other (e.g. two types of steel), we can understand how their mechanical properties compare to each other.

2.3.6 True Stress–Strain Diagram

Another way to represent the stress–strain relationship of a material is to depict the true stress versus true strain instead of the engineering stress versus engineering strain shown in Figure 2.3. Equation (2.1) shows that the engineering stress is based on the original cross-section A_0 of the dog-bone specimen, before we apply any load. However, in reality, the cross-section is a function of the axial force P (see Figure 2.2), i.e. the actual cross-sectional area $A = f(P)$. In the case of a tensile force, $A \leqslant A_0$ if the Poisson coefficient of the material $\nu > 0$. Indeed, if we stretch the dog-bone specimen, its cross-sectional area decreases because its volume must remain constant. This effect is minimal for elastic deformation but becomes noticeable when plastic deformation occurs. Furthermore, when necking occurs, we observe a significant difference between the actual (local) cross-sectional area A and the original cross-sectional area A_0.

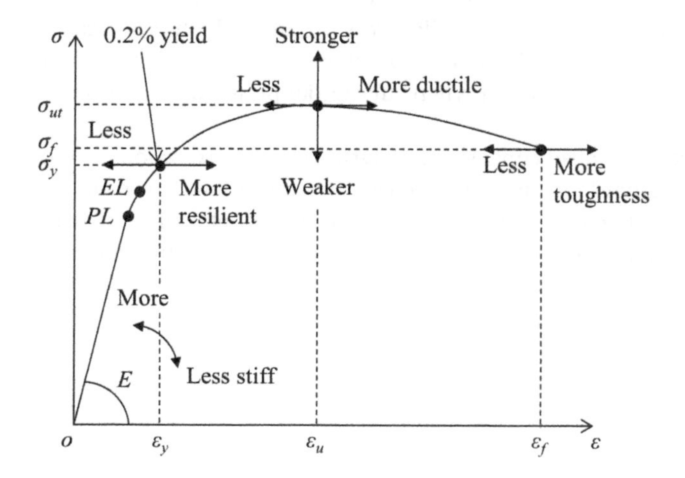

Figure 2.7 Engineering stress–strain diagram, graphically indicating how its shape relates to the mechanical properties of the material.

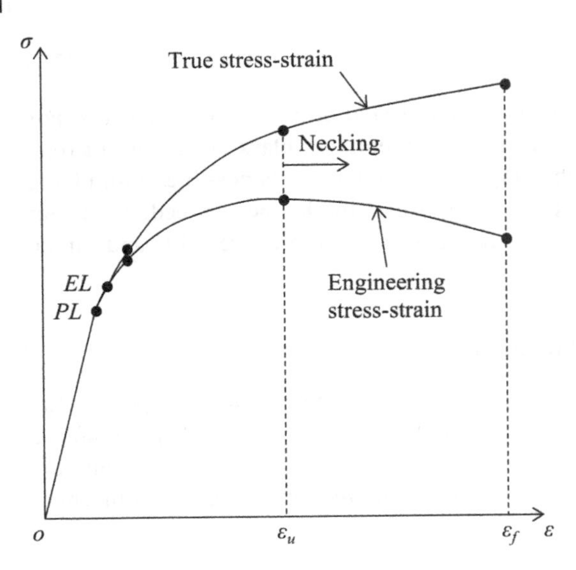

Figure 2.8 Comparison of the true versus engineering stress–strain diagram of a ductile material.

Thus, the true stress based on A is larger than the engineering stress based on A_0. Hence, we define the true stress as

$$\sigma = \frac{P}{A}. \tag{2.7}$$

Similarly, the true strain or logarithmic strain is based on the actual length L rather than the original length L_0 of the dog-bone specimen. Specifically, we define the true strain as the sum of the incremental elongations of the dog-bone specimen, divided by the actual length of the specimen. Thus,

$$\varepsilon = \int_{L_0}^{L} \frac{dL}{L} = \ln \frac{L}{L_0}. \tag{2.8}$$

Figure 2.8 shows a graphical comparison of the true stress–strain and engineering stress–strain diagram of the same ductile material. We observe that the true stress increases monotonically with increasing strain because we account for the actual cross-sectional area A. In contrast, the engineering stress reaches a maximum at the ultimate tensile stress σ_{ut} after which it decreases with increasing strain because it is based on the original cross-sectional area A_0, which is not a good approximation of the actual cross-sectional area A when substantial plastic deformation or necking occurs.

Other relevant mechanical properties may need consideration during material selection for the design of a mechanical element, such as hardness, impact, or creep, to name a few. However, we do not provide a discussion of these mechanical properties here. Instead, we refer the reader to other texts on materials science or mechanical behavior of materials, for a detailed treatment of this topic.[2]

2.4 Materials Processing

2.4.1 Hot Versus Cold Processing

Processing techniques alter the mechanical properties of a material by creating plastic deformation within the material. Depending on whether the plastic deformation is created at elevated or room

2 See, e.g. Meyers, M.A. and Chawla, K. (2008). *Mechanical Behavior of Materials*, 2e. Cambridge University Press.

temperature, we refer to a processing technique as *hot working* (e.g. rolling, forging, hot extrusion) or *cold working* (e.g. rolling, drawing), respectively.

The appearance of a hot and cold worked specimen is distinctly different. A cold worked specimen will show a bright surface finish, sharp corners and edges, and have tight tolerances. In contrast, a hot worked specimen typically shows a dull surface finish because an oxide layer formed on its surface at high temperature. Furthermore, tolerances are typically larger than for a cold worked specimen. For those reasons, cold worked materials tend to be more expensive than similar hot worked materials.

2.4.2 Hot Working

Hot working requires creating plastic deformation at a temperature above the re-crystallization temperature of the material. Hot rolling typically creates a bar of specific shape and dimensions, whereas extrusion requires forcing a heated metal blank through a restricted orifice. Forging involves hot working of metals using hammers or presses.

2.4.3 Cold Working

2.4.3.1 Process

Cold working requires creating plastic deformation of the material at room temperature. Plastic deformation below the recrystallization temperature distorts the grain structure, and changes the mechanical properties of the material, including

- Increase of the yield stress
- Increase of the ultimate tensile stress
- Increase of the hardness
- Decrease of ductility, i.e. the material behaves more brittle.

Table 2.2 shows a comparison of the yield stress and ultimate tensile stress of selected examples of steel types both hot rolled and cold worked. We observe that σ_y and σ_{ut} are indeed higher for the cold worked than for the corresponding hot rolled specimens.

Table 2.2 Comparison of material properties of selected hot rolled and cold worked steels.

Material	Hot rolled		Cold worked	
Steel	**σ_y (MPa)**	**σ_{ut} (MPa)**	**σ_y (MPa)**	**σ_{ut} (MPa)**
AISI 1006	170	300	280	330
AISI 1010	180	320	300	370
AISI 1018	220	400	370	440
AISI 1020	210	380	350	420
AISI 1030	260	470	440	520
AISI 1035	270	500	460	560
AISI 1040	290	520	490	590
AISI 1045	310	530	570	630
AISI 1050	340	620	580	690
AISI 1080	420	770	—	—
AISI 1095	460	830	—	—

Source: Data from *ASM Handbook* (2000), 10e. American Society for Metals.

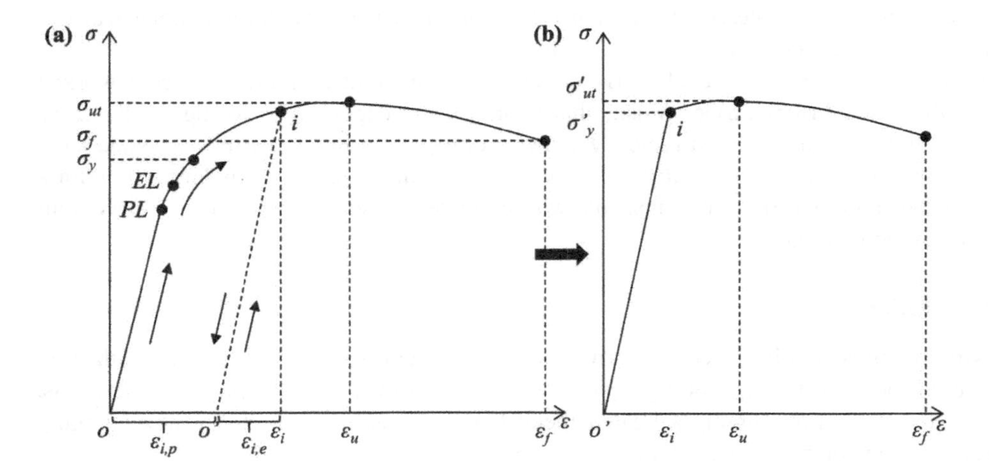

Figure 2.9 (a) Engineering stress–strain diagram of a ductile material before cold work and (b) new engineering stress–strain diagram of the same material after cold work, showing increased yield stress, ultimate tensile stress, and reduced ductility.

The mechanics of cold work are as follows. First, we load the material specimen into the plastic region of the engineering stress–strain diagram ($\sigma_i > \sigma_y$), as indicated in Figure 2.9a with arrows. Subsequently removing the load relieves the elastic strain, whereas plastic strain remains and a new stress–strain diagram exists with origin in o', as shown in Figure 2.9b. This new stress–strain diagram displays a higher yield stress σ_y' and ultimate tensile stress σ_{ut}' than the original stress–strain diagram. Indeed, this cold worked material may be loaded and unloaded elastically any number of times along the straight line of the proportional region of the new stress–strain diagram with origin in o'.

2.4.3.2 Reduction in Area

Figure 2.10 shows load P versus cross-sectional area A, rather than stress versus strain. Note that the horizontal axis shows decreasing cross-sectional area (which we indicate by a downward pointing arrow next to the axis label A). The origin represents the original cross-sectional area A_0 of the

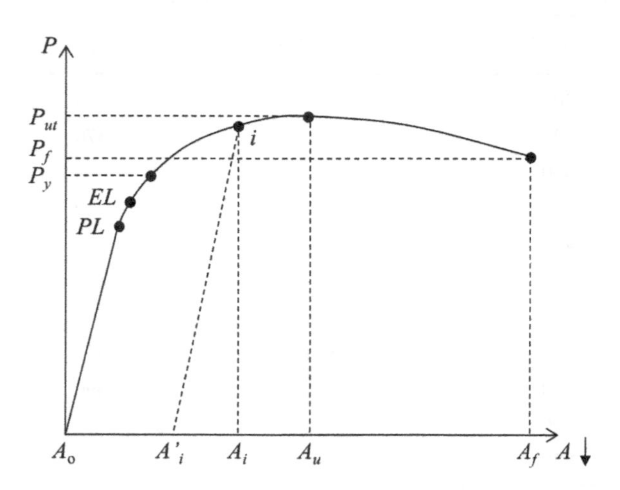

Figure 2.10 Load versus (decreasing) cross-sectional area.

material specimen. We apply a force P_i to load the specimen to an arbitrary point i in the plastic region of the diagram, where it experiences a combination of elastic and plastic strain. The corresponding cross-sectional area is A_i. When we subsequently remove the force P_i, the elastic strain disappears and only the plastic strain remains, and the corresponding cross-sectional area is A_i'. Ultimately, when fracture occurs, the cross-sectional area is A_f.

Thus, we define the reduction-in-area R, corresponding to the load at fracture P_f, as

$$R = \frac{A_0 - A_f}{A_0}. \tag{2.9}$$

Physically, R is a measure of ductility and, thus, the ability of a material to absorb overload and be cold worked.

2.4.3.3 Cold Work Factor

We define the cold work factor as

$$W = \frac{A_0 - A_i'}{A_0} \cong \frac{A_0 - A_i}{A_0}, \tag{2.10}$$

and we approximate $A_i' \cong A_i$ because the difference in cross-sectional area resulting from elastic deformation only is small.

2.4.3.4 Modifying Material Properties Using Cold Work

To calculate how cold work modifies the yield stress of a ductile material from σ_y to σ_y' and its ultimate tensile stress from σ_{ut} to σ_{ut}', we must know the mathematical relationship between true stress and true strain in the plastic region of the stress–strain diagram. Indeed, this relationship allows determining the true stress in the material that corresponds to the true (plastic) strain, i.e. the cold work applied to the material. Figure 2.11 shows a true stress–strain diagram, and we emphasize the plastic region with a dotted line. Within this region, the relationship between true stress and true strain is given as

$$\sigma = \sigma_0 \varepsilon^m. \tag{2.11}$$

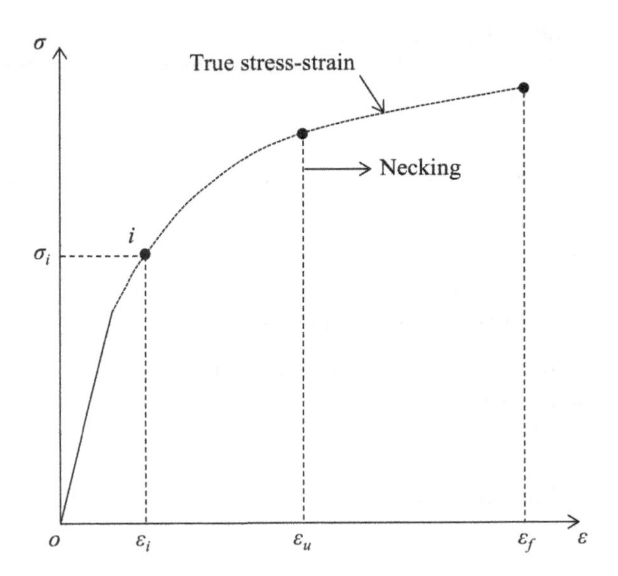

Figure 2.11 True stress–strain diagram, showing the region of plastic deformation as a dotted line.

In Eq. (2.11), σ_0 is the strain-strengthening coefficient and m is the strain-strengthening exponent, with $m = \varepsilon_u$. These parameters are tabulated for commonly used materials (see, e.g. Table 2.3). Using Eq. (2.11), we calculate the modified yield stress after cold work as

$$\sigma'_y = \frac{P_i}{A'_i} = \sigma_0 \varepsilon_i^m, \tag{2.12}$$

where P_i is the load that causes a strain of ε_i in the material, and A'_i is the cross-sectional area of the material specimen after applying strain ε_i and removing the load P_i so that only plastic strain remains. We also refer back to Figure 2.10, where we indicate these parameters.

Similarly, we calculate the modified ultimate tensile stress after cold work as

$$\sigma'_{ut} = \frac{P_u}{A'_i}. \tag{2.13}$$

We emphasize that the ultimate load P_u is constant for a material specimen and remains constant throughout a tensile test regardless whether we load and unload the specimen in the plastic region. However, when loading the material in the plastic region, the cross-sectional area of the material decreases from A_0 to A'_i, increasing the ultimate tensile stress that corresponds to the ultimate load P_u. Thus,

$$P_u = A_0 \sigma_{ut}. \tag{2.14}$$

Combining Eqs. (2.13) and (2.14), we write that

$$\sigma'_{ut} = \frac{A_0}{A'_i} \sigma_{ut}. \tag{2.15}$$

Considering Eq. (2.10), then Eq. (2.15) reduces to

$$\sigma'_{ut} = \frac{\sigma_{ut}}{1 - W}. \tag{2.16}$$

It is important to point out that the calculation of σ'_{ut} is valid only if $\varepsilon_i < \varepsilon_u = m$. If $\varepsilon_i > \varepsilon_u$, necking occurs and we typically approximate $\sigma'_{ut} \cong \sigma'_y$, as there is no practical difference anymore between the yield stress and ultimate tensile stress.

Example problem 2.1

Calculate the yield stress and the ultimate tensile stress of 1144 annealed steel after applying 5% cold work.

Solution

The material properties of 1144 annealed steel are (from Table 2.3):

$\sigma_y = 358$ MPa
$\sigma_{ut} = 646$ MPa
$\sigma_0 = 992$ MPa (strain-strengthening coefficient)
$m = 0.14$ (strain-strengthening exponent).

Starting from the relationship between true stress and true strain in the plastic region, we write that

$$\sigma'_y = \sigma_0 \varepsilon_i^m.$$

Volume conservation requires that $V = A_0 l_0 = A'_i l'_i$ and, thus,

$$\varepsilon_i = \ln(l'_i / l_0) = \ln(A_0 / A'_i).$$

Table 2.3 Mechanical properties of selected materials.

Material	Condition	σ_y (MPa)	σ_{ut} (MPa)	σ_0 (MPa)	m
1018 steel	Annealed	220	341	620	0.25
1020 steel	Hot rolled	210	380	793	0.22
1144 steel	Annealed	358	646	992	0.14
1212 steel	Hot rolled	193	424	758	0.24
1045 steel	Q and T 600 °F	1520	1580	1880	0.041
4142 steel	Q and T 600 °F	1720	1930	1760	0.048
303 stainless steel	Annealed	241	601	1410	0.51
304 stainless steel	Annealed	276	568	1270	0.45
2011 aluminum	T6	169	324	620	0.28
2024 aluminum	T4	296	446	689	0.15
7075 aluminum	T6	542	593	882	0.13

Source: Shigley, J.E. and Mischke, C.R. (1996). *Standard Handbook of Machine Design*, 2e. McGraw-Hill.

Taking into account Eq. (2.10), we the write that

$$\varepsilon_i = \ln(\tfrac{1}{1-W}) = \ln(\tfrac{1}{1-0.05}) = 5.13 \cdot 10^{-2}.$$

Thus, $\sigma_y^{'} = \sigma_0 \ln(\frac{1}{1-W})^m = 992(5.13 \cdot 10^{-2})^{0.14} = 654.50$ MPa.
Because $\varepsilon_i < m$, we determine the ultimate tensile stress after cold work as

$$\sigma_{ut}^{'} = \sigma_{ut}/(1-W) = 646/(1-0.05) = 680 \text{ MPa}.$$

Example problem 2.2

Calculate the yield stress and the ultimate tensile stress of 1212 HR steel after applying 25% cold work.

Solution

The material properties of 1212 HR steel are (from Table 2.3):

$\sigma_y = 193$ MPa
$\sigma_{ut} = 424$ MPa
$\sigma_0 = 758$ MPa (strain-strengthening coefficient)
$m = 0.24$ (strain-strengthening exponent).

Starting from the relationship between true stress and true strain in the plastic region, we write that

$$\sigma_y^{'} = \sigma_0 \varepsilon_i^m.$$

Volume conservation requires that $V = A_0 l_0 = A_i^{'} l_i^{'}$ and, thus,

$$\varepsilon_i = \ln(l_i^{'}/l_0) = \ln(A_0/A_i^{'}).$$

Taking into account Eq. (2.10), we the write that

$$\varepsilon_i = \ln(\tfrac{1}{1-W}) = \ln(\tfrac{1}{1-0.25}) = 2.88 \cdot 10^{-1}$$

Thus, $\sigma_y^{'} = \sigma_0 \ln(\frac{1}{1-W})^m = 758(2.88 \cdot 10^{-1})^{0.24} = 555.14$ MPa.

Because $\varepsilon_i > m$, we determine the ultimate tensile stress after cold work as

$$\sigma'_{ut} \cong \sigma'_y = 555.14 \text{ MPa}.$$

2.5 Alloys

An alternative method to change the properties of steels is to modify its alloying elements. Strictly, every steel is an alloy because it contains both iron (Fe) and carbon (C). We refer to plain carbon steels as steels that contain carbon as the principal alloying element, with only small amounts of other elements. Plain carbon steels are not considered *alloy steels*. In contrast, *alloy steels* contain a variety of alloying elements in amounts between 1% and 50%. Common alloying elements include manganese, nickel, chromium, molybdenum, vanadium, silicon, and boron.

2.5.1 Numbering Systems

With so many possible steel alloys, it is important to have a method to classify the different alloys and their corresponding composition (amounts of alloying elements). Several classification systems exist in North America and Europe. In North America, the Society of Automotive Engineers (SAE) steel grades system is a standard alloy numbering system in which carbon and alloy steels have a four-digit designation, shown in Table 2.4. The first digit designates the main alloying element, the second digit identifies top-grade alloying elements, and the last two digits indicate the amount of carbon in 1/100s of a percent by weight.

For instance, a 1040 steel is a plain carbon steel with 0.40 weight percent carbon.

Another widely used numbering system is the SAE/ASTM Unified Numbering System (UNS), which relates a specific alloy to its corresponding chemical composition (not mechanical properties). A UNS number comprises a prefix letter that designates the material (not limited to steels), followed by five digits that designate the chemical composition. The first two digits refer to the composition of the alloy, whereas the next two digits refer to the carbon content in 1/100s of a percent by weight. The last digit is reserved for special designations.

Table 2.4 SAE steel grades (major classification only).

SAE designation	Steel type
1xxx	Carbon steels
2xxx	Nickel steels
3xxx	Nickel–chromium steels
4xxx	Molybdenum steels
5xxx	Chromium steels
6xxx	Chromium–vanadium steels
7xxx	Tungsten steels
8xxx	Nickel–chromium–molybdenum steels
9xxx	Silicon–manganese steels

Table 2.5 SAE/ASTM unified numbering system (steel classification only).

SAE UNS	Steel type
G10	Plain carbon
G11	Free-cutting carbon steel with more sulfur or phosphorus
G13	Manganese
G23	Nickel
G25	Nickel
G31	Nickel–chromium
G33	Nickel–chromium
G40	Molybdenum
G41	Chromium–molybdenum
G43	Nickel–chromium–molybdenum
G46	Nickel–molybdenum
G48	Nickel–molybdenum
G50	Chromium
G51	Chromium
G52	Chromium
G61	Chromium–vanadium
G86	Chromium–nickel–molybdenum
G87	Chromium–nickel–molybdenum
G92	Manganese–silicon
G94	Nickel–chromium–molybdenum

The prefix letter G refers to plain and alloy steels, A refers to aluminum, S is stainless steel, N refers to nickel alloys, to name a few. Table 2.5 shows examples of the UNS numbering system for plain carbon and alloy steels, showing the prefix and the first two digits that indicate the chemical composition.

For instance, G10400 is a plain carbon steel with 0.40 weight percent carbon.

2.5.2 Plain Carbon Steels

Plain carbon steels contain carbon as the principal alloying element and only contain small amounts of other elements. The strength of plain carbon steels increases with increasing carbon content because carbon and iron atoms form cementite (Fe_3C), which impedes dislocation movement. Low-carbon steels contain less than 0.20% C and are widely used for, e.g. sheet metal applications. Medium-carbon steels contain between 0.20% and 0.50% C and are used in applications that require higher strength than low-carbon steels, such as machine elements. High-carbon steels contain more than 0.50% carbon and are used in applications that require hardness, such as cutting tools.

2.5.3 Alloy Steels

Alloying elements modify the properties of steel. In low amounts (<5%), they increase strength or hardenability, whereas in large amounts (>5%) they achieve special properties, such as corrosion resistance or extreme temperature stability. For instance, we define stainless steels as steels that contain more than 12% chromium (Cr). Chromium improves the corrosion resistance and the toughness of the steel alloy, but it also makes machining and welding more difficult compared to plain carbon steels. Stainless steels typically also contain nickel (Ni) to increase strength and molybdenum (Mo) to improve toughness. Thus, so-called "Cr-Ni-Mo steels" are corrosion resistant and display high strength and toughness. For that reason, they are often used as high-quality spindles in machinery.

The effect of a few common alloying elements on the material properties of the resulting steel alloy is as follows:

- **Chromium**: Increases tensile stress, hardness, hardenability, toughness, resistance to wear and abrasion, and resistance to corrosion.
- **Manganese**: Deoxidizes and degasifies and reacts with sulfur to improve forgeability of the alloy. It increases tensile stress, hardness, hardenability, and resistance to wear.
- **Molybdenum**: Increases strength, hardness, hardenability, and toughness, as well as creep resistance and strength at elevated temperatures. It improves machinability and resistance to corrosion.
- **Nickel**: Increases strength and hardness without sacrificing ductility and toughness. It also increases resistance to corrosion at elevated temperatures when used in high-chromium (stainless) steels.
- **Silicon**: Deoxidizes and degasifies. It increases tensile and yield stress, hardness, forgeability, and magnetic permeability.
- **Tungsten**: Increases strength, wear resistance, hardness, and toughness.
- **Vanadium**: Increases strength, hardness, wear resistance, and resistance to impact.

We refer the reader to specialized textbooks and literature for detailed information on this topic.[3]

2.6 Key Takeaways

1. We can determine mechanical properties from the stress–strain diagram of a material, including yield stress, ultimate tensile stress, Young's modulus, fracture stress, and modulus of resilience and toughness. Thus, comparing stress–strain diagrams of different materials allows comparing their suitability for a specific mechanical design.
2. We can modify the mechanical properties of materials by means of hot and cold processing or adding alloy elements.
3. When selecting a material (steel) for a specific mechanical design, we start by considering a plain carbon steel. If the properties of the plain carbon steel are insufficient to obtain the desired outcome and functionality of the mechanical design, then we consider material processing methods and alloy steels.

---

3 See, e.g. (2000). *ASM Handbook*, 10e. American Society for Metals.

2.7 Problems

2.1 Look up the yield stress σ_y and the ultimate tensile stress σ_{ut} for the following materials:

Material	σ_y	σ_{ut}
AISI 1018		
AISI 1040		
303 stainless steel		
2024 aluminum		
1212 HR steel		

2.2 Figure 2.12 shows the engineering stress–strain diagram of a ductile metal. The original length of the test specimen $L_0 = 250$ mm and diameter $D_0 = 30$ mm. Estimate the elastic modulus E, the ultimate tensile stress σ_{ut}, the yield stress σ_y, the uniform elongation ε_u, the modulus of toughness U_T, and the modulus of resilience U_R.

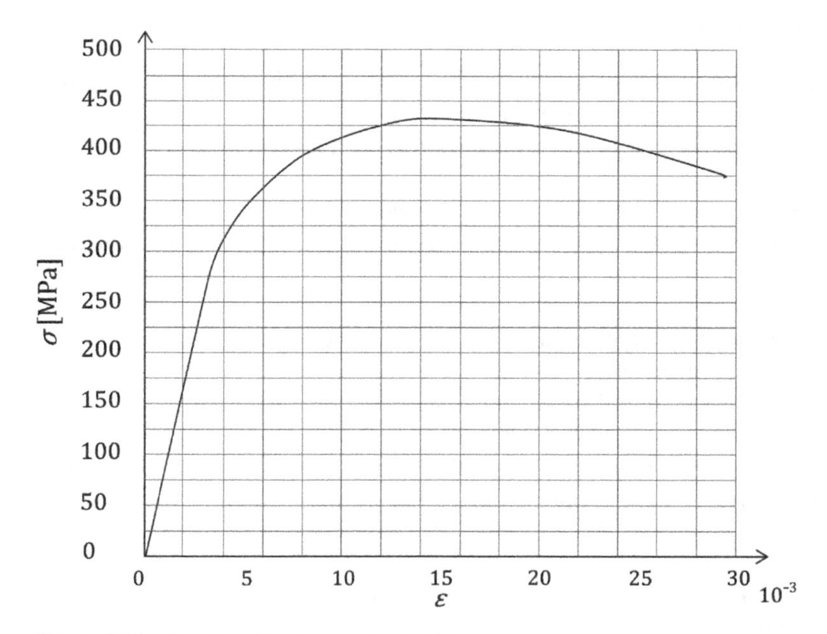

Figure 2.12 Engineering stress–strain diagram.

2.3 Figure 2.13 shows the engineering load-extension diagram of an aluminum alloy. The original length of the test specimen $L_0 = 200$ mm and diameter $D_0 = 5$ mm. Estimate the elastic modulus E, the ultimate tensile stress σ_{ut}, the yield stress σ_y, the uniform elongation ε_u, the modulus of toughness U_T, and the modulus of resilience U_R.

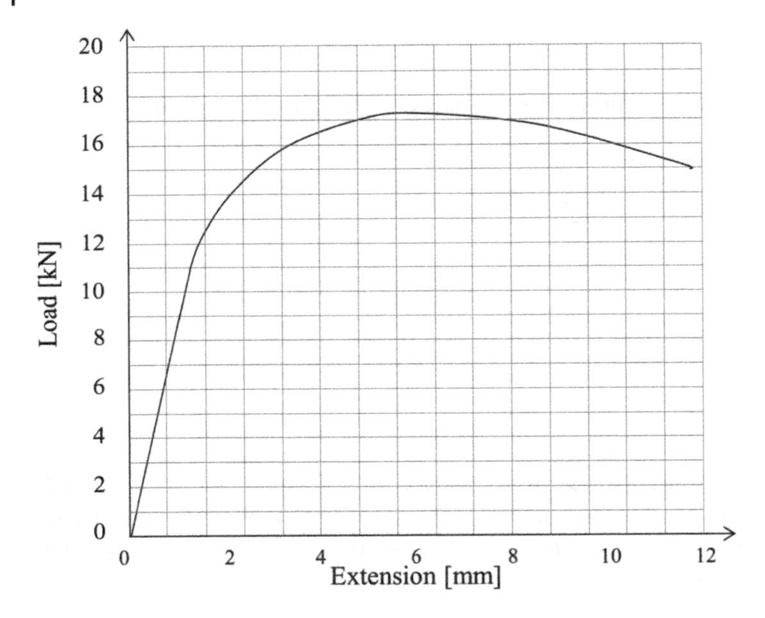

Figure 2.13 Engineering load-extension diagram.

2.4 Consider a plate of 1212 HR steel, with length $L = 500$ mm, and rectangular cross-section with width $w = 100$ mm and thickness $t = 10$ mm.

(a) Calculate the yield stress σ_y' and ultimate tensile stress σ_{ut}' after applying $W = 2\%$ cold work to the material.

(b) Graph the yield stress after cold work σ_y' versus the amount of cold work W, for $\varepsilon_i \leq m$.

(c) Calculate the yield stress σ_y' after applying cold work until the inception of necking, i.e. $\varepsilon_i = m$.

2.5 Consider a plate of 1144 annealed steel, with length $L = 1$ m, and rectangular cross-section with width $w = 50$ mm and thickness $t = 10$ mm.

(a) Calculate the yield stress σ_y' and ultimate tensile stress σ_{ut}' after applying $W = 5\%$ cold work to the material.

(b) Graph the yield stress after cold work σ_y' versus the amount of cold work W, for $\varepsilon_i \leq m$.

(c) Calculate the yield stress σ_y' after applying cold work until the inception of necking, i.e. $\varepsilon_i = m$.

2.6 Consider a rod of 1144 annealed steel with length $L = 500$ mm, and circular cross-section with diameter $D = 20$ mm.

(a) Calculate the yield stress σ_y' and the ultimate tensile stress σ_{ut}' after applying $W = 2.5\%$ cold work to the material.

(b) Calculate the length of the rod after applying $W = 2.5\%$ cold work to the material.

2.7 Figure 2.14 shows a 304 stainless steel rod with circular cross-section, attached to a rigid frame. The diameter and length after cold work are $D = 10$ mm and $L = 0.5$ m, respectively. The rod is subject to an axial load $F = 20$ kN. Calculate the amount of cold work W that was performed on the rod prior to attaching it to the rigid frame, such that a safety factor of $n = 2$ was achieved after cold work?

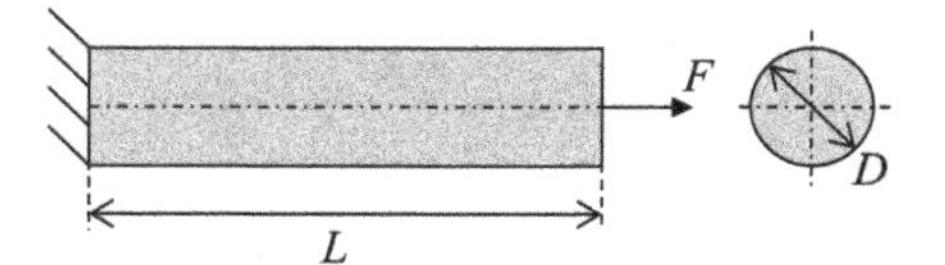

Figure 2.14 Stainless steel rod attached to a rigid frame.

2.8 Figure 2.15 shows a 303 stainless steel bar with square cross-section, attached to a rigid frame. The size and length after cold work are $b = 20$ mm and $L = 0.7$ m, respectively. The bar is subject to an axial load $F = 40$ kN. Calculate the amount of cold work W that was performed on the bar prior to attaching it to the rigid frame, such that a safety factor of $n = 3$ was achieved after cold work?

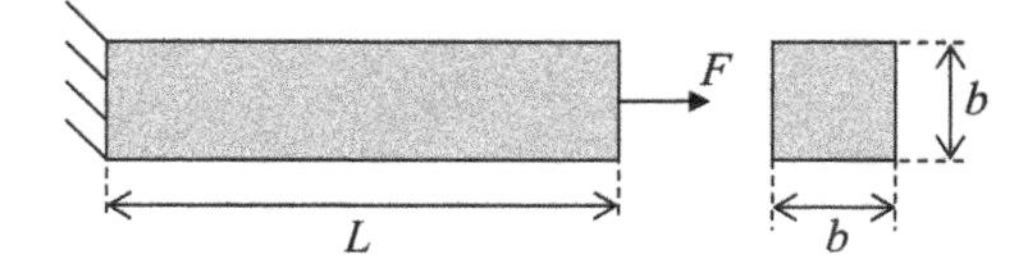

Figure 2.15 Stainless steel bar attached to a rigid frame.

2.9 Consider a shaft machined from 1018 annealed steel with diameter $D = 20$ mm.

(a) Calculate the amount of cold work W that must be applied to the material to increase its yield stress σ_y' by 20% compared to its original value σ_y.

(b) Calculate the diameter of the shaft D' after cold work.

(c) Calculate the safety factor against yielding of the cold-worked shaft, when subject to an axial load $F = 25$ kN.

2.10 Consider a plate of 304 stainless steel with width $w = 20$ mm, length $L = 40$ mm, and thickness $t = 5$ mm.

(a) Calculate the safety factor against yielding of this design, if the plate is subject to an axial load $F = 15$ kN.

(b) Calculate the amount of cold work W that must be applied to this material to double the safety factor against yielding.

3

Statistical Considerations

3.1 Introduction

When designing a mechanical element, we must consider that most dimensions, geometry elements, and physical properties show variability as a result of, e.g. a manufacturing process. For instance, if we manufacture 100 mechanical elements with length L, diameter D, and yield stress σ_y, we expect that not all values of L, D, and σ_y are identical because of imperfections of the stock material and finite accuracy of the manufacturing process. Thus, it is important to understand probability theory and statistics to account for the uncertainty associated with the actual values of L, D, and σ_y.

3.2 Random Variables and Distributions

Probability theory and statistics require the use of random variables. A random variable, or stochastic variable, is a variable whose value depends on the outcome of a random experiment. For instance, randomly selecting one mechanical element from the 100 elements mentioned in the introductory example above, and measuring L, D, and σ_y of this element, assigns a value to random variables L, D, and σ_y. A random variable can be discrete, which means that the outcome of a random experiment is limited to a discrete number of possibilities, such as rolling a dice. Alternatively, a random variable can be continuous, which means that the outcome of a random experiment is not limited to a discrete number of possibilities, but it can be any value between an upper and lower limit. Note that those limits could extend to ∞.

Furthermore, each possible outcome of the random experiment has a probability associated with it. In probability theory and statistics, we use distributions or density functions to describe the probability that the different possible outcomes of a random experiment occur. For instance, if we roll a dice, the discrete possible outcomes of that random experiment are 1, 2, 3, 4, 5, or 6. The probability that each possible outcome of the random experiment occurs is $1/6$ because each possible outcome is equally likely to occur. We describe the distribution of a random variable by means of the *probability density function* and the *cumulative density function*, which are related to each other.

Design of Mechanical Elements: A Concise Introduction to Mechanical Design Considerations and Calculations, First Edition. Bart Raeymaekers.

Companion website: www.wiley.com/go/raeymaekers/designofmechanicalelements

3.3 Density Functions

3.3.1 Probability Density Function

The probability density function (PDF) of random variable x shows the possible outcomes of a random experiment on the horizontal axis and the corresponding frequency or probability of occurrence p of each of the possible outcomes on the vertical axis, i.e. $p = f(x)$. Figure 3.1 shows an example of an arbitrary PDF for (a) a continuous random variable x and (b) a discrete random variable x. Thus, the PDF shows a "picture" of the distribution of random variable x and assigns a probability to each measurable subset of the possible outcomes of a random experiment. Conceptually, we may, e.g. consider the probability p that a random experiment assigns a value to random variable x that remains below value 1 ($p(x \leq \text{value } 1)$), above value 2 ($p(x \geq \text{value } 2)$), or between values 1 and 2 ($p(\text{value } 1 \leq x \leq \text{value } 2)$), to give a few examples of how we can define a measurable subset of the possible outcomes of a random experiment.

We calculate the probability as the area under the PDF and between the limits that define the measurable subset that we evaluate. Note that the area under the entire PDF must always be equal to 1, as the probability that a random experiment assigns a value to the random variable x that falls within the upper and lower limit of the possible outcomes of a random experiment is always 100%. While theoretically one could interpret the PDF as showing the probability that the random variable x is exactly equal to a specific value, this probability is small for a continuous random variable because it is unlikely to occur and, thus, this information is of limited practical importance.

3.3.2 Cumulative Density Function

The cumulative density function (CDF) of random variable x shows all possible outcomes of a random experiment on the horizontal axis, and the probability that a random experiment assigns to the random variable x an outcome less than or equal to a specific outcome, on the vertical axis. Thus, the CDF shows the probability that a random experiment will assign to the random variable x a value smaller than or equal to a specific outcome of the random experiment. Conceptually, the CDF represents the area under the PDF to the left of the specific value we evaluate. Thus, for a

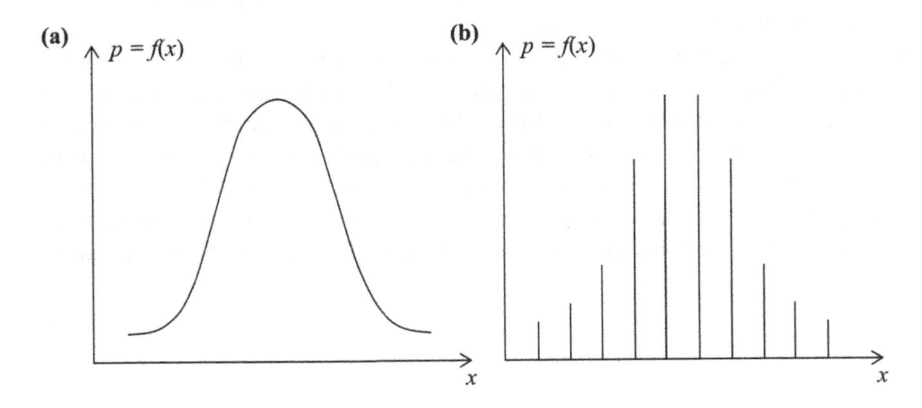

Figure 3.1 Example of a probability density function of (a) a continuous random variable x and (b) a discrete random variable x.

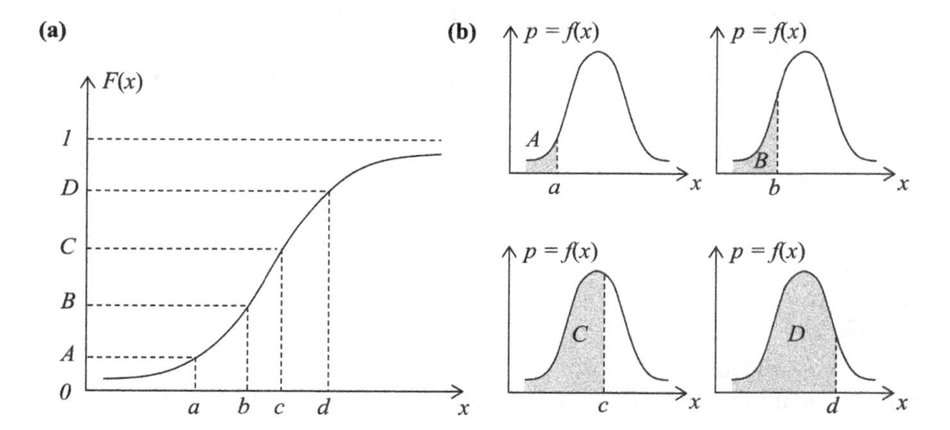

Figure 3.2 (a) Example of a cumulative density function of random variable x, and (b) corresponding probability density function from which we derive the cumulative density function.

continuous random variable x, it is the integral of the PDF from its lower limit to the value for which we evaluate the CDF, i.e.

$$p(x \leq d) = F(d) = \int_{-\infty}^{d} f(x)dx = D, \tag{3.1}$$

where d is the value for which we evaluate the CDF and D is the corresponding probability.

Similarly, for a discrete random variable x, we sum the discrete probability values $f(x_j)$ associated with each possible discrete outcome of the random experiment x_j, smaller or equal than the value d for which we evaluate the CDF. Hence,

$$p(x \leq d) = F(d) = \sum_{x_j \leq d} f(x_j) = D. \tag{3.2}$$

Figure 3.2 illustrates the relationship between the CDF and PDF for a continuous random variable x. Figure 3.2a shows the CDF that corresponds to the arbitrary PDF, repeated four times in Figure 3.2b. Furthermore, Figure 3.2 illustrates that we determine the area under the PDF between its lower limit and each possible outcome of a random experiment by calculating Eq. (3.1), to construct the CDF. We depict four graphical examples in Figure 3.2b, where we calculate the area underneath the PDF between its lower limit and $x = a$, $x = b$, $x = c$, and $x = d$, resulting in A, B, C, and D, respectively.

3.4 Metrics to Describe a Distribution

We can use several metrics to quantify different aspects of the distribution of a random variable. If we consider random variable x, with $a \leq x \leq b$, then we define the arithmetic mean as

$$\mu_x = \int_{a}^{b} xf(x)dx, \tag{3.3}$$

if x is a continuous random variable, and

$$\mu_x = \frac{1}{M}\sum_{i=1}^{M} x_i, \tag{3.4}$$

if x is a discrete random variable. M is the number of discrete data points (or outcomes of a random experiment) we consider for x.

We also define the variance of random variable x, which represents how far the data points spread around its arithmetic mean μ_x.

$$s_x^2 = \int_a^b (x - \mu_x)^2 f(x)dx, \tag{3.5}$$

if x is a continuous random variable, and

$$s_x^2 = \frac{1}{M-1}\sum_{i=1}^{M}(x_i - \mu_x)^2, \tag{3.6}$$

if x is a discrete random variable.

We calculate the standard deviation s_x of random variable x as the positive square root of the variance s_x^2. Thus, s_x is always a positive number, and represents the mean deviation from the arithmetic mean.

Finally, we define the covariance between two continuous random variables x and y as

$$s_{xy} = \int_a^b (x - \mu_x)(y - \mu_y)f(x)dx, \tag{3.7}$$

where μ_x and μ_y are the arithmetic mean of random variables x and y, respectively. Additionally, if x and y are discrete random variables, then

$$s_{xy} = \frac{1}{M-1}\sum_{i=1}^{M}(x_i - \mu_x)(y_i - \mu_y). \tag{3.8}$$

Note that Eq. (3.8) requires that we consider the same number of discrete datapoints M for both random variables x and y. Physically, the covariance between two random variables represents how "similar" these two datasets are. Thus, the covariance between random variables x and y and that between y and x is identical and, therefore, $s_{xy} = s_{yx}$.

The physical meaning of the covariance is further illustrated by converting the covariance s_{xy} between random variable x and y to the correlation coefficient C_{xy}. Indeed, it might be difficult to interpret the physical meaning of the covariance because its units are the square of the units of both random variables, i.e. identical to the units of the variance s_x^2. However, the correlation coefficient represents a nondimensional covariance, which we calculate by dividing the covariance with the standard deviation of each random variable x and y and, thus,

$$C_{xy} = \frac{s_{xy}}{s_x s_y}, \tag{3.9}$$

which ranges between −1 and 1. If the data sets associated with two random variables are identical, i.e. $x = y$, then Eq. (3.8) shows that the covariance is equal to the variance, $s_{xx} = s_x^2$. Correspondingly, Eq. (3.9) shows that $C_{xx} = s_{xx}/s_x s_x = s_x^2/s_x^2 = 1$. Thus, if two random variables are identical, their correlation coefficient is equal to 1.

Example problem 3.1

Table 3.1 shows experimental yield stress data of AISI 1040 HR steel specimens that we procure from two different suppliers; Supplier 1 and Supplier 2. We compare both data sets by calculating their respective arithmetic mean, variance, standard deviation, and the covariance and correlation coefficient between both data sets.

Table 3.1 Comparison of the yield stress of AISI 1040 HR steel specimens from two different suppliers.

Supplier 1, σ_y (MPa)	Supplier 2, σ_y (MPa)
520	530
530	527
515	526
505	519
501	513
523	521
527	523

Solution

We consider the yield stress as a random variable for each supplier, and we calculate the arithmetic mean of each random variable using Eq. (3.4).

$$\mu_1 = (520 + 530 + 515 + 505 + 501 + 523 + 527)/7 = 517.28 \text{ MPa},$$
$$\mu_2 = (530 + 527 + 526 + 519 + 513 + 521 + 523)/7 = 522.71 \text{ MPa}.$$

Next, we calculate the variance of each random variable using Eq. (3.6).

$$s_1^2 = \frac{1}{6}\left[(520 - 517.28)^2 + (530 - 517.28)^2 + \cdots + (527 - 517.28)^2\right] = 119.57 \text{ MPa}^2,$$

$$s_2^2 = \frac{1}{6}\left[(530 - 522.71)^2 + (527 - 522.71)^2 + \cdots + (523 - 522.71)^2\right] = 32.24 \text{ MPa}^2.$$

From the variance, we determine the standard deviation, i.e.

$$s_1 = \sqrt{119.57} = 10.93 \text{ MPa},$$
$$s_2 = \sqrt{32.24} = 5.67 \text{ MPa}.$$

We also calculate the covariance between the yield stress data of Supplier 1 and Supplier 2 using Eq. (3.8), i.e.

$$s_{12} = \frac{1}{6}[(520 - 517.28)(530 - 522.71) + (530 - 517.28)(527 - 522.71) + \cdots$$
$$+ (527 - 517.28)(523 - 522.71)] = 43.93 \text{ MPa}^2.$$

Finally, we calculate the correlation coefficient using Eq. (3.9) as $C_{12} = 43.94/(10.93 \times 5.67) = 0.71$.

3.5 Linear Combination of Random Variables

We extend the understanding of random variables, distributions, and metrics to characterize distributions, to linear combinations of random variables. Indeed, if we consider several random variables of which we know the mean and standard deviation, we can calculate the mean and standard deviation of any linear combination of those variables. This can help to understand error propagation or tolerance stacking in a mechanical design or manufacturing context.

For instance, we consider three random variables x, y, and z, where z is a linear combination of x and y, i.e.

$$z = Ax + By, \tag{3.10}$$

with A and B coefficients that describe the linear combination. If we know μ_x, s_x, and μ_y, s_y, then we calculate μ_z and s_z as

$$\mu_z = A\mu_x + B\mu_y \tag{3.11}$$

and

$$s_z = \left(A^2 s_x^2 + B^2 s_y^2 + 2ABs_{xy}\right)^{1/2} \tag{3.12}$$

or, alternatively, when substituting Eq. (3.9) in Eq. (3.12),

$$s_z = \left(A^2 s_x^2 + B^2 s_y^2 + 2ABs_x s_y C_{xy}\right)^{1/2}. \tag{3.13}$$

We note that $s_{ii} = s_i^2$ because the covariance between a random variable and itself is the variance, and $s_{ij} = s_{ji}$ because the covariance between random variables x and y is identical to the covariance between y and x. Hence, we generalize Eq. (3.12) to determine the mean μ and standard deviation s of N random variables as

$$\mu = \sum_{i=1}^{N} K_i \mu_i \tag{3.14}$$

and,

$$s = \sum_{i=1}^{N} \sum_{j=1}^{N} K_i K_j s_{ij}. \tag{3.15}$$

For instance, if $N = 3$, Eq. (3.15) expands to

$$\begin{aligned} s = {} & K_1K_1s_{11} + K_1K_2s_{12} + K_1K_3s_{13} + K_2K_1s_{21} + K_2K_2s_{22} \\ & + K_2K_3s_{23} + K_3K_1s_{31} + K_3K_2s_{32} + K_3K_3s_{33}. \end{aligned} \tag{3.16}$$

Rearranging and simplifying the terms gives

$$s = K_1^2 s_{11} + K_2^2 s_{22} + K_3^2 s_{33} + 2K_1K_2s_{12} + 2K_1K_3s_{13} + 2K_2K_3s_{23}. \tag{3.17}$$

Example problem 3.2

x, y, and z are random variables with $\mu_x = 3$, $\mu_y = 5$, $s_x = 0.3$, and $s_y = 0.5$. $C_{xy} = 0.3$. If $z = 0.2x + 0.6y$, determine μ_z and s_z.

Solution

We first calculate the arithmetic mean of the linear combination of random variables using Eq. (3.14):

$$\mu_z = 0.2\mu_x + 0.6\mu_y = (0.2 \times 3) + (0.5 \times 5) = 3.1.$$

Next, we calculate the standard deviation of the linear combination of random variables using Eq. (3.15):

$$s_z^2 = (0.2^2 \times 0.3^2) + (0.6^2 \times 0.5^2) + (2 \times 0.2 \times 0.6 \times 0.3 \times 0.5 \times 0.3) = 10.44 \cdot 10^{-2}.$$

Thus, $s_z = 0.32$.

Example problem 3.3

A company produces cylinders with length 100 cm and diameter 2 cm, using two production lines. Production line A produces cylinders with a mean length of 100 cm and a standard deviation of 0.50 cm, whereas production line B produces cylinders with a mean length of 100 cm and a standard deviation of 0.35 cm. Production line A accounts for 60% of the company's total production. Furthermore, experimental data shows that the length of the cylinders manufactured on both production lines is correlated with a correlation coefficient $C = 0.3$. If quality control is performed on the total production of cylinders, then determine the mean and standard deviation of the lengths of all cylinders.

Solution

Random variable L_A: Length of a cylinder from production line A (cm).
Random variable L_B: Length of a cylinder from production line B (cm).
Random variable L_T: Length of a cylinder from the entire production (cm).

Production line A and B produce 60% and 40%, respectively, of the total cylinder production.

We first calculate the arithmetic mean of the linear combination of random variables using Eq. (3.14):

$$\mu_{L_T} = 0.6\mu_{L_A} + 0.4\mu_{L_B} = (0.6 \times 100) + (0.4 \times 100) = 100 \text{ cm}.$$

Next, we calculate the standard deviation of the linear combination of random variables using Eq. (3.15):

$$s_{L_T} = \left[(0.60^2 \times 0.50^2) + (0.40^2 \times 0.35^2) + (2 \times 0.60 \times 0.40 \times 0.50 \times 0.35 \times 0.3)\right]^{1/2},$$

$$s_{L_T} = 0.37 \text{ cm}.$$

3.6 Types of Distributions

Many different types of distributions exist to describe the probability associated with each measurable subset of possible outcomes of a random experiment. Here, we discuss three specific types of distributions, relevant to the design of mechanical elements.

1. **Uniform distribution**: This is the most simple distribution.
2. **Normal distribution**: This is probably the most common distribution.
3. **Weibull distribution**: This distribution is often used to describe failure of mechanical elements.

3.6.1 Uniform Distribution

In a uniform distribution, each possible outcome of a random experiment is equally likely to occur. Figure 3.3 shows (a) the PDF and (b) the CDF of a uniformly distributed random variable x. The lower and upper limits of the random variable are $x = a$ and $x = b$, respectively. The area underneath the PDF is 1 because the probability that a random experiment assigns a value to the random variable between the lower and upper limit is 100% (= certainty).

We express the uniform distribution of random variable x as

$$f(x) = 0, \text{ if } x < a \text{ or } x > b, \text{ and } f(x) = c, \text{ if } a \leqslant x \leqslant b. \tag{3.18}$$

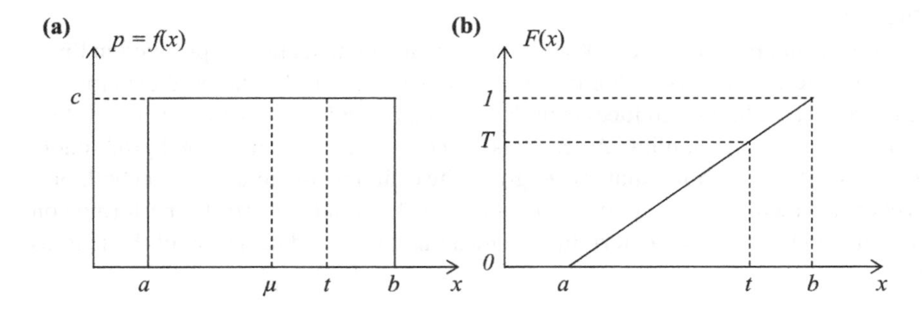

Figure 3.3 (a) Probability density function and (b) cumulative density function of a uniformly distributed random variable $a \leq x \leq b$.

The CDF shows an upward sloping line from 0 to 1, corresponding to $x = a$ and $x = b$. The probability that a random experiment assigns a value to the random variable smaller than a, $p(x < a) = 0$, or smaller than b, $p(x \leqslant b) = 1$, respectively. The CDF represents the incremental integral from $x = a$ to $x = b$ of the PDF. Since the PDF is constant, i.e. $f(x) = c$ for $a \leqslant x \leqslant b$, its integral is a linear function with respect to the random variable x.

In general terms, we express the probability that a random experiment assigns a value to the random variable x that is smaller than a specific value t as

$$p(x \leqslant t) = \int_a^t f(x)dx = \int_a^t c\,dx = c(t - a). \tag{3.19}$$

We evaluate the CDF at its upper limit $x = b$ and, thus, $F(b) = 1$, because the area underneath the PDF must be equal to 1. Hence, using Eq. (3.19), we write that

$p(x \leqslant b) = 1 = c(b - a)$ and, thus,

$c = 1/(b - a)$, with $a \leqslant x \leqslant b$.

Finally, we write the CDF as

$$F(t) = \int_a^t f(x)dx = \int_a^t c\,dx = \frac{t - a}{b - a}. \tag{3.20}$$

Example problem 3.4

The probability that a machine element must be replaced within a machine assembly is uniformly distributed between one and eight years. Calculate the probability that the machine element needs replacement before six years have passed. Calculate the probability that the machine element does not need replacement before six years have passed.

Solution

Define random variable x: time until the machine element needs replacement (years).

We express the probability that a machine element requires replacement before six years have passed as

$p(x \leqslant 6) = \int_1^6 c\,dx$, with $c = 1/(b - a) = 1/(8 - 1)$.

Using the expression for the CDF of a uniform distribtion (Eq. (3.20)), we write that

$p(x \leqslant 6) = \int_1^6 \frac{dx}{8-1} = \frac{6-1}{8-1} = 0.71$.

Hence, a 71% probability exists that the machine element needs replacement before six years have passed.

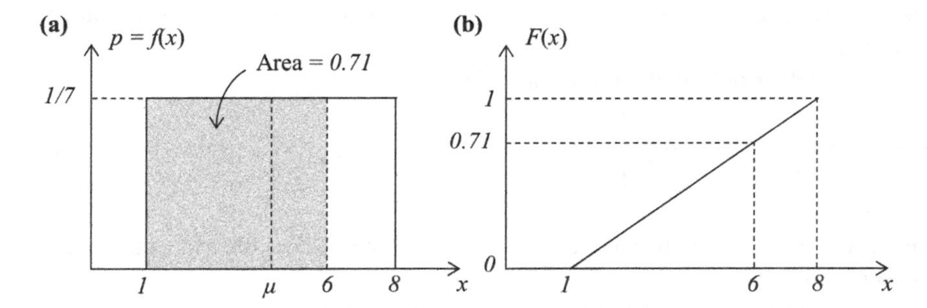

Figure 3.4 (a) Probability density function and (b) cumulative density function of a uniformly distributed random variable x, showing the probability that a machine element needs replacement before six years have passed.

Consequently, because the area under the PDF is always equal to 1, we express the probability that the machine element does not need replacement before six years have passed as

$$p(x > 6) = 1 - p(x \leqslant 6) = 1 - 0.71 = 0.29.$$

Hence, a 29% probability exists that the machine element does not need replacement before six years have passed.

Figure 3.4 shows the relationship between the calculation of the probability and its graphical representation in the PDF (Figure 3.4a) and CDF (Figure 3.4b), respectively. The shaded area in the PDF corresponds to the probability that the machine element requires replacement before six years have passed. The total area under the PDF minus the shaded area then represents the probability that the machine element does not require replacement before six years have passed. The CDF shows the same information.

3.6.2 Normal Distribution

A normal or Gaussian distribution is probably the most common type of distribution because many natural effects follow a normal distribution. Figure 3.5 shows (a) the PDF and (b) the CDF of a random variable $-\infty \leq x \leq +\infty$ that follows a normal distribution. The area underneath the PDF is 1 because the probability that a random experiment assigns a value to the random variable between the lower and upper limit is 100% (= certainty). A normal distribution of a

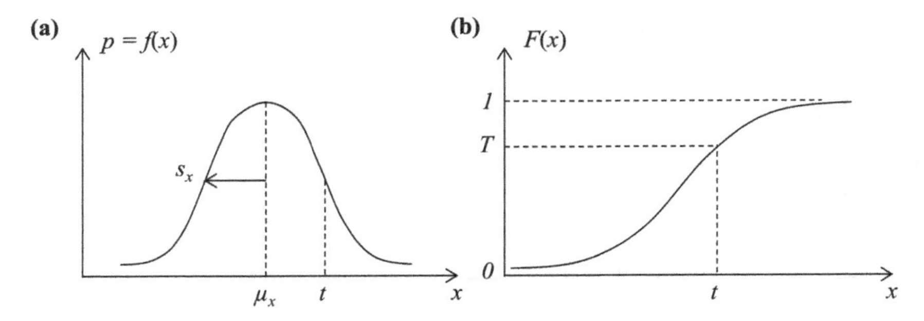

Figure 3.5 (a) Probability density function and (b) cumulative density function of a normally distributed random variable $-\infty \leq x \leq +\infty$.

random variable x is entirely defined by its mean μ_x and its standard deviation s_x or, using compact notation $x \sim N(\mu_x, s_x)$.

We express the normal distribution of random variable x as

$$f(x) = \frac{1}{\sqrt{2\pi}s_x} \exp\left[-\frac{1}{2}\left(\frac{x-\mu_x}{s_x}\right)^2\right], \tag{3.21}$$

with $-\infty \leq x \leq +\infty$.

The CDF represents the incremental integral from $x = -\infty$ to $x = +\infty$ of the PDF and changes nonlinearly between 0 and 1.

In general terms, we express the probability that a random experiment assigns a value to the random variable x that is smaller than a specific value t as

$$p(x \leqslant t) = F(t) = \int_{-\infty}^{t} f(x)dx. \tag{3.22}$$

Solving the integral of Eq. (3.21) is not straightforward. For that reason, we transform the normal distribution of random variable x to the standard normal distribution with standard normal random variable z, where

$$z = \frac{x-\mu_x}{s_x}. \tag{3.23}$$

Essentially, we nondimensionalize the normal distribution of variable x, which may have any unit (meter, kg, volt, etc.) to the standard normal distribution where the random variable z is dimensionless with mean $\mu_z = 0$ and standard deviation $s_z = 1$. Thus, we transform $x \sim N(\mu_x, s_x)$ to $z \sim N(0,1)$.

Figure 3.6 shows (a) the PDF and (b) the CDF of a standard normally distributed random variable $-\infty \leq z \leq +\infty$.

We express the standard normal distribution as

$$f(z) = \frac{1}{\sqrt{2\pi}} \exp^{-\frac{z^2}{2}} \tag{3.24}$$

with $-\infty \leq z \leq +\infty$.

We can transform any (dimensional) normal distribution into a (nondimensional) standard normal distribution by transforming its random variable x into z using Eq. (3.23). Then,

$$p(z \leqslant k) = F(z) = \int_{-\infty}^{k} f(z)dz. \tag{3.25}$$

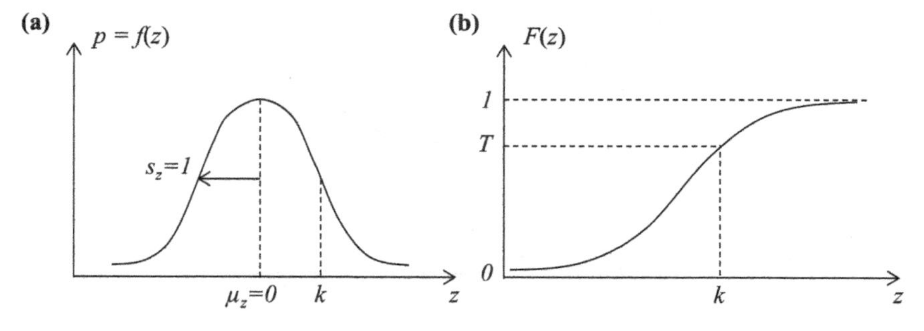

Figure 3.6 (a) Probability density function and (b) cumulative density function of a standard normally distributed random variable z.

We calculate the CDF as the incremental integral from $z = -\infty$ to $z = +\infty$ of Eq. (3.24). The standard normal distribution table (see Table 3.2) shows the results of this integral for different values of z, with $0 \leqslant z \leqslant 3$. Thus, when performing a probability calculation that involves a random variable that follows a normal distribution, we first transform that normal distribution into the standard normal distribution. Rather than calculating complex integrals, we can then use the standard normal distribution table to perform probability calculations.

For instance, we calculate the probability that a normally distributed random variable x is smaller than a specific value t, i.e. $p(x \leqslant t)$. First, we transform the normally distributed random variable x to the standard normally distributed random variable z, with $z = (x - \mu_x)/s_x$. Correspondingly, we transform value t in the normal distribution to value k in the standard normal distribution, with $k = (t - \mu_x)/s_x$. Then, we rewrite the original probability calculation as

$$p(x \leqslant t) = p\left(\frac{x - \mu_x}{s_x} \leqslant \frac{t - \mu_x}{s_x}\right) = p(z \leqslant k) = F(k). \tag{3.26}$$

We look up the value of $z = k$ in the standard normal distribution table (see Table 3.2) to find the answer to $p(z \leqslant k) = F(k)$.

Example problem 3.5

The probability that a machine element needs replacement within a machine assembly follows a normal distribution with mean five years and standard deviation two years. Calculate the probability that the machine element needs replacement before six years have passed. Calculate the probability that the machine element does not need replacement before six years have passed.

Solution

Define random variable x: time until the machine element needs replacement (years).

We express the probability that a machine element requires replacement before six years have passed as follows:

$p(x \leqslant 6) = p(\frac{x-\mu_x}{s_x} \leqslant \frac{6-5}{2.0}) = p(z \leqslant 0.5) = F(0.5)$.

The standard normal distribution table shows that $p(z \leqslant 0.50) = 0.6915$.

Hence, a 69.15% probability exists that the machine element requires replacement before six years have passed.

Because the area under the PDF is always 1, we express the probability that the machine element does not need replacement before six years have passed as

$p(x > 6) = 1 - p(x \leqslant 6) = 1 - p(z \leqslant 0.50) = 1 - 0.6915 = 0.3075$.

Hence, a 30.15% probability exists that the machine element does not need replacement before six years have passed.

Figure 3.7 shows the relationship between the calculation of the probability and its graphical representation. Figure 3.7a,b show the PDF and CDF, respectively, of the normally distributed random variable x with units of years. We first transform the normal distribution to the standard normal distribution. Figure 3.7c,d shows the PDF and CDF, respectively, of the standard normally distributed random variable z. The shaded area in the PDF corresponds to the probability that the machine element requires replacement before six years have passed. The difference between the total area under the PDF and the shaded area represents the probability that the machine element does not require replacement before six years have passed. The CDF contains the same information.

Table 3.2 Standard normal distribution table.

The table shows the cumulative probability of the standard normal distribution variable $z \leq k$

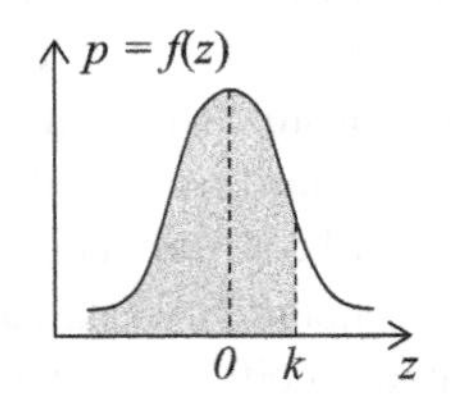

z	0.00	0.01	0.02	0.03	0.04	0.05	0.06	0.07	0.08	0.09
0.0	0.5000	0.5040	0.5080	0.5120	0.5159	0.5199	0.5239	0.5279	0.5319	0.5359
0.1	0.5398	0.5438	0.5478	0.5517	0.5557	0.5596	0.5636	0.5675	0.5714	0.5753
0.2	0.5793	0.5832	0.5871	0.5910	0.5948	0.5987	0.6026	0.6064	0.6102	0.6141
0.3	0.6179	0.6217	0.6255	0.6293	0.6331	0.6368	0.6406	0.6443	0.6480	0.6517
0.4	0.6554	0.6591	0.6628	0.6664	0.6700	0.6736	0.6772	0.6808	0.6844	0.6879
0.5	0.6915	0.6950	0.6985	0.7019	0.7054	0.7088	0.7123	0.7157	0.7190	0.7224
0.6	0.7257	0.7291	0.7324	0.7357	0.7389	0.7422	0.7454	0.7486	0.7517	0.7549
0.7	0.7580	0.7511	0.7642	0.7673	0.7704	0.7734	0.7764	0.7794	0.7823	0.7854
0.8	0.7881	0.7910	0.7939	0.7967	0.7995	0.8023	0.8051	0.8078	0.8106	0.8133
0.9	0.8159	0.8186	0.8212	0.8238	0.8264	0.8289	0.8315	0.8340	0.8365	0.8389
1.0	0.8413	0.8438	0.8461	0.8485	0.8508	0.8531	0.8554	0.8577	0.8599	0.8621
1.1	0.8643	0.8665	0.8686	0.8708	0.8729	0.8749	0.8770	0.8790	0.8804	0.8830
1.2	0.8849	0.8869	0.8888	0.8907	0.8925	0.8944	0.8962	0.8980	0.8997	0.9015
1.3	0.9032	0.9049	0.9066	0.9082	0.9099	0.9115	0.9131	0.9147	0.9162	0.9177
1.4	0.9192	0.9207	0.9222	0.9236	0.9251	0.9265	0.9279	0.9292	0.9306	0.9319
1.5	0.9332	0.9345	0.9357	0.9370	0.9382	0.9394	0.9406	0.9418	0.9429	0.9441
1.6	0.9452	0.9463	0.9474	0.9484	0.9495	0.9505	0.9515	0.9525	0.9535	0.9545
1.7	0.9554	0.9564	0.9573	0.9582	0.9591	0.9599	0.9608	0.9616	0.9625	0.9633
1.8	0.9641	0.9649	0.9656	0.9664	0.9671	0.9678	0.9686	0.9693	0.9699	0.9706
1.9	0.9713	0.9719	0.9726	0.9732	0.9738	0.9744	0.9750	0.9756	0.9761	0.9767
2.0	0.9773	0.9778	0.9783	0.9788	0.9793	0.9798	0.9803	0.9808	0.9812	0.9817
2.1	0.9821	0.9826	0.9830	0.9834	0.9838	0.9842	0.9846	0.9850	0.9854	0.9857
2.2	0.9861	0.9865	0.9868	0.9871	0.9874	0.9878	0.9881	0.9884	0.9887	0.9890
2.3	0.9893	0.9896	0.9898	0.9901	0.9904	0.9906	0.9909	0.9911	0.9913	0.9916
2.4	0.9918	0.9920	0.9922	0.9924	0.9927	0.9929	0.9931	0.9932	0.9934	0.9936
2.5	0.9938	0.9940	0.9941	0.9943	0.9945	0.9946	0.9948	0.9949	0.9951	0.9952
2.6	0.9953	0.9955	0.9956	0.9957	0.9959	0.9960	0.9961	0.9962	0.9963	0.9964
2.7	0.9965	0.9966	0.9967	0.9968	0.9969	0.9970	0.9971	0.9972	0.9973	0.9974
2.8	0.9974	0.9975	0.9976	0.9977	0.9977	0.9978	0.9979	0.9980	0.9980	0.9981
2.9	0.9981	0.9982	0.9982	0.9983	0.9984	0.9984	0.9985	0.9985	0.9986	0.9986
z	3.00	3.10	3.20	3.30	3.40	3.50	3.60	3.70	3.80	3.90
p	0.9986	0.999 02	0.9993	0.9995	0.9997	0.9998	0.9998	0.9999	0.9999	1.0000

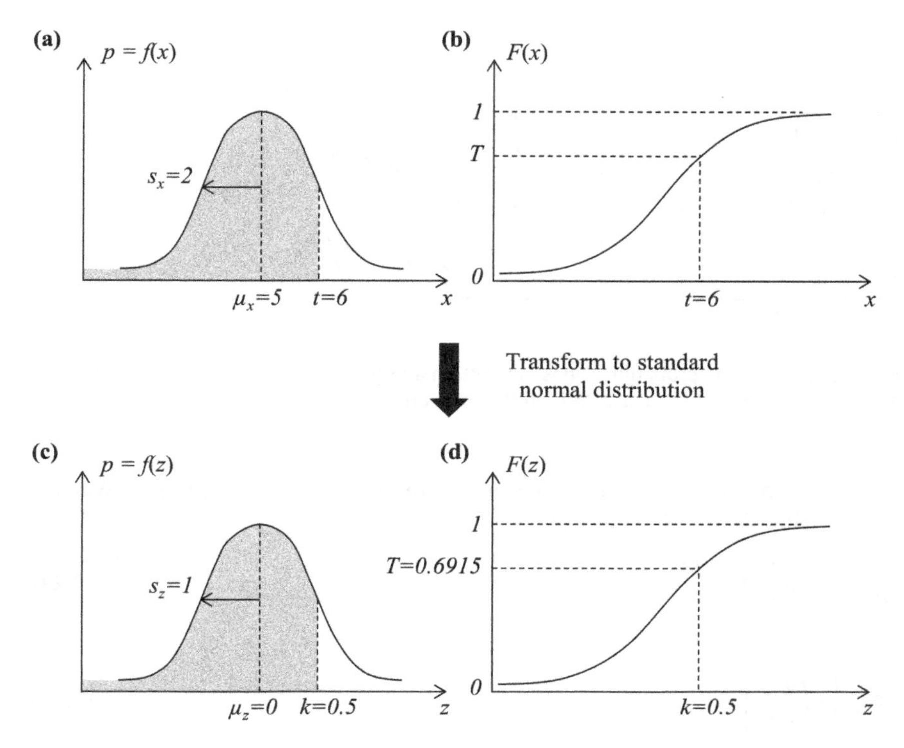

Figure 3.7 (a) Probability density function and (b) cumulative density function of normally distributed variable x, (c) probability density function, and (d) cumulative density function of standard normally distributed variable z, indicating the probability that a machine element needs replacement before six years have passed.

3.6.3 Weibull Distribution

A Weibull distribution of a random variable x is entirely defined by three parameters: the minimum expected value x_0, a scale factor θ, and a Weibull slope b. For that reason, it is often referred to as a "family" of distributions. We often use the Weibull distribution to describe the "time-to-failure" of a mechanical element because its three parameters allow for flexibility when fitting a Weibull PDF to a cloud of experimental data points that for instance represent the time-to-failure of a mechanical element. As with all types of distributions, the area underneath the PDF is 1 because the probability that a random experiment assigns a value to the random variable between the lower and upper limit is 100% (= certainty), independent of the value of each of the three parameters x_0, θ, and b.

We express the Weibull distribution as

$$f(x) = \frac{b}{\theta - x_0}\left(\frac{x - x_0}{\theta - x_0}\right)^{b-1} \exp\left[-\left(\frac{x - x_0}{\theta - x_0}\right)^{b}\right]. \tag{3.27}$$

The CDF represents the incremental integral from $x = x_0$ to $x = +\infty$ of the PDF and changes nonlinearly between 0 and 1, i.e.

$$F(t) = \int_{x_0}^{t} f(x)dx = 1 - \exp\left[-\left(\frac{x - x_0}{\theta - x_0}\right)^{b}\right]. \tag{3.28}$$

Figure 3.8 shows three examples of PDFs and their corresponding CDFs of example Weibull distributions, indicating the three parameters x_0, θ, and b.

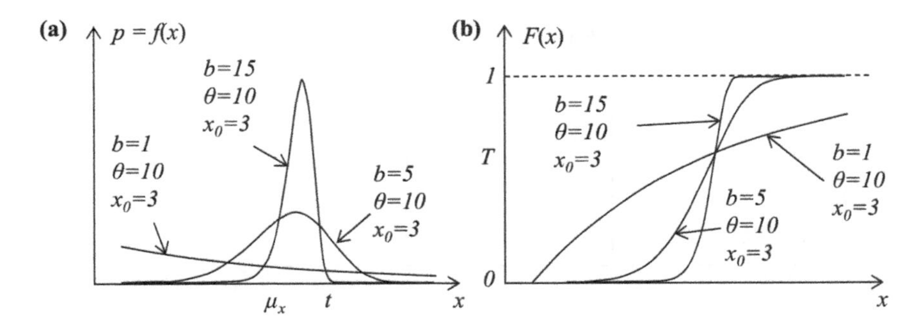

Figure 3.8 Three examples of (a) the probability density function and (b) the cumulative density function of a random variable $x_0 \leq x \leq +\infty$ that follows a Weibull distribution.

If $x_0 = 0$, then the three-parameter Weibull distribution reduces to a two-parameter Weibull distribution, and we rewrite Eqs. (3.27) and (3.28) as

$$f(x) = \frac{b}{\theta}\left(\frac{x}{\theta}\right)^{b-1} \exp\left[-\left(\frac{x}{\theta}\right)^{b}\right], \tag{3.29}$$

and,

$$F(t) = 1 - \exp\left[-\left(\frac{t}{\theta}\right)^{b}\right]. \tag{3.30}$$

The concept of performing a probability calculation with a random variable that follows a Weibull distribution is identical to that of a uniform or normal distribution, as demonstrated in the following example problem.

Example problem 3.6

The probability that a machine element needs replacement within a machine assembly follows a Weibull distribution with $x_0 = 0$, $\theta = 10$, and $b = 3$. Calculate the probability that the machine element needs replacement before six years have passed. Calculate the probability that the machine element does not need replacement before six years have passed.

Solution

Define the random variable x: time until the machine element replacement (years).

We express the probability that a machine element requires replacement before six years have passed as follows:

$$p(x \leqslant 6) = F(6) = 1 - \exp\left[-\left(\frac{6}{10}\right)^{3}\right] = 0.1943.$$

Hence, a 19.43% probability exists that the machine element requires replacement before six years have passed.

Because the area under the PDF is always 1, we express the probability that the machine element does not need replacement before six years have passed as follows

$$p(x > 6) = 1 - p(x \leqslant 6) = 1 - 0.1943 = 0.8057.$$

Hence, a 80.57% probability exists that the machine element does not need replacement before six years have passed.

Note that while provided in the problem statement of Example problem 3.6, we typically have to determine the Weibull parameters from an experimental data set, for instance, historical data of the time-to-failure of a mechanical element. Finding the best fit of the Weibull PDF to the experimental data by varying the three parameters x_0, θ, and b, yields the value that we use in the probability calculations.

We also point out that independent of the type of distribution (uniform, normal, or Weibull), the concept of probability calculations is identical. We first identify and define the random variable we will use. Then, we mathematically express the probability we will calculate. When the random variable follows a uniform or Weibull distribution, we may use the equation for the CDF to calculate probability values. When the random variable follows a normal distribution, we first transform the normal distribution to a standard normal distribution, after which we use the standard normal distribution table to calculate probability values.

Example problem 3.7

A Mars rover is designed to grasp and analyze rocks in a crater. The rocks are spherical and vary in size. Hence, the rover is outfitted with a grasper that adapts its width k between k_{min} and k_{max} (Figure 3.9). Design k_{max} such that 95% of all Mars rocks will fit within the grasper, if $k_{min} = 1.5$ cm and,

(a) If the diameter D of the rocks is uniformly distributed between 2 and 6 cm.
(b) If the diameter D of the rocks follows a normal distribution with mean 4 cm and standard deviation 1.5 cm.
(c) Assume that the Mars soil comprises two distinctly different types of spherical granular media. The particle size of medium 1 follows a normal distribution with mean diameter 1 mm and standard deviation 0.1 mm. The particle size of medium 2 follows a normal distribution with mean diameter 0.8 mm and standard deviation 0.2 mm. Calculate the expected mean and standard deviation of a granular medium that consists of a 50/50 mix of medium 1 and 2, if the distributions of the particle size of both types of granular media are uncorrelated.

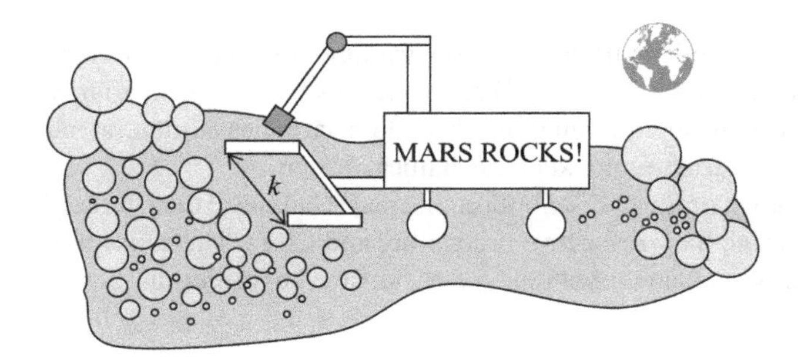

Figure 3.9 Mars rover grasping rocks.

Solution

Define random variable D: diameter of a Mars rock (cm).

We express the probability that the diameter of a Mars rock is between k_{min} and k_{max} as

$$p(k_{min} \leqslant D \leqslant k_{max}) = 0.95.$$

We recognize that a probability represents part of the area under the PDF and, thus, we rewrite this expression as follows:

$$p(D \leqslant k_{max}) - p(D \leqslant k_{min}) = 0.95.$$

(a) Considering a uniform distribution, we evaluate the CDF for k_{max} and k_{min} as

$$F(k_{max}) - F(k_{min}) = (k_{max} - 2)/(6 - 2) - 0 = 0.95.$$

Thus, $k_{max} = (0.95 \times 4) + 2 = 5.80$ cm

(b) Considering a normal distribution, we first convert the normal distribution to a standard normal distribution, with the standard normal distribution random variable $z = (D - \mu_D)/s_D$. Hence,

$$p(z \leqslant (k_{max} - 4)/1.5) - p(z \leqslant (k_{min} - 4)/1.5) = 0.95, \text{ or}$$
$$p(z \leqslant (k_{max} - 4)/1.5) = 0.95 + p(z \leqslant (k_{min} - 4)/1.5), \text{ and}$$
$$p(z \leqslant (k_{max} - 4)/1.5) = 0.95 + p(z \leqslant -1.67).$$

Finally, $p(z \leqslant (k_{max} - 4)/1.5) = 0.95 + [1 - p(z \leqslant 1.67)] = 0.95 + [1 - 0.9525]$.
Thus, $p(z \leqslant (k_{max} - 4)/1.5) = 0.998$.
Performing a reverse look-up in the standard normal distribution table (see Table 3.2), we find that $p(z \leqslant 2.87) = 0.998$.
Thus, $(k_{max} - 4)/1.5 = 2.87$ and $k_{max} = 8.31$ cm.

(c) We first calculate the mean of the 50/50 mix, using Eq. (3.14), i.e.

$$\mu_{mix} = (0.5 \times 1) + (0.5 \times 0.8) = 0.900 \text{ mm}.$$

We then calculate the standard deviation of the 50/50 mix, using Eq. (3.15), i.e.

$$s_{mix} = ((0.5^2 \times 0.1^2) + (0.5^2 \times 0.2^2))^{1/2} = 0.112 \text{ mm}.$$

3.7 Key Takeaways

1. We use probability theory and statistics to account for uncertainty in the design and manufacturing of mechanical elements.
2. We characterize distributions of discrete and continuous random variables using the arithmetic mean, variance, and standard deviation. We compare different data sets using the covariance and the correlation coefficient. We use linear combinations of random variables to understand error propagation and stacking of errors in the context of mechanical design.
3. We describe three types of distributions: uniform distribution, normal (Gaussian) distribution, and Weibull distribution, and determine their respective probability and CDFs. The concept of performing probability calculations is independent of the type of statistical distribution.

3.8 Problems

3.1 A manufacturing process yields metal brackets with nominal length L_o that follows a normal distribution with mean $\mu_{L_o} = 10$ cm and standard deviation $s_{L_o} = 1$ mm. If we add a manufacturing line to double the production volume (i.e. each production line produces half of the total production), and the new production line yields metal brackets with nominal length L_n that follows a normal distribution with $\mu_{L_n} = 10$ cm and $s_{L_n} = 0.5$ mm, determine and graph the standard deviation of the nominal length of all brackets (total production) as a function of the correlation coefficient between the lengths of the brackets manufactured on both manufacturing lines L_o and L_n, with $-1 \leq C_{L_oL_n} \leq 1$.

3.2 A CNC machining center is programmed to drill two holes 4 cm apart in a mass produced part.

(a) Calculate the probability that the distance between the two holes of a randomly selected part is between 4.01 and 4.04 cm, if the distance between the holes in the part is uniformly distributed between 3.9 and 4.1 cm.

(b) Calculate the probability that the distance between the two holes of a randomly selected part is between 4.01 and 4.04 cm, if the distance between the holes in the part follows a normal distribution with mean 4 cm and standard deviation 1.0 mm.

(c) Calculate the probability that the distance between the two holes of a randomly selected part is smaller than 3.85 cm, if the distance between the holes in the part follows a normal distribution with mean 4 cm and standard deviation 1.0 mm.

(d) A second CNC machining center is available that is more accurate than the first one. The distance between the holes it drills in the mass produced part follows a normal distribution with mean 4 cm and standard deviation 0.1 mm. The distance between the holes produced with the first CNC machine follows a normal distribution with mean 4 cm and standard deviation 1.0 mm. The outcomes of both machines are uncorrelated. Calculate the fraction of the total production volume that must be directed to the new CNC machining center to ensure that the distance between the two holes is smaller than 4.03 cm for 98% of all parts.

3.3 A company produces covers for smartphones. To achieve a snug fit between the cover and the smartphone, the width of the cover must be between 4.99 and 5.01 cm.

(a) Calculate the probability that the width of the smartphone cover is between these limits, if the width is uniformly distributed between 4.98 and 5.02 cm.

(b) Calculate the probability that the width of the smartphone cover is between these limits, if the width follows a normal distribution with mean 5 cm and standard deviation 0.01 cm.

(c) Calculate the standard deviation of the width of the total production of smartphone covers, required to ensure that the width of 98% of all covers is smaller than 5.01 cm, if the width of the smartphone covers follows a normal distribution with mean 5.00 cm.

(d) A new manufacturing process is available that manufactures smartphone covers such that their width follows a normal distribution with mean 5 cm and standard deviation 0.05 mm. This contrasts the original process that produces smartphone covers of which the width follows a normal distribution with mean 5.00 cm and standard deviation 0.10 mm. The outcomes of both manufacturing processes are uncorrelated. Calculate the fraction of the total production volume that must be directed to the new manufacturing process to ensure that the width of 98% of all smartphone covers is smaller than 5.05 cm.

(e) How does the calculation of (d) change when considering that the width of the smartphone covers manufactured with either process are correlated? Graph the fraction of the total production of smartphone covers that must be manufactured using the new process as a function of the correlation coefficient $-1 \leq C \leq 1$ between the width of the smartphone covers manufactured with both processes.

3.4 A company produces cylinders with length $L = 10$ cm and diameter $D = 1$ cm.

(a) Calculate the probability that the yield stress of a randomly selected cylinder $\sigma_y \leq 540$ MPa, if the yield stress follows a normal distribution with mean $\mu_{\sigma_y} = 510$ MPa and standard deviation $s_{\sigma_y} = 10$ MPa.

(b) Calculate the probability that the yield stress of a randomly selected cylinder is between 500 and 520 MPa, if the yield stress follows a normal distribution with mean $\mu_{\sigma_y} = 510$ MPa and standard deviation $s_{\sigma_y} = 10$ MPa.

(c) Calculate the probability that the yield stress of a randomly selected cylinder exceeds 500 MPa, if the yield stress follows a Weibull distribution with $x_0 = 420$ MPa, $b = 3.64$, and $\theta = 550$.

(d) Calculate the standard deviation of the length of the cylinders to guarantee that 95% of all cylinders is shorter than 10.60 cm, if the length of the cylinders follows a normal distribution with mean 10.00 cm.

3.5 An AISI 1020 HR shaft is an intricate part of a machine assembly. When the shaft deforms plastically, the machine can no longer perform its function, and the customer will file a warranty claim to replace the shaft.

(a) Calculate the fraction of machines that will require a shaft replacement under warranty if the maximum stress in the shaft σ during operation is uniformly distributed between 180 and 230 MPa.

(b) Calculate the fraction of machines that will require a shaft replacement under warranty if the maximum stress in the shaft σ during operation follows a normal distribution with mean $\mu_\sigma = 190$ MPa and standard deviation $s_\sigma = 20$ MPa.

(c) Determine which steel to select from Table 2.2 to reduce the fraction of warranty replacements to 4%, if the maximum stress in the shaft σ during operation follows a normal distribution with mean $\mu_\sigma = 190$ MPa and standard deviation $s_\sigma = 20$ MPa.

3.6 Tensile tests of 1038 heat-treated steel bolts show that the yield stress σ_y follows a Weibull distribution with $x_0 = 720$ MPa, $\theta = 920$, $b = 3.64$.

(a) Calculate the probability that a randomly selected bolt meets the minimum required yield stress of $\sigma_y = 900$ MPa.

(b) Graph the PDF and the CDF of the Weibull distribution of σ_y with $x_0 = 720$ MPa, $\theta = 920$, $b = 3.64$.

3.7 Experiments with bicycle wheels show that failure of the rim occurs after many thousands of kilometers and that the longevity of the wheel follows a Weibull distribution with $x_0 = 20 \cdot 10^6$ m, $\theta = 40 \cdot 10^6$, $b = 3.64$.

(a) Calculate the probability that the longevity of a randomly selected wheel exceeds $30 \cdot 10^6$ m.

(b) Calculate the probability that the longevity of a randomly selected wheel is between $25 \cdot 10^6$ m and $45 \cdot 10^6$ m.

(c) Graph the PDF and the CDF of the Weibull distribution of the longevity of the wheel with $x_0 = 20 \cdot 10^6$ m, $\theta = 40 \cdot 10^6$, $b = 3.64$.

3.8 Consider a production plant that packages concrete mix. The mix comes in bags of 50 kg, and consist of 50 weight% cement, 25 weight% sand, and 25 weight% gravel. A bag of concrete mix is filled by sequentially adding the three components to the bag and then mixing it. The weight of each component added to the bag follows a normal distribution with mean 25 kg and standard deviation 0.2 kg for cement, mean 12.5 kg and standard deviation 0.2 kg for sand, and mean 12.5 kg and standard deviation 0.4 kg for gravel. The correlation coefficient between the weights of any two materials is 0.5.

(a) Calculate the mean and standard deviation of the weight of the bags of concrete mix.
(b) The production plant considers expanding its product line by adding a concrete mix that consists of 60 weight% cement, and 40 weight% sand. This new product is packaged in bags of which the weight follows a normal distribution with mean 25 kg. The weight of the cement portion follows a normal distribution with mean 15 kg and standard deviation 0.2 kg. The correlation coefficient between the weights of cement and sand in a bag is 0.5. Graph the standard deviation of the weight of the entire bag as a function of the standard deviation of the weight of the sand, ranging between 0 and 0.5 kg.
(c) How does the answer to (b) change, when considering that the correlation coefficient between the weights of cement and sand can vary, but the standard deviation of the weight of sand remains 0.2 kg? Graph the standard deviation of the total concrete mix weight of (b) as a function of the correlation coefficient $-1 \leq C \leq 1$ between the weights of cement and sand.

4

Tolerances

4.1 Introduction

Manufacturing processes are imperfect, which inevitably leads to variations of the geometry, dimensions, and properties of a mechanical element. Thus, prescribing, controlling, and verifying the allowable variation of the design specifications of a mechanical element is important to ensure that (i) it will perform its function, and (ii) it will fit and operate together with other mechanical elements in a mechanical assembly, such as a device or a machine. In mechanical design, we explicitly control the allowable variation of design specifications by prescribing tolerances.

In this chapter, we discuss how to consider tolerances for design specifications of mechanical elements such that they fit with other elements and reliably perform their function within an assembly with other mechanical elements. We specifically focus on tolerances of dimensions of a mechanical element. Furthermore, we discuss tolerance stacks, i.e. multiple tolerances "in series" or "stacked," to understand how we determine the tolerance of a dependent variable that is a function of any number of independent variables, each with a prescribed tolerance. Finally, we describe and calculate the worst-case and statistical error of a dependent variable that relates to any number of independent variables.

4.2 Terminology

A tolerance is defined as the allowable limit of variation of a design specification. We always use the following informal *engineering rule*: "As coarse as possible, as fine as necessary." Indeed, fine or more restrictive tolerances are more expensive to manufacture than coarse or less-restrictive tolerances because they require advanced equipment and tools, more time, or operator skill. However, sometimes fine tolerances are required to ensure functionality or fit of the mechanical element.

Figure 4.1 shows a block of size L (design specification). We refer to the dimension indicated on the drawing as the nominal size. Additionally, Figure 4.1 indicates the tolerance limits, which define the allowable variation of the design specification L. The upper and lower tolerance limits bound the maximum and minimum value of the design specification, respectively. As long as the actual dimension falls within the tolerance limits, the part meets the design specification, is "acceptable," and will be able to perform its function or fit with other elements in an assembly. In the example of Figure 4.1, the upper tolerance limit is $+0.002L$, whereas the lower tolerance limit is $-0.003L$. Hence, we quantify the tolerance of a design specification as the difference between the

Design of Mechanical Elements: A Concise Introduction to Mechanical Design Considerations and Calculations,
First Edition. Bart Raeymaekers.

Companion website: www.wiley.com/go/raeymaekers/designofmechanicalelements

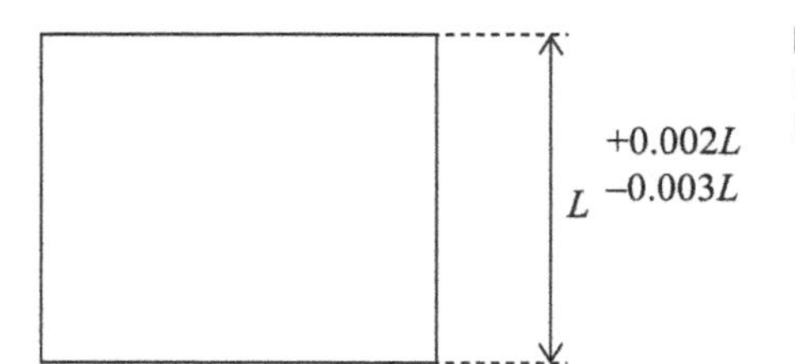

Figure 4.1 Mechanical element with nominal size L, indicating the upper ($0.002L$) and lower ($-0.003L$) tolerance limits.

upper and lower tolerance limit, i.e. $0.002L - (-0.003L) = 0.005L$, which covers its allowable limit of variation.

We define a bilateral tolerance as a tolerance of which the absolute value of the upper and lower tolerance limits is equal, e.g. $L \pm 0.002L$. In contrast, a unilateral tolerance requires the nominal size to be equal to one of the tolerance limits.

Tolerances define how two mechanical elements fit together. Figure 4.2 shows a block of size $L \pm \Delta L$ that fits into a bracket of size $Z \pm \Delta Z$. Depending on the actual dimensions of the block and the bracket, which fall within the respective tolerance limits, the fit between these two elements may change. Specifically,

- If $Z - \Delta Z > L + \Delta L$, the minimum size of the bracket is larger than the maximum size of the block. Thus, we obtain a *clearance fit* because in this scenario, the block is always smaller than the bracket, and an air gap always exists between both mechanical elements.
- If $Z + \Delta Z < L - \Delta L$, the maximum size of the bracket is smaller than the minimum size of the block. Thus, we obtain an *interference fit* because in this scenario, the block is always larger than the bracket, and an air gap never exists between both mechanical elements. An interference fit is sometimes also referred to as a "press fit" because assembling two mechanical elements with an interference fit requires exerting force to "press" them together. Large interference fits sometimes even require heating one element and using thermal expansion to facilitate assembly.
- Anything in between the two previous scenarios results in an *intermediate* or *transition fit*, i.e. until we measure both mechanical elements, we cannot predict whether their assembly will result in a slight interference or slight clearance fit. We use a transition fit when accuracy of location is important, but a small amount of clearance or interference is permissible.

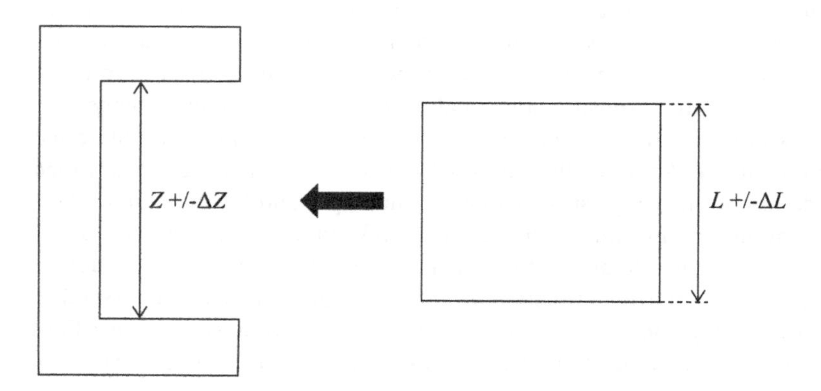

Figure 4.2 Two mechanical elements that fit together in an assembly. The fit between the mechanical elements depends on their actual dimensions, which are bound by tolerance limits.

4.3 Preferred Fits and Tolerances

4.3.1 ISO 286 Method

We may define any tolerance for a design specification. However, this leads to an infinite number of possibilities, and many different combinations of tolerances to obtain the same fit between two mechanical elements. Alternatively, using a standard way of describing tolerances is beneficial because it decreases the number of possible outcomes of the design process and eliminates redundant tolerances that lead to the same fit between two mechanical elements. The ISO 286 standard[1] describes a method to define tolerances for dimensions of mechanical elements in a methodical fashion.

Figure 4.3 illustrates the ISO 286 concept. We show the nominal size as a horizontal dash-dot line, and we represent the tolerance as a box. We define a tolerance by means of (i) the size of the tolerance, i.e. the difference between the upper and lower tolerance limit, represented by the (vertical) size of the box (double arrows in Figure 4.3), and (ii) the location of the tolerance with respect to the nominal size, represented by the orthogonal distance between the nominal size and the location of the box (single arrows in Figure 4.3). We show three example tolerances, numbered 1, 2, and 3. Tolerance 1 straddles the nominal size, whereas tolerance 2 and 3 are located above and below the nominal size, respectively. While intuitively, we might expect the nominal size to always be in between the upper and lower tolerance limits, this is not the case. Often, we use a round number for the nominal size of a dimension of a mechanical element, but we may choose both tolerance limits either larger or smaller than the nominal size to obtain the desired fit with another mechanical element.

The ISO method assigns an *IT-grade*, indicated by an integer number, to define the size of the tolerance. The size of the tolerance increases with increasing IT-grade. The ISO method also uses a letter to define the location of the tolerance with respect to the nominal size. Furthermore, it uses a lowercase letter when referring to the location of a tolerance of an external dimension, such as a shaft diameter, but it uses a capital letter when referring to the location of a tolerance of an internal dimension, such as the diameter of a hole. By definition, h has an upper tolerance limit equal to zero, i.e. it coincides with the nominal size, whereas H has a lower tolerance limit equal to zero, i.e. it coincides with the nominal size. Figure 4.3 illustrates both h and H tolerances, and their location with respect to the nominal size.

The IT-grade values are tabulated, typically in different columns, with rows representing different brackets of nominal sizes (see Table 4.1). An IT-grade represents the size of the tolerance, which also depends on the nominal size of the dimension because, for instance, a 1 μm tolerance with respect to a 1 mm nominal size is much less accurate than that same tolerance with respect to a 1 m nominal size. Thus, we could consider the absolute value of the IT-grade a "constant level of accuracy" across different nominal sizes.

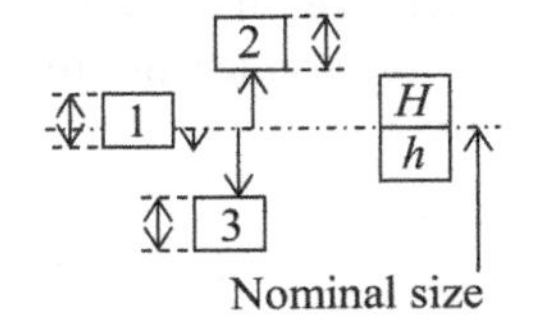

Figure 4.3 ISO 286 tolerance method, showing the size of the tolerance (double arrows) and the location of the tolerance with respect to the nominal size (single arrows).

1 See ISO 286-1:2010. *Geometrical Product Specifications (GPS) – ISO Code System for Tolerances on Linear Sizes – Part 1: Basis of Tolerances, Deviations and Fits*. International Organization for Standardization.

Table 4.1 International tolerance (IT) grades for metric dimensions in (mm).

Nominal	IT6	IT7	IT8	IT9	IT10	IT11
0–3	0.006	0.010	0.014	0.025	0.040	0.060
3–6	0.008	0.012	0.018	0.030	0.048	0.075
6–10	0.009	0.015	0.022	0.036	0.058	0.090
10–18	0.011	0.018	0.027	0.043	0.070	0.110
18–30	0.013	0.021	0.033	0.052	0.084	0.130
30–50	0.016	0.025	0.039	0.062	0.100	0.160
50–80	0.019	0.030	0.046	0.074	0.120	0.190
80–120	0.022	0.035	0.054	0.087	0.140	0.220
120–180	0.025	0.040	0.063	0.100	0.160	0.250
180–250	0.029	0.046	0.072	0.115	0.185	0.290
250–315	0.032	0.052	0.081	0.130	0.210	0.320
315–400	0.036	0.057	0.089	0.140	0.230	0.360

Size ranges are for *over* the lower limit and *including* the upper limit.
Source: Modified from ISO 286-1:2010. *Geometrical Product Specifications (GPS) – ISO Code System for Tolerances on Linear Sizes – Part 1: Basis of Tolerances, Deviations and Fits.* International Organization for Standardization.

Similarly, the values that correspond to letters describing the location of the tolerance with respect to the nominal size are also tabulated. Figure 4.4 illustrates that letters refer to different locations of the tolerance with respect to the nominal size. First, Figure 4.4 shows the h and H tolerances as previously defined. Lowercase letters (tolerances for external dimensions) start from a below the nominal size and continue to z above the nominal size. They cross over the nominal size with tolerance h. Tolerance j, js, and k straddle the nominal size, whereas every other tolerance is located either above or below the nominal size. Uppercase letters (tolerances for internal dimensions) start from A above the nominal size and continue to Z below the nominal size. They cross over the nominal size with tolerance H. Tolerance J, JS, and K straddle the nominal size, whereas every other tolerance is located either above or below the nominal size. Note that the tolerance letter always indicates the location of the tolerance limit that is closest to the nominal size. Adding or subtracting the size of the tolerance, defined by the IT-grade, determines the tolerance limit farthest from the nominal size.

Table 4.2 shows the values corresponding to each letter in different columns, and for different brackets of nominal sizes in different rows. Note that j/J and js/JS are not included in Table 4.2. We account for the nominal size for the same reason we discussed in the context of the IT-grade. The value in Table 4.2 for each letter is the distance between the nominal size and the closest tolerance limit. Thus, from Figure 4.4, we observe that for tolerances A–H and k–z the table shows the lower tolerance limit, whereas for a–h and K–Z it shows the upper-tolerance limit. The absolute

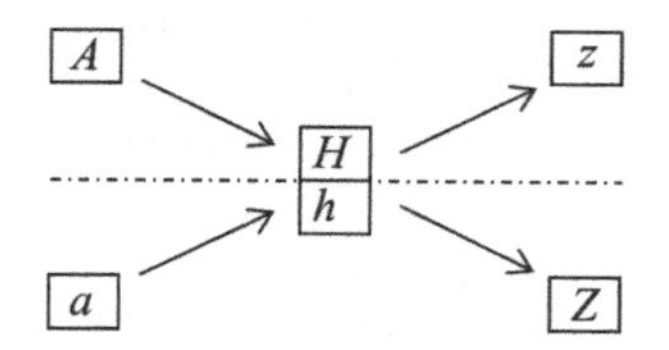

Figure 4.4 ISO 286 tolerance method, showing the locations of h and H tolerances, and the locations of other tolerances with respect to the nominal size. We use this diagram to determine graphically whether any specific tolerance is located above or below the nominal size.

Table 4.2 Fundamental deviations for metric dimensions in (mm).

Nominal	c/C	d/D	f/F	g/G	h/H	k/K	n/N	p/P	s/S	u/U
0–3	0.060	0.020	0.006	0.002	0	0	0.004	0.006	0.014	0.018
3–6	0.070	0.030	0.010	0.004	0	0.001	0.008	0.012	0.019	0.023
6–10	0.080	0.040	0.013	0.005	0	0.001	0.010	0.015	0.023	0.028
10–14	0.095	0.050	0.016	0.006	0	0.001	0.012	0.018	0.028	0.033
14–18	0.095	0.050	0.016	0.006	0	0.001	0.012	0.018	0.028	0.033
18–24	0.110	0.065	0.020	0.007	0	0.002	0.015	0.022	0.035	0.041
24–30	0.110	0.065	0.020	0.007	0	0.002	0.015	0.022	0.035	0.048
30–40	0.120	0.080	0.025	0.009	0	0.002	0.017	0.026	0.043	0.060
40–50	0.130	0.080	0.025	0.009	0	0.002	0.017	0.026	0.043	0.070
50–65	0.140	0.100	0.030	0.010	0	0.002	0.020	0.032	0.053	0.087
65–80	0.150	0.100	0.030	0.010	0	0.002	0.020	0.032	0.059	0.102
80–100	0.170	0.120	0.036	0.012	0	0.003	0.023	0.037	0.071	0.124
100–120	0.180	0.120	0.036	0.012	0	0.003	0.023	0.037	0.079	0.144
120–140	0.200	0.145	0.043	0.014	0	0.003	0.027	0.043	0.092	0.170
140–160	0.210	0.145	0.043	0.014	0	0.003	0.027	0.043	0.100	0.190
160–180	0.230	0.145	0.043	0.014	0	0.003	0.027	0.043	0.108	0.210
180–200	0.240	0.170	0.050	0.015	0	0.004	0.031	0.050	0.122	0.236
200–225	0.260	0.170	0.050	0.015	0	0.004	0.031	0.050	0.130	0.258
225–250	0.280	0.170	0.050	0.015	0	0.004	0.031	0.050	0.140	0.284
250–280	0.300	0.190	0.056	0.017	0	0.004	0.034	0.056	0.158	0.315
280–315	0.330	0.190	0.056	0.017	0	0.004	0.034	0.056	0.170	0.350
315–355	0.360	0.210	0.062	0.018	0	0.004	0.037	0.062	0.190	0.390
355–400	0.400	0.210	0.062	0.018	0	0.004	0.037	0.062	0.208	0.435

Size ranges are for *over* the lower limit and *including* the upper limit.
Source: Modified from ISO 286-1:2010. *Geometrical Product Specifications (GPS) – ISO Code System for Tolerances on Linear Sizes – Part 1: Basis of Tolerances, Deviations and Fits*. International Organization for Standardization.

values of the locations of the tolerance is identical for upper and lowercase letters. Thus, by considering Figure 4.4, we only must know the absolute value of the location of the tolerance, as we can determine graphically whether this tolerance will be located above the nominal size (positive tolerance limits), or below the nominal size (negative tolerance limits).

Example problem 4.1
Consider a shaft with a 45 mm nominal diameter. Determine the upper and lower limit of the following tolerance: $45h6$ mm. The shaft fits in a hole with the same nominal diameter. Determine the upper and lower limit of the following tolerances of the hole, and the type of fit with the shaft:

(a) $45H6$
(b) $45C6$
(c) $45S9$

Calculate the minimum and maximum clearance between $45h6/C6$.

Solution

45*h*6: We know that *h* refers to a shaft because *h* is a lowercase letter, and that the upper tolerance limit coincides with the nominal size because of the definition of *h*. Thus,

Upper limit: +0.000 mm.

The IT-grade indicates the size of the tolerance, from which we determine the lower tolerance limit. The nominal size of 45 mm falls within the 30 – 50 mm bracket in the IT-grade table (see Table 4.1) and, thus, the size of the tolerance is 0.016 mm. Hence,

Lower limit: −0.016 mm

We summarize this tolerance as $45^{+0.000}_{-0.016}$ mm.

(a) **45*H*6**: We know that *H* refers to a hole because *H* is an uppercase letter, and that the lower tolerance limit coincides with the nominal size because of the definition of *H*. Thus,

Lower limit: +0.000 mm.

The IT-grade indicates the size of the tolerance from which we determine the upper tolerance limit. The nominal size of 45 mm falls within the 30–50 mm bracket in the IT-grade table (see Table 4.1) and, thus, the size of the tolerance is 0.016 mm. Hence,

Upper limit: +0.016 mm.

We summarize this tolerance as $45^{+0.016}_{+0.000}$ mm.

(b) **45*C*6**: We know that *C* refers to a hole because *C* is an uppercase letter, and that it is located above the nominal size because *C* comes before *H* in the alphabet, i.e. both tolerance limits are positive. Thus, the letter *C* indicates the distance between the lower tolerance limit and the nominal size, as the lower tolerance limit is closest to the nominal size. The nominal size of 45 mm falls within the 40–50 mm bracket in Table 4.2 and, thus,

Lower limit: +0.130 mm

The IT-grade indicates the size of the tolerance from which we determine the upper tolerance limit. The nominal size of 45 mm falls within the 30–50 mm bracket in the IT-grade table (see Table 4.1) and, thus, the size of the tolerance is 0.016 mm. Hence,

Upper limit: Lower limit + IT grade = +0.130 + 0.016 = +0.146 mm.

We summarize this tolerance as $45^{+0.146}_{+0.130}$ mm.

(c) **45*S*9**: We know that *S* refers to a hole because *S* is an uppercase letter, and that it is located below the nominal size because *S* comes after *H* in the alphabet, i.e. both tolerance limits are negative. Thus, the letter *S* indicates the distance between the upper tolerance limit and the nominal size. The nominal size of 45 mm falls within the 40–50 mm bracket in Table 4.2 and, thus,

Upper limit: −0.043 mm.

The IT-grade indicates the size of the tolerance from which we determine the lower tolerance limit. The nominal size of 45 mm falls within the 30–50 mm bracket in the IT-grade table (see Table 4.1) and, thus, the size of the tolerance is 0.062 mm. Hence,

Lower limit: Upper limit–IT grade = −0.043 − 0.062 = −0.105 mm.

We summarize this tolerance as $45^{-0.043}_{-0.105}$ mm.

Figure 4.5 schematically illustrates the different tolerances (a), (b), and (c) with respect to 45*h*6. We observe that (a) $45h6/H6$ results in a clearance or transition fit, (b) $45h6/C6$ results in a clearance fit because the respective tolerances do not overlap, which indicates that the diameter of the hole is always larger than the diameter of the shaft. Thus, an air gap always exists between the shaft and hole assembly. Finally, (c) $45h6/S9$ results in an interference fit because the diameter of the hole is always smaller than the diameter of the shaft and, thus, they always interfere when assembled.

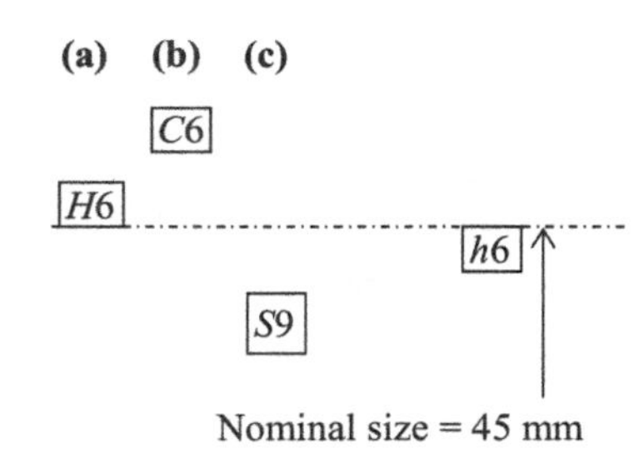

Figure 4.5 Schematic illustration of the different tolerances with respect to 45*h*6.

Finally, we calculate the minimum and maximum clearance between 45$h6/C6$.

The minimum clearance physically occurs when the hole has its smallest possible dimension and the shaft has its largest possible dimension. This will minimize the air gap between both mechanical elements.

Minimum clearance = smallest hole–largest shaft = 45.130 − 45.000 = 0.130 mm.

The maximum clearance physically occurs when the hole has its largest possible dimension, whereas the shaft has its smallest possible dimension. This will maximize the air gap between both mechanical elements.

Maximum clearance = largest hole–smallest shaft = 45.146 − 44.984 = 0.162 mm.

Note that *interference* and *clearance* are equal in magnitude but with opposite sign, i.e. a clearance of 0.162 mm is equal to an interference of −0.162 mm.

4.3.2 Unit Shaft and Unit Hole System

The ISO 286 tolerance system reduces the number of possible tolerances and tolerance combinations from an infinite to a finite number of possibilities. However, it still allows a large number of different tolerance combinations to define the fit between two mechanical elements. Furthermore, different tolerance combinations can still lead to an identical or very similar fit between two mechanical elements. To address that problem and further reduce the number of tolerance combinations that are possible (or recommended), we introduce the unit shaft and unit hole system.

In the *unit hole system*, we assign all internal dimensions (hole) a tolerance H. Then, by changing the tolerances of the external dimensions, we create the desired fit between two mechanical elements. However, fixing the tolerance of the internal dimensions drastically reduces the number of possible tolerance combinations and also unambiguously defines the tolerance combinations required to obtain a specific fit between two mechanical elements. Figure 4.6 visualizes the unit hole system.

In the *unit shaft system*, we assign all external dimensions (shaft) a tolerance h. Then, by changing the tolerances of the internal dimensions, we create the desired fit between two mechanical elements. However, fixing the tolerance of the external dimensions drastically reduces the number of possible tolerance combinations and also unambiguously defines the tolerance combinations required to obtain a specific fit between two mechanical elements. Figure 4.7 visualizes the unit shaft system.

Finally, other standards in the United States, such as the ANSI B4.2 standard,[2] define a number of preferred tolerance combinations to obtain a specific fit between two mechanical elements, thus

---

2 See ANSI B4.2: 1978(R2020). *Preferred Metric Limits and Fits.* American National Standards Institute.

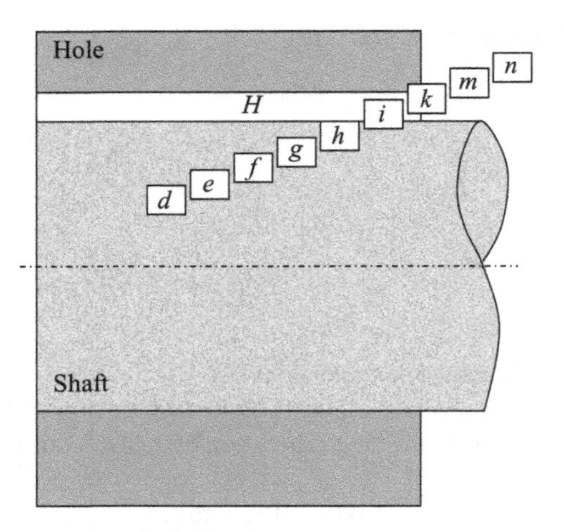

Figure 4.6 Unit hole system: we fix the tolerance of all internal dimensions to H and obtain the desired fit by changing the tolerance of the external dimension.

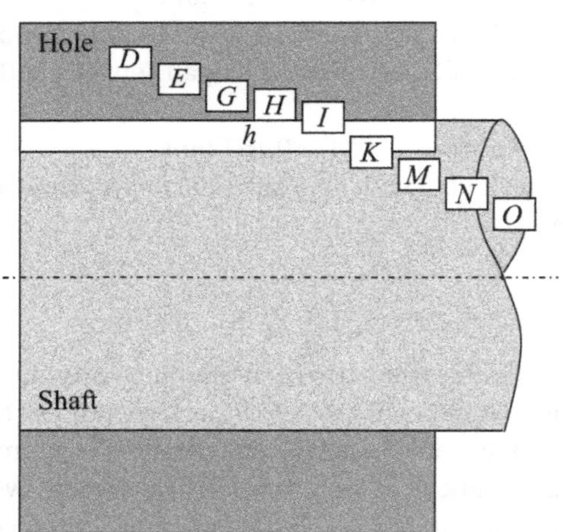

Figure 4.7 Unit shaft system: we fix the tolerance of all external dimensions to h and obtain the desired fit by changing the tolerance of the internal dimension.

even further reducing the possible tolerance combinations. These preferred combinations are the following:

- **Clearance fit**: $H11/c11, H9/d9, H8/f7, H7/g6, H7/h6$
- **Transition fit**: $H7/k6, K7/h6, N7/h6$
- **Interference fit**: $H7/p6, H7/S6, H7/u6$.

4.4 Tolerance Stacks

In mechanical design, we routinely must calculate the error or tolerance of a dimension or design specification that depends on a number of other dimensions or design specifications, each of which also have an error or a tolerance associated with it. We refer to this as *tolerance stacks* because one could think of it as individual tolerances stacked on top of each other, and we attempt to calculate the resulting tolerance of the entire stack.

We define two types of error:

1. Worst-case error
2. Statistical error.

Furthermore, we define two types of variables:

1. **Independent variables**: Variables/design specifications whose value does not depend on other variables.
2. **Dependent variables**: Variables/design specifications whose value is a function of a set of independent variables/design specifications.

In general terms, we consider the dependent variable $y = f(x_1, x_2, \dots, x_n)$, which is a function of a number of independent variables $(x_1, x_2, \dots, x_n)$, each with a tolerance associated with them, i.e. $x_i = \overline{x_i} \pm \delta x_i$. Here, $\overline{x_i}$ represents the nominal value, whereas δx_i represents a bilateral tolerance of the independent variable. We restrict this analysis to bilateral tolerances, i.e. the nominal size divides the tolerance in half.

We define the worst-case error as

$$(\Delta y)_{WCE} = \left|\frac{\delta f}{\delta x_1}\right| \delta x_1 + \left|\frac{\delta f}{\delta x_2}\right| \delta x_2 + \dots + \left|\frac{\delta f}{\delta x_n}\right| \delta x_n. \tag{4.1}$$

Correspondingly, we define the statistical error as

$$(\Delta y)_{STAT} = \left[\left(\frac{\delta f}{\delta x_1}\right)^2 \delta x_1^2 + \left(\frac{\delta f}{\delta x_2}\right)^2 \delta x_2^2 + \dots + \left(\frac{\delta f}{\delta x_n}\right)^2 \delta x_n^2\right]^{1/2}. \tag{4.2}$$

The worst-case error is often too conservative and results in costly tolerances because it implicitly assumes that all tolerance will combine or “stack” in the worst possible way, which is possible, but unlikely. In contrast, using the statistical error is less conservative and leads to less restrictive tolerances.

Example problem 4.2

Calculate the worst-case and statistical error for the following function of two variables: $C = \sqrt{A^2 + B^2}$, with $A = 3.0 \pm 0.3$, and $B = 4.0 \pm 0.4$.

Solution

We first calculate the partial derivatives of the dependent variable with respect to each independent variable.

$\delta C/\delta A = A/\sqrt{A^2 + B^2}$, and

$\delta C/\delta B = B/\sqrt{A^2 + B^2}$.

Then, using Eq. (4.1), we find

$$(\Delta C)_{WCE} = \left|\frac{A}{\sqrt{A^2+B^2}}\right| \delta A + \left|\frac{B}{\sqrt{A^2+B^2}}\right| \delta B, \text{ and,}$$

$$(\Delta C)_{WCE} = \left|\frac{3.0}{\sqrt{3.0^2+4.0^2}}\right| 0.3 + \left|\frac{4.0}{\sqrt{3.0^2+4.0^2}}\right| 0.4 = 0.50.$$

Correspondingly, using Eq. (4.2), we find

$$(\Delta C)_{STAT} = \left[\left(\frac{A}{\sqrt{A^2+B^2}}\right)^2 \delta A^2 + \left(\frac{B}{\sqrt{A^2+B^2}}\right)^2 \delta B^2\right]^{1/2}, \text{ and,}$$

$$(\Delta C)_{STAT} = \left[\left(\frac{3.0}{\sqrt{3.0^2+4.0^2}}\right)^2 0.3^2 + \left(\frac{4.0}{\sqrt{3.0^2+4.0^2}}\right)^2 0.4^2\right]^{1/2} = 0.37.$$

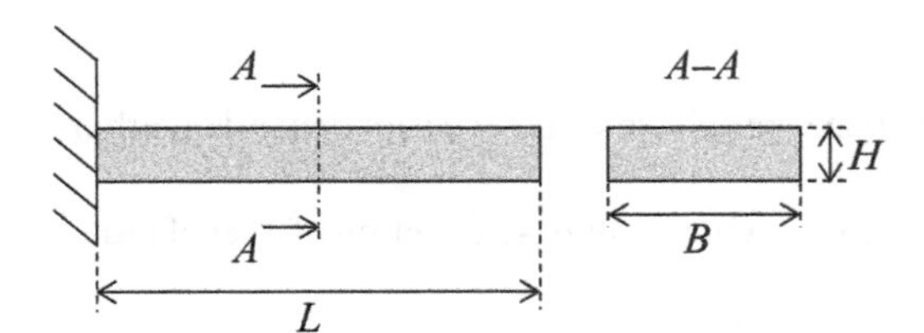

Figure 4.8 Microscale cantilever beam with rectangular cross-section.

Example problem 4.3

Figure 4.8 schematically shows a microscale silicon cantilever beam with rectangular cross-section. The first resonance frequency (bending mode) is given as $f_{res} = \frac{0.08H}{L^2}\sqrt{E/\rho}$.

$L = 3 \cdot 10^{-4} \pm 0.02 \cdot 10^{-4}$ m and $H = 2 \cdot 10^{-6} \pm 0.1 \cdot 10^{-6}$ m. The density of silicon $\rho = 2330 \pm 1$ kg/ m^3, and its elastic modulus $E_{Si} = 100$ GPa.

(a) Calculate the nominal resonance frequency of the cantilever beam.
(b) Calculate the worst-case error and the statistical error of the resonance frequency.
(c) Calculate the bilateral tolerance of H that the fabrication process must achieve to limit the statistical error of the resonance frequency to 200 Hz.

Solution

(a) $f_{res} = \frac{0.08H}{L^2}\sqrt{E/\rho} = \frac{0.08 \times 2 \cdot 10^{-6}}{(3 \cdot 10^{-4})^2}\sqrt{100 \cdot 10^9/2330} = 11646.6$ Hz.

(b) We first calculate the partial derivatives of the dependent variable with respect to each independent variable, i.e.

$$\delta f_{res}/\delta L = \frac{-0.16H}{L^3}\sqrt{E/\rho},$$

$$\delta f_{res}/\delta \rho = \frac{-0.04H}{L^2}E^{1/2}/\rho^{3/2},$$

$$\delta f_{res}/\delta H = \frac{0.08}{L^2}\sqrt{E/\rho}.$$

Then, using Eq. (4.1), we write

$$\left(\Delta f_{res}\right)_{WCE} = \left|\frac{\delta f_{res}}{\delta L}\right|\Delta L + \left|\frac{\delta f_{res}}{\delta \rho}\right|\Delta \rho + \left|\frac{\delta f_{res}}{\delta H}\right|\Delta H, \text{ and}$$

$$\left(\Delta f_{res}\right)_{WCE} = \left|-7.76 \cdot 10^7\right| \times 2 \cdot 10^{-6} + |2.50| \times 1 + \left|5.82 \cdot 10^9\right| \times 1 \cdot 10^{-7} = 740 \text{ Hz}.$$

Correspondingly, using Eq. (4.2), we write

$$\left(\Delta f_{res}\right)_{STAT} = \left[\left(\frac{\delta f_{res}}{\delta L}\right)^2\Delta L^2 + \left(\frac{\delta f_{res}}{\delta \rho}\right)^2\Delta \rho^2 + \left(\frac{\delta f_{res}}{\delta H}\right)^2\Delta H^2\right]^{1/2}, \text{ and}$$

$$\left(\Delta f_{res}\right)_{STAT} = [(-7.76 \cdot 10^7)^2 \times (2 \cdot 10^{-6})^2 + (2.5)^2 \times (1)^2 + (5.82 \cdot 10^9)^2 \times (1 \cdot 10^{-7})^2]^{1/2} = 602 \text{ Hz}.$$

$$\text{(c) } \Delta H = \left[\left(\Delta f_{res}\right)^2_{STAT} - \left(\frac{\delta f_{res}}{\delta L}\right)^2\Delta L^2 - \left(\frac{\delta f_{res}}{\delta \rho}\right)^2\Delta \rho^2\right]^{1/2}\left(\frac{\delta f_{res}}{\delta H}\right)^{-1} = 4.33 \cdot 10^{-8} \text{ m}.$$

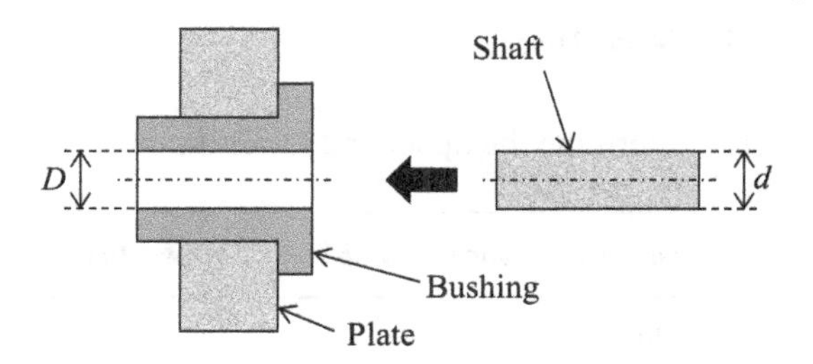

Figure 4.9 Shaft and bushing assembly.

Example problem 4.4

Figure 4.9 shows a blank drawn shaft with diameter 10h9 mm. Using a micrometer gauge, we measure its exact diameter as $d = 9.985$ mm. The shaft fits in a bushing pressed into a metal plate. The inner diameter of the bushing is $D = 10N7$. Assume the actual inner diameter of the bushing is a random variable that follows a uniform distribution between the upper and lower limit of the tolerance of D. Calculate the likelihood that the assembly of these two elements results in a clearance fit?

Solution

10N7: We know that N refers to a hole because N is a capital letter and that it is located below the nominal size because N comes after H in the alphabet, i.e. both tolerance limits are negative. Thus, the letter indicates the distance between the upper tolerance limit and the nominal size. The nominal size of 10 mm falls within the 6–10 mm bracket in Table 4.2 and, thus,

Upper limit: −0.010 mm

The IT-grade indicates the size of the tolerance from which we determine the lower tolerance limit. The nominal size of 10 mm falls within the 6–10 mm bracket in the IT-grade table (see Table 4.1) and, thus, the size of the tolerance is 0.015 mm. Hence,

Lower limit: Upper limit–IT grade $= -0.010 - 0.015 = -0.025$ mm.

We summarize this tolerance as $10^{-0.025}_{-0.010}$ mm.

We define random variable D, diameter of the bushing.

A clearance fit requires the diameter of the bushing to be strictly larger than the diameter of the shaft. Thus,

$p(D > 9.985) = 1 - p(D \leq 9.985) = 1 - (9.985 - 9.975)/(9.990 - 9.975)$,

$p(D > 9.985) = 0.333 \cong 33.333\%$.

4.5 Key Takeaways

1. We use tolerances to define the allowable limits of variation of design specifications.
2. We use the statistical and worst-case error to determine the tolerance of a dependent variable that is a function of a number of independent variables.
3. When a tolerance describes the allowable limit of variation of a dimension, then combining tolerances of two mechanical elements allows tailoring the fit of those mechanical elements, including clearance, interference, or transition fits.
4. ISO and ANSI standards help reducing the infinite number of tolerance combinations to a finite number of preferred tolerance combinations and fits.

4.6 Problems

4.1 Determine the upper and lower limits of the following tolerances:

Nominal size and tolerance	Lower limit	Upper limit
$140h6$		
$15p6$		
$260H7$		
$260n8$		
$40D9$		
$60h6$		
$80H7$		
$15P7$		
$130S8$		
$65H11$		
$50c9$		

4.2 Determine the minimum and maximum clearance of the following tolerance pairs, and identify the resulting fit:

Tolerance pair	Minimum clearance	Maximum clearance	Fit
$70H7/u6$			
$150K7/h6$			
$20H7/h6$			
$75H8/f7$			
$150H7/n6$			
$40P7/h6$			

4.3 Figure 4.10 shows four mechanical elements that fit together in an assembly, with
$a = 30 \pm 0.01$ mm
$b = 40 \pm 0.02$ mm
$c = 50 \pm 0.02$ mm

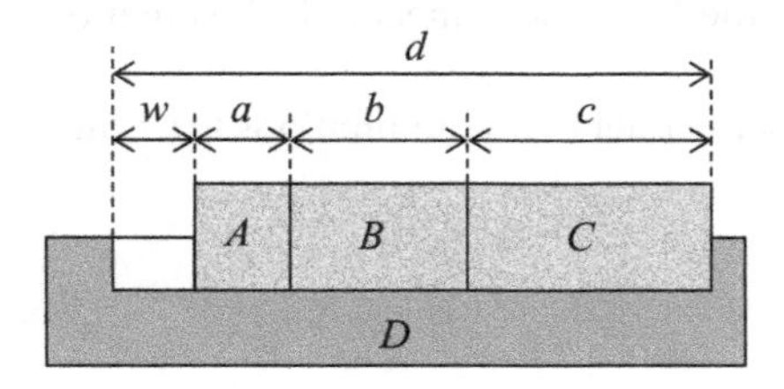

Figure 4.10 Several elements fitting together in an assembly.

$d = 141 \pm 0.02$ mm

(a) Determine the nominal value of the gap $\overline{w}$ and its bilateral tolerance Δw.

(b) Calculate the worst-case and statistical error of the gap w.

4.4 The acceleration of a swinging pendulum $g = 4\pi^2 L/T^2$ depends on a measurement of the length of the pendulum L and its swinging period T.

(a) Determine the worst-case and statistical error of g if the length measurement results in 100 ± 1 mm and the time measurement is 3.5 ± 0.1 seconds.

(b) Express the worst-case and statistical error as a fractional error by normalizing the error with the nominal value of g. Determine whether to invest in a more accurate length or time measurement device to reduce the fractional error of the acceleration.

4.5 Figure 4.11 shows a microscale silicon (Si) cantilever beam with rectangular cross-section. The first resonance frequency (bending mode) of the cantilever beam is $f_{res} = \frac{0.08}{L^2}\sqrt{\frac{E}{\rho}}$. The elastic modulus $E_{Si} = 100$ GPa. Table 4.3 shows the parameters that determine the resonance frequency of the cantilever beam and their corresponding tolerance.

Figure 4.11 Cantilever beam with rectangular cross-section.

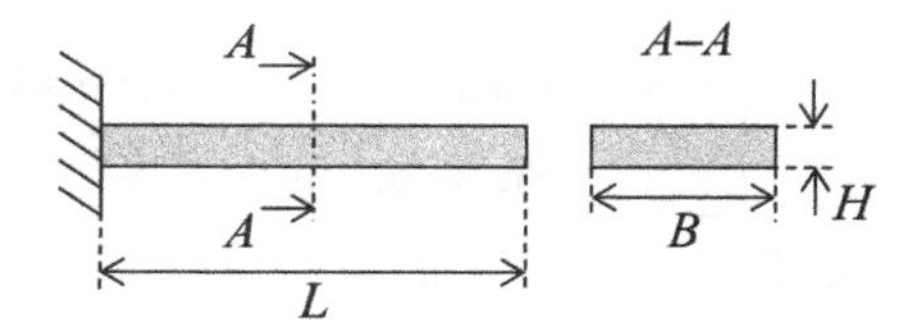

Table 4.3 Cantilever parameters.

Parameter	Symbol	Nominal value	Bilateral tolerance
Length	L (m)	$3 \cdot 10^{-4}$	$1.0 \cdot 10^{-6}$
Height	H (m)	$3 \cdot 10^{-6}$	$0.2 \cdot 10^{-6}$
Density	ρ (kg/m^3)	2330	3

(a) Calculate the nominal resonance frequency of the cantilever beam.

(b) Calculate the worst-case and statistical error of f_{res}.

(c) Calculate the bilateral tolerance of H, required to constrain the statistical error of the resonance frequency to 200 Hz.

4.6 Figure 4.12 shows a microscale aluminum (Al) cantilever beam with circular cross-section. The first resonance frequency (bending mode) of the cantilever beam is $f_{res} = \frac{0.56}{L^2}\sqrt{\frac{ED^2}{16\rho}}$. The elastic modulus $E_{Al} = 80$ GPa. Table 4.4 shows the parameters that determine the resonance frequency of the cantilever beam, and their corresponding tolerance.

(a) Calculate the nominal resonance frequency of the cantilever beam.

(b) Calculate the worst-case and statistical error of f_{res}.

(c) Calculate the bilateral tolerance of H, required to constrain the statistical error of the resonance frequency to 400 Hz.

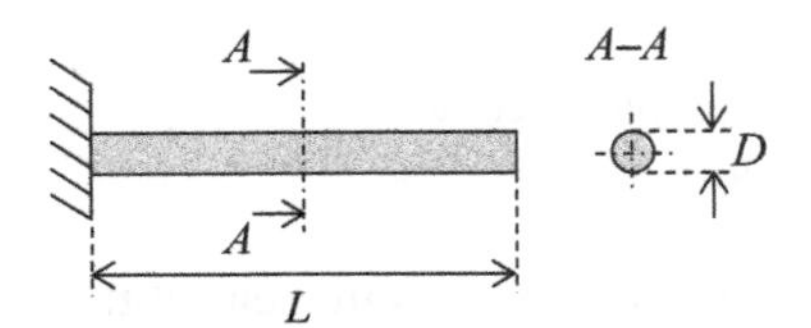

Figure 4.12 Cantilever beam with circular cross-section.

Table 4.4 Cantilever parameters.

Parameter	Symbol	Nominal value	Bilateral tolerance
Length	L (m)	$1 \cdot 10^{-2}$	$2.0 \cdot 10^{-4}$
Diameter	D (m)	$1 \cdot 10^{-3}$	$0.1 \cdot 10^{-4}$
Density	ρ (kg/m^3)	2700	1

4.7 Figure 4.13 shows a shaft that runs through a sleeve bearing pressed into a frame. The tolerance of the outer diameter of the sleeve bearing is 20*n*6, and the tolerance of the hole diameter is 20*H*7.

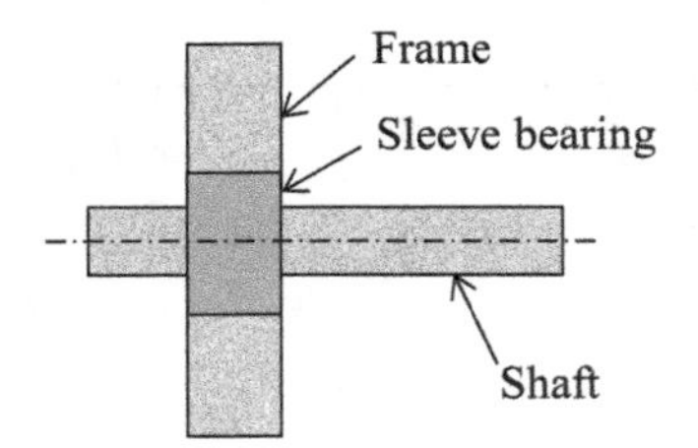

Figure 4.13 Shaft that runs through a sleeve bearing pressed into a frame.

(a) Determine the minimum and maximum clearance between the outer diameter of the sleeve bearing and the hole in the frame.
(b) Determine the fit between the outside diameter of the sleeve bearing and the hole in the frame.
(c) Determine the probability that an interference fit occurs between the outer diameter of the bearing and the hole in the frame, if the outer diameter of the bearing follows a uniform distribution between the lower and upper limit of the tolerance. The actual diameter of the hole in the frame measures 20.016 mm.
(d) How does the answer to (c) change if the outer diameter of the bearing follows a normal distribution with mean 19.995 mm and standard deviation 0.015 mm?

4.8 Figure 4.14 shows a bushing that fits into a hole in a bracket. The tolerance of the OD of the bushing is 30*s*6. The actual diameter of the hole, determined using a micrometer gauge, is

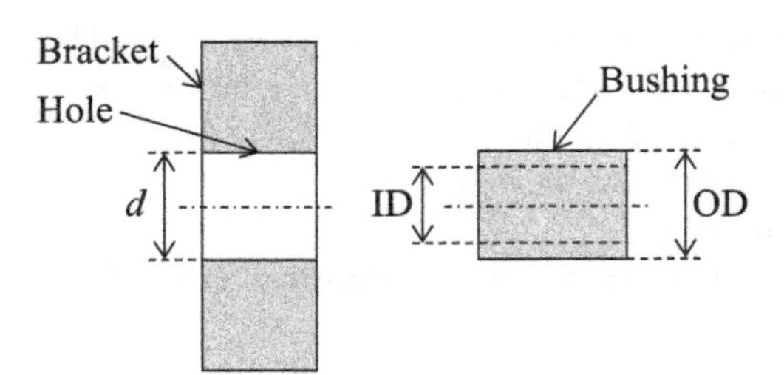

Figure 4.14 Bushing pressed into a bracket.

$d = 30.036$ mm. Calculate the probability of an interference fit between the OD of the bushing and the hole, if the OD of the bushing follows a uniform distribution between the tolerance limits.

4.9 Consider a shaft with diameter $d = 25h6$ mm. Its exact diameter, determined with a micrometer gauge, is $d = 24.993$ mm, which falls within the tolerance limits. The shaft fits in a bore with diameter $D = 25K7$ mm. D is uniformly distributed between the upper and lower limit of the tolerance. Calculate the probability of a press fit between these two elements.

5

Design for Static Strength

5.1 Introduction

We define a static load as an external force or moment that remains constant as a function of time. This contrasts a dynamic load, which changes as a function of time.

When designing or verifying a mechanical element for static strength, we analyze the physical connection between the mechanical element and its environment and replace this connection with boundary conditions and reaction forces. Together, the external forces and the reaction forces describe the free-body diagram of the mechanical element. From the free-body diagram, we determine the internal forces in the mechanical element as an axial force, a shear force, a bending moment, and a torque moment diagram. The internal force diagrams display each internal force component as a function of the location in the mechanical element and, thus, they enable determining which section of the mechanical element is subject to the highest load. Subsequently, when considering the internal force diagrams together with the (local) geometry of the mechanical element, we convert internal forces to internal stresses. Figure 5.1 schematically shows that we relate the external forces acting on a mechanical element to the corresponding internal forces by means of the free-body diagram, and subsequently to the internal stress by considering the local geometry of the mechanical element.

Ultimately, we compare the internal stress to the yield stress (ductile materials) or the ultimate tensile stress (brittle materials) to determine whether the external load causes failure of the mechanical element. Alternatively, if the geometry of the mechanical element is unknown, we use this approach to determine an unknown dimension of a mechanical element, by prescribing a design factor to which the design must adhere. Thus, we define two different types of calculations.

1. **Design calculation:** We design an unknown dimension of the mechanical element to meet a prescribed design factor n_D, considering the external loading and boundary conditions.
2. **Verification calculation:** We determine the safety factor n of the mechanical design, considering the dimensions of the mechanical element, which are known *a priori*, and the external loading and boundary conditions.

Design of Mechanical Elements: A Concise Introduction to Mechanical Design Considerations and Calculations, First Edition. Bart Raeymaekers.

Companion website: www.wiley.com/go/raeymaekers/designofmechanicalelements

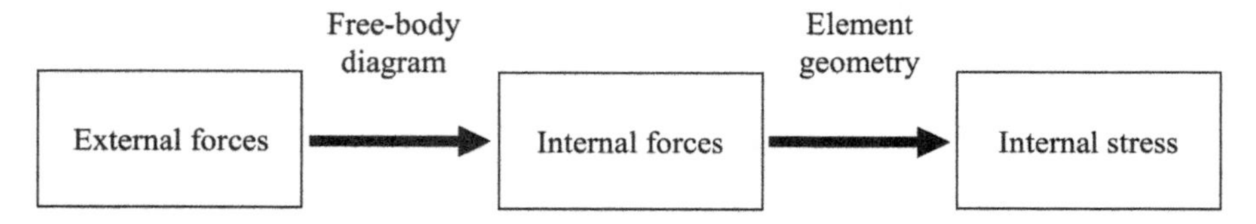

Figure 5.1 Relating external forces to internal forces and stress.

5.2 Simple Loading

We briefly review the basic concepts and equations that describe different types of simple, static loading.[1] Here, we refer to "simple" as situations in which the mechanical element is subject to only one type of static load.

5.2.1 Axial Loading

Figure 5.2 illustrates a simple axial load, which results in a uniform (a) normal stress σ or (b) shear stress τ. We calculate the normal stress as

$$\sigma = \frac{F}{A}, \tag{5.1}$$

with $A = b^2$ for the square cross-section with side b of Figure 5.2a. We calculate the shear stress as

$$\tau = \frac{F}{A}, \tag{5.2}$$

with A the cross-sectional area of the element subject to the shear force F.

Furthermore, stress and strain are linearly related through Hooke's law for small, elastic deformations, i.e.

$$\sigma = E\varepsilon, \tag{5.3}$$

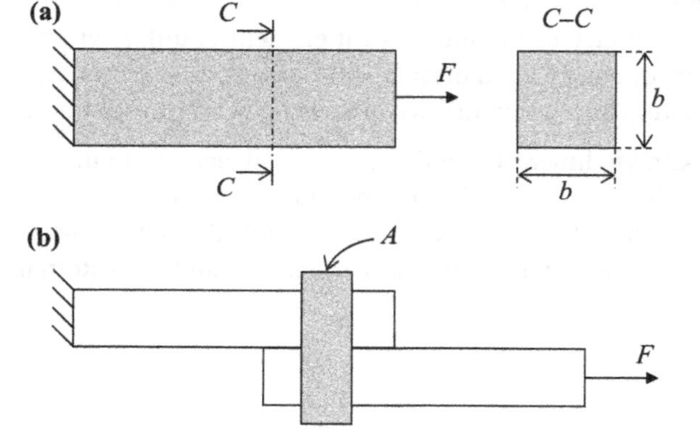

Figure 5.2 Simple axial loading, showing uniform (a) normal and (b) shear stress.

1 For a comprehensive presentation see, e.g. Beer, F.P. and Johnston, E.R. Jr. (2014). *Mechanics of Materials*, 7e. McGraw-Hill.

where E is the elastic modulus. Equivalently, for shear stress,

$$\tau = G\gamma, \tag{5.4}$$

where γ is the shear strain and G is the shear modulus. We also define the Poisson ratio as

$$\nu = -\frac{\varepsilon_{lateral}}{\varepsilon_{axial}}, \tag{5.5}$$

i.e. the ratio of the strain in the lateral and axial directions. Indeed, under tensile axial strain, the lateral strain is negative (for most ductile materials), and vice versa, because volume must be conserved. To avoid that the Poisson coefficient would be a negative number, we use a negative sign in front of the strain ratio. Finally, the three elastic constants of a material E, G, and ν relate to each other as

$$E = 2G(1 + \nu). \tag{5.6}$$

5.2.2 Bending

Figure 5.3 illustrates simple bending, which results in a normal stress. Here, we only discuss "Euler–Bernouilli" or "thin-beam" theory. Thus, we only consider long, thin beams, subject to small, elastic deformation resulting from bending, in which plane cross-sections remain plane because the shear stress resulting from bending is negligibly small.

The normal stress resulting from bending is a function of the orthogonal distance from the neutral axis. The neutral axis is the location that experiences zero stress and strain from bending, and it runs through the centroid of the mechanical element. We calculate the normal stress from bending using the *flexure formula*, i.e.

$$\sigma = -\frac{My}{I_{zz}}, \tag{5.7}$$

where M is the bending moment, y is the distance from the neutral axis in the y-direction (orthogonal to the neutral axis, see Figure 5.3), and I_{zz} is the area moment of the cross-section of the mechanical element around its bending axis (z in Figure 5.3 is oriented orthogonal to the page). Equation (5.7) shows that the magnitude of the normal stress changes linearly with respect to the orthogonal distance from the neutral axis y, which we illustrate in the inset image of Figure 5.3. Note that material on one side of the neutral axis will experience tensile stress, whereas the other side will experience compressive stress. We determine tension and compression from the sign in

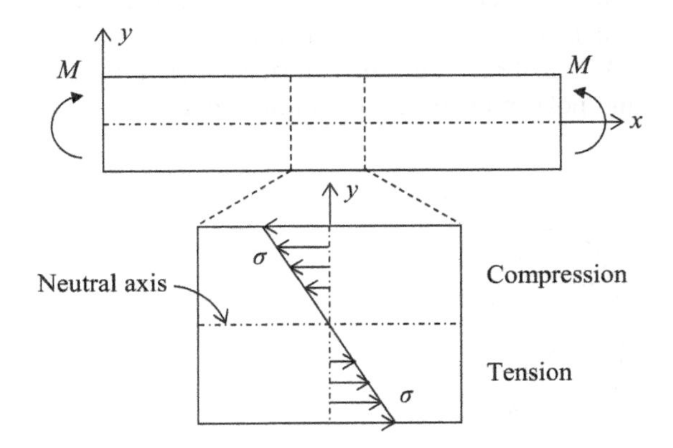

Figure 5.3 Uniaxial bending of a prismatic beam, showing the normal stress and its linear relationship to the orthogonal distance from the neutral axis in the inset image.

front of the stress value resulting from the flexure formula, or from a physical interpretation of the bending moment.

Thus, the maximum tensile and compressive stress occur on the surface of the mechanical element, where the distance from the neutral axis is maximum, i.e. we calculate

$$\sigma_{max} = \frac{My_{max}}{I_{zz}}. \tag{5.8}$$

We point out that this concept also applies to beams with a nonsymmetric cross-section. The neutral axis always runs through the centroid of the cross-section of the beam and, thus, while still creating a normal stress in the beam that changes linearly with the orthogonal distance from the neutral axis, the portion of the beam that is subject to tensile and compressive stress might not be equal, as opposed to in a beam with a symmetric cross-section.

5.2.3 Torsion

Figure 5.4 illustrates simple torsion, which results in a shear stress. A torque moment M_t is applied to a cylinder with diameter $2R$ and length L. The torque moment twists the cylinder around its center axis. The angle of twist ψ that results from the torque moment is given as

$$\psi = \frac{M_t L}{G I_p}, \tag{5.9}$$

with G the shear modulus and I_p the polar moment of inertia around the center axis of the cylinder. We observe that the angle of twist is proportional to the torque moment and the length of the cylinder. Indeed, if we increase the torque moment applied to the cylinder, it increases the angle over which the cylinder twists. Additionally, the twist is amplified by the length of the cylinder. In contrast, the angle of twist is inversely related to the shear modulus, which expresses the stiffness of the material in the shear direction, and the polar moment of inertia, which represents the resistance of the cylinder against torsion.

We calculate the shear stress resulting from the torque moment M_t at any point in the cylinder as

$$\tau = \frac{M_t r}{I_p}, \tag{5.10}$$

where r represents the radial coordinate with origin at the center of the cylinder (see Figure 5.4). The shear stress increases with increasing distance from the center axis of the cylinder, where $\tau = 0$, and with increasing torque moment. It is inversely proportional to the polar moment of inertia, which again represents the resistance of the cylinder against torsion.

When designing mechanical elements, we routinely calculate the area moment of inertia and polar moment of inertia of a solid and hollow cylinder, as many shaft designs involve such

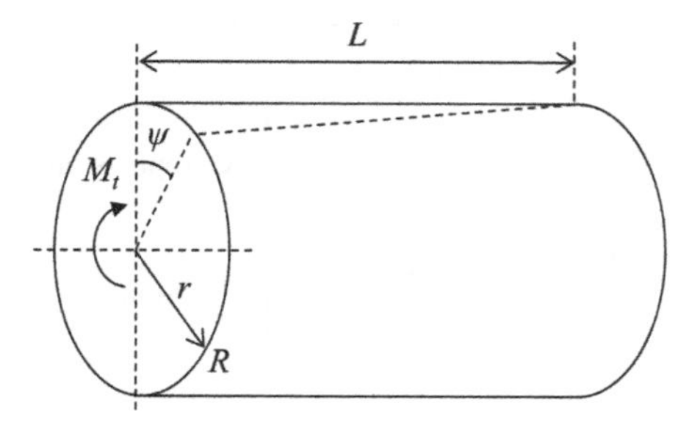

Figure 5.4 Torsion of a cylinder, illustrating the angle of twist, and the different parameters that are important in the context of torsion.

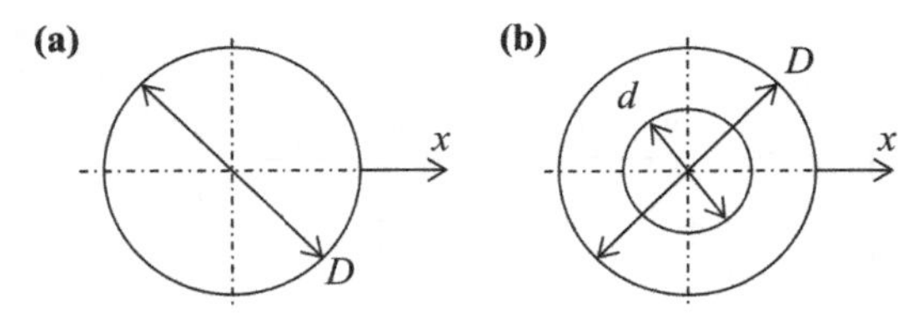

Figure 5.5 (a) Solid and (b) hollow cylinder cross-section.

geometries. Therefore, we illustrate these two geometries in Figure 5.5, showing (a) a solid cylinder with diameter D, and (b) a hollow cylinder with outer diameter D and inner diameter d. The area moment of inertia around the x-axis for the hollow cylinder is given as

$$I_{xx} = \frac{\pi\left(D^4 - d^4\right)}{64}, \tag{5.11}$$

which reduces to

$$I_{xx} = \frac{\pi D^4}{64}, \tag{5.12}$$

when $d = 0$ for a solid cylinder. Similarly, the polar moment of inertia of a hollow cylinder is given as

$$I_o = \frac{\pi\left(D^4 - d^4\right)}{32}, \tag{5.13}$$

which reduces to

$$I_o = \frac{\pi D^4}{32}, \tag{5.14}$$

when $d = 0$ for a solid cylinder.

5.3 Stress Concentrations

The equations and concepts in Section 5.2 apply to mechanical elements (or parts thereof) without geometric irregularities or discontinuities. In reality, this is difficult because mechanical elements, such as shafts, have shoulders to seat bearings, or slots to seat keys for pulleys, gears, and other components that mount on the shaft. Irregularities could also include oil grooves, notches, or any other local geometry change of the mechanical element.

An irregularity or discontinuity in the geometry of the mechanical element alters the local stress in the immediate neighborhood of that discontinuity. Thus, elementary stress equations like those discussed in the previous section, no longer accurately describe the local stress. To account for the stress around a discontinuity, we use a stress concentration factor.

A stress concentration factor relates the actual maximum stress σ_{max} or τ_{max} at the discontinuity, to the nominal stress σ_0 or τ_0 away from the discontinuity (or when the discontinuity would not be present). Thus, we define the *theoretical stress concentration factor* for a normal stress as

$$K_t = \frac{\sigma_{max}}{\sigma_0}, \tag{5.15}$$

and the *theoretical shear stress concentration factor* for a shear stress as

$$K_{ts} = \frac{\tau_{max}}{\tau_0}. \tag{5.16}$$

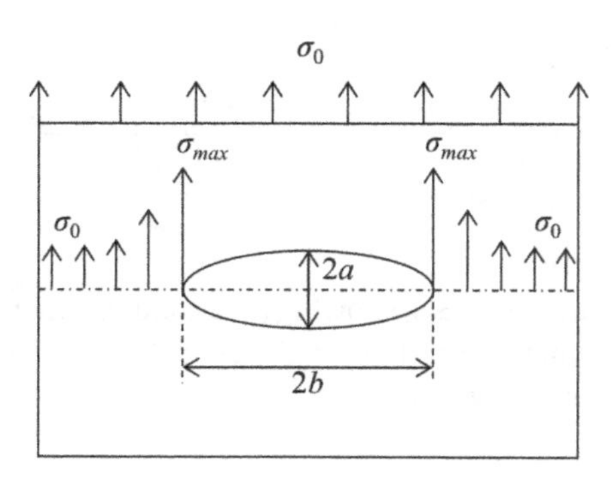

Figure 5.6 Stress concentration around an elliptical hole in a plate, with a uniform axial load.

It is important to emphasize that the stress concentration factors are only dependent on the geometry of the mechanical element, specifically the geometry of the discontinuity, and the type of loading (axial, bending, or torsion). However, the stress concentration factor is independent of the material type.

A typical example of a discontinuity in a mechanical element that causes a stress concentration is an elliptical hole in a plate. Figure 5.6 illustrates this geometry subject to a uniform axial load. If the elliptical hole would not exist, the uniform axial load would create a nominal stress σ_0 in the plate. However, the elliptical hole creates a discontinuity, which leads to a stress concentration around the hole. The maximum stress σ_{max} occurs immediately adjacent to the discontinuity and decreases to the nominal stress σ_0 away from the discontinuity, as schematically illustrated by the stress vectors in Figure 5.6. We can show that for this specific geometry and type of loading, the maximum normal stress is given as

$$\sigma_{max} = \sigma_0 \left(1 + \frac{2b}{a}\right). \tag{5.17}$$

In the limit case that the elliptical hole approaches a circular hole, i.e. when $a = b$, Eq. (5.17) reduces to

$$\sigma_{max} = 3\sigma_0. \tag{5.18}$$

Thus, a circular hole in a plate (e.g. to accommodate a bolt) increases the local stress around that hole by a factor 3 compared to the nominal stress in the plate, if the hole would not exist. Thus, the theoretical stress concentration factor $K_t = \sigma_{max}/\sigma_0 = 3$.

Several reference works exist that tabulate stress concentration factors. Perhaps the most common reference work is *Peterson's stress concentration factors*.[2] Experimentally determining stress concentration factors is time-consuming. However, several methods exist, including photoelasticity, where translucent models of the discontinuity geometry we evaluate, in conjunction with interferometry, allow visualizing and quantifying the local stress around the discontinuity. Furthermore, we can use strain gauges to quantify stress concentrations. However, strain gauges can be too large for some mechanical elements. Spray paints that are sensitive to mechanical stress also exist. We apply the spray paint to a model of the mechanical element before loading the element. The spray paint discolors according to the local stress magnitude, which we can then decipher using a color scale. Finally, finite element simulations provide an economical alternative to determining stress concentrations, as they avoid physical experiments altogether.

The following figures display examples of stress concentration factors (Figures 5.7–5.13):

2 See, Pilkey, W.D. and Pilkey, D.F. (2008). *Peterson's Stress Concentration Factors*, 3e. Wiley.

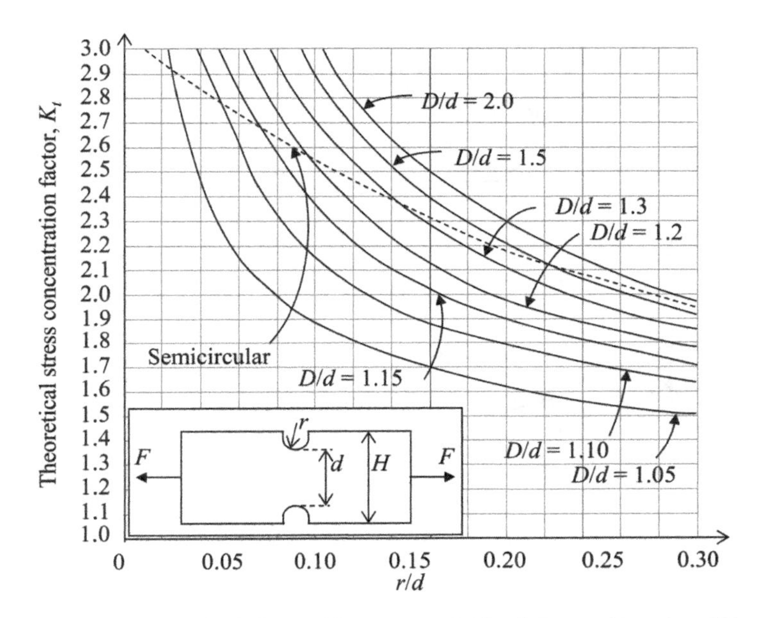

Figure 5.7 Notched plate subject to an axial load. Source: Based on Pilkey, W.D. and Pilkey, D.F. (2008). *Peterson's Stress Concentration Factors*, 3e. Wiley.

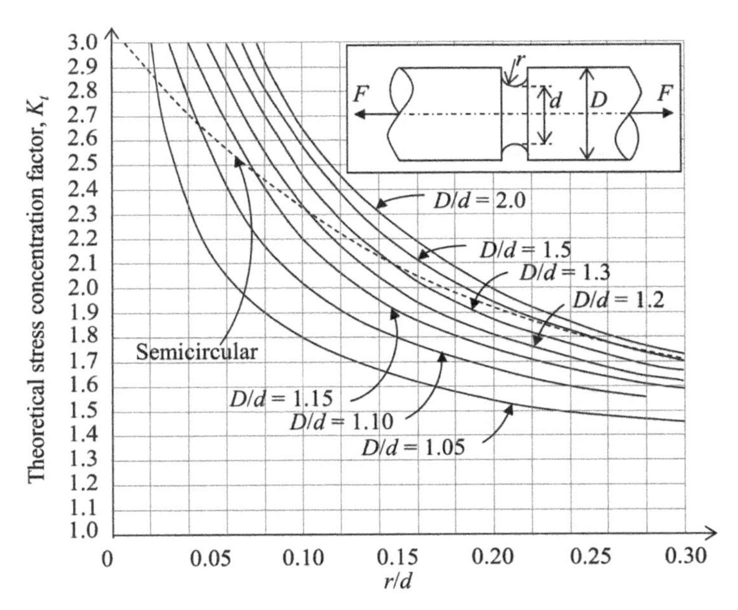

Figure 5.8 Grooved shaft subject to an axial load. Source: Based on Pilkey, W.D. and Pilkey, D.F. (2008). *Peterson's Stress Concentration Factors*, 3e. Wiley.

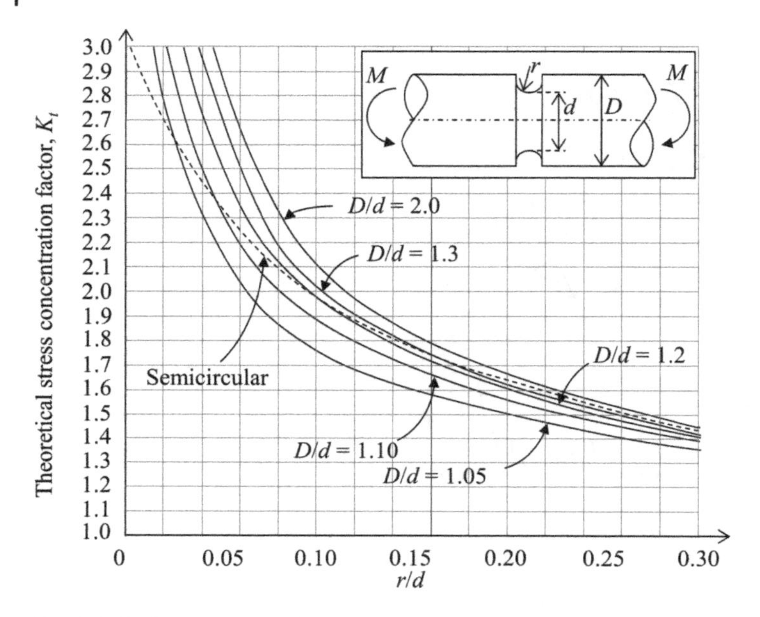

Figure 5.9 Grooved shaft subject to bending. Source: Based on Pilkey, W.D. and Pilkey, D.F. (2008). *Peterson's Stress Concentration Factors*, 3e. Wiley.

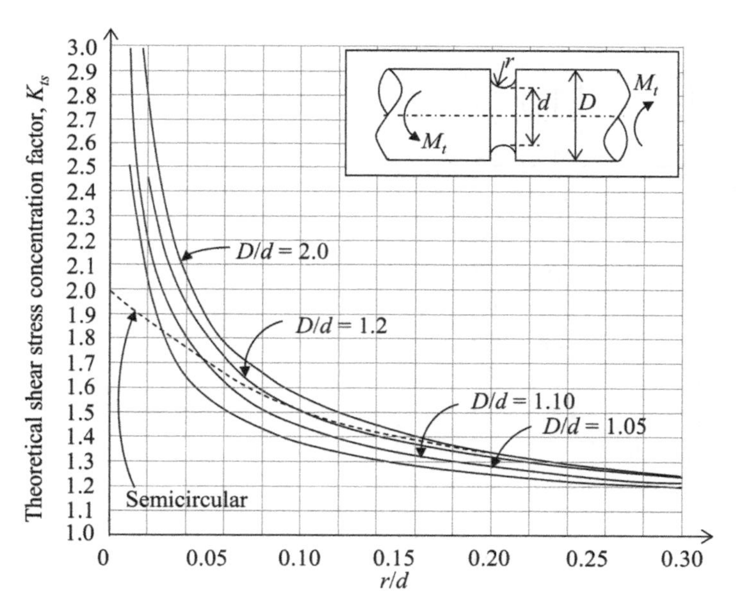

Figure 5.10 Grooved shaft subject to torque. Source: Data from Pilkey, W.D. and Pilkey, D.F. (2008). *Peterson's Stress Concentration Factors*, 3e. Wiley.

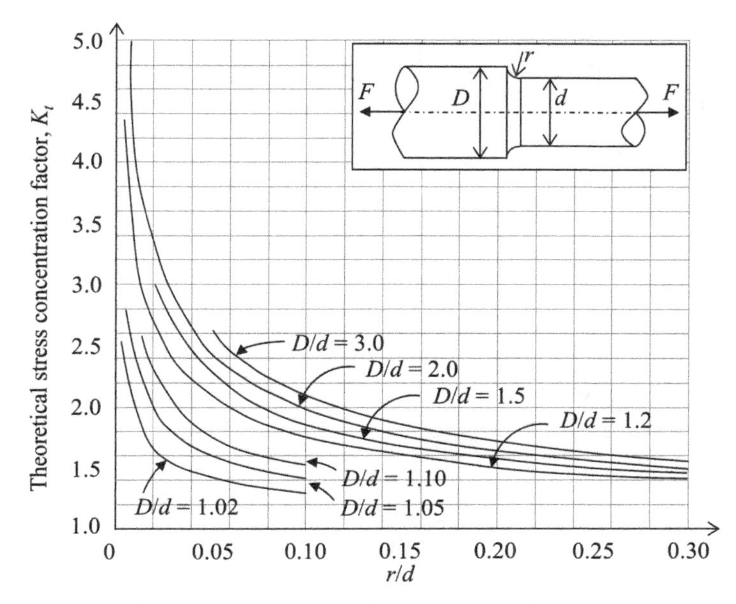

Figure 5.11 Shoulder fillet subject to an axial load. Source: Based on Pilkey, W.D. and Pilkey, D.F. (2008). *Peterson's Stress Concentration Factors*, 3e. Wiley.

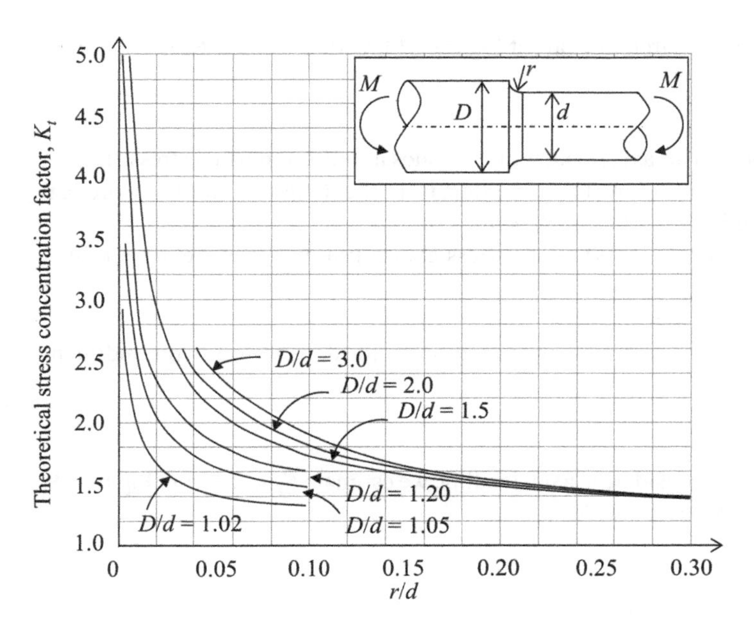

Figure 5.12 Shoulder fillet subject to bending. Source: Based on Pilkey, W.D. and Pilkey, D.F. (2008). *Peterson's Stress Concentration Factors*, 3e. Wiley.

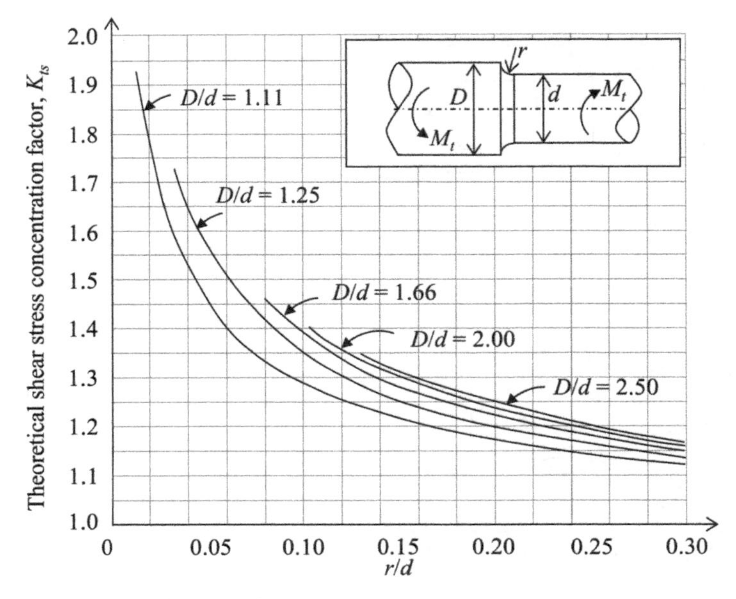

Figure 5.13 Shoulder fillet subject to torque. Source: Based on Pilkey, W.D. and Pilkey, D.F. (2008). *Peterson's Stress Concentration Factors*, 3e. Wiley.

Example problem 5.1

Calculate the maximum stress in the circular shaft of Figure 5.14, with $F = 100$ kN, $R = 2.5$ mm, $D = 50$ mm, and $d = 45$ mm.

Solution

The shaft of Figure 5.14 is subject to an axial load, which creates a uniform normal stress. The oil groove locally creates a stress concentration. Thus, we start from the definition of the theoretical stress concentration factor given in Eq. (5.15), i.e. $K_t = \sigma_{max}/\sigma_0$.

We first calculate the nominal stress σ_0, when no stress concentration exists, i.e. $\sigma_0 = F/A = 4F/\pi d^2$.

Thus, $\sigma_0 = (4 \times 100 \cdot 10^3)/0.045^2\pi = 62.88$ MPa.

We determine K_t from Figure 5.8. The stress concentration factor depends on two nondimensional parameters:

$D/d = 50/45 = 1.11$,

$R/d = 2.5/45 = 5.60 \cdot 10^{-3}$.

Thus, based on D/d and R/d, and considering axial loading, we estimate $K_t \cong 2.4$ from Figure 5.8.

Finally, $\sigma_{max} = K_t\sigma_0 = 2.4 \times 62.88 = 150.90$ MPa.

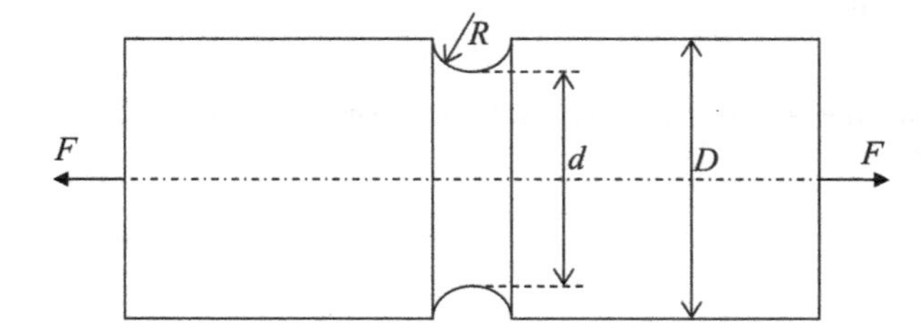

Figure 5.14 A circular shaft with an oil groove subject to an axial load.

5.4 Failure Criteria

When designing a mechanical element, we always ensure that the internal stress in the mechanical element, which results from the external loading, is smaller than the strength of the material. This raises two questions. What is the strength of the material? What is the internal stress in the mechanical element? The former question is relatively straightforward to answer. Typically, we use the yield stress σ_y and the ultimate tensile stress σ_{ut} as the strength of the material for ductile and brittle materials, respectively. The latter question is also straightforward to answer in cases of simple loading, when we use the concepts and equations discussed earlier in this chapter.

However, determining the internal stress in the context of combined loading, where multiple cases of simple loading occur simultaneously, is not straightforward. There is no single unified theory to quantify the internal stress in the mechanical element but, instead, several *failure criteria* exist. Failure criteria attempt to capture the physics of how materials fail. However, each failure criterion may be accurate in certain situations, but not in others. For that reason, we need to understand the most common failure criteria, how we use them, and what their shortcomings are, so we can apply the most appropriate failure criterion in each combined loading calculation.

5.4.1 Failure Criteria for Ductile Materials

5.4.1.1 Maximum Normal Stress Theory (Rankine)

The maximum normal stress theory predicts that failure occurs when the largest principal stress in the mechanical element equals the strength of the material. Note that we always rank the principal stresses σ_1, σ_2, and σ_3 in descending order, i.e. $\sigma_1 > \sigma_2 > \sigma_3$. Thus, failure occurs when

$$\sigma_1 > \sigma_y \tag{5.19}$$

for tension, and

$$\sigma_3 < -\sigma_y \tag{5.20}$$

for compression.

5.4.1.2 Maximum Shear Stress Theory (Tresca)

The maximum shear stress theory predicts that yielding occurs when the maximum shear stress in the mechanical element is equal to the maximum shear stress in a uniaxial tensile test of a specimen of the same material, when that specimen starts to yield.

The maximum shear stress in a uniaxial tensile test is

$$\tau_{max} = \frac{\sigma_1 - \sigma_3}{2}, \tag{5.21}$$

which is the radius of the Mohr circle diagram that represents this stress-state. Figure 5.15 shows the Mohr circle diagram for a uniaxial tensile test, where $\sigma_2 = \sigma_3 = 0$. The specimen starts to yield when $\sigma_1 = \sigma_y$ during the uniaxial tensile test. Thus, the maximum shear stress at the inception of yielding during a uniaxial tensile test, $\tau_{max} = \sigma_y/2$.

Hence, according to the maximum shear stress criterion, failure occurs when

$$\tau_{max} = \frac{\sigma_1 - \sigma_3}{2} = \frac{\sigma_y}{2}. \tag{5.22}$$

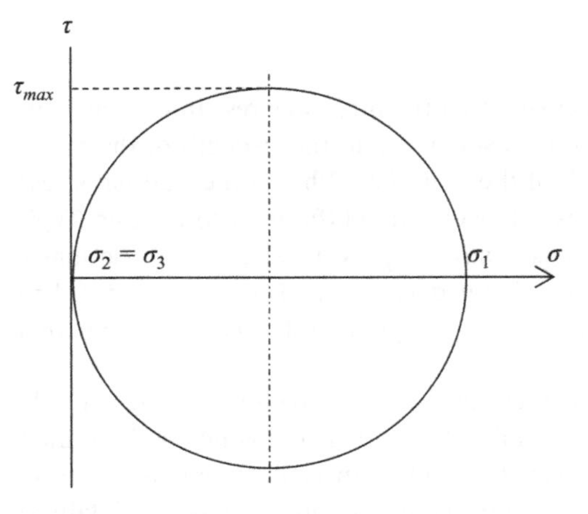

Figure 5.15 Mohr circle of a uniaxial tensile test.

We define an equivalent normal stress σ_{eq}, which is the normal stress that causes the same stress in the material than the combination of different cases of simple loading (according to the maximum shear stress criterion), and is given as

$$\sigma_{eq} = \sigma_1 - \sigma_3. \tag{5.23}$$

Failure occurs when $\sigma_{eq} > \sigma_y$.

5.4.1.3 Distortion Energy Theory (Von Mises)

The distortion energy theory predicts that yielding occurs when the distortion energy stored in the stressed mechanical element is equal to the distortion energy stored in a uniaxial tensile test specimen of the same material, when that specimen starts to yield.

Figure 5.16 illustrates the distortion energy concept. We show three stress elements. Figure 5.16a shows a stress element subject to its three principal stresses σ_1, σ_2, and σ_3, acting on mutually orthogonal planes, and representing the total strain energy associated with this stress state. The strain energy causes a volume change as well as angular distortion of the stress element. Figure 5.16b shows a stress element subject to the average stress (average of three principal stress components), applied to each of the coordinate directions, and representing the strain energy that only creates a volume change of the stress element. Indeed, since the three stress components are identical and equal to σ_{avg}, the stress element expands and contracts uniformly in the three coordinate directions. Finally, Figure 5.16c shows a stress element subject to the difference

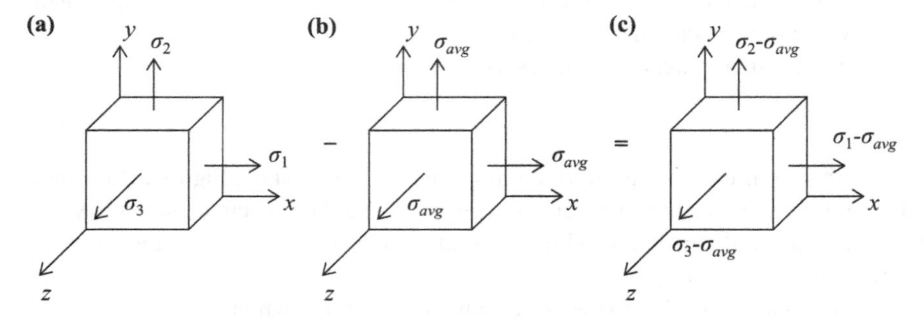

Figure 5.16 Distortion energy concept, showing (a) total strain energy, (b) strain energy creating a volume change, and (c) strain energy creating distortion.

between the respective principal stress components and the average stress, which represents the strain energy that only creates distortion of the stress element.

We define an equivalent stress σ_{eq}, which is the normal stress that causes the same stress in the material than the combination of different cases of simple loading (according to the distortion energy stress criterion), and is given as

$$\sigma_{eq} = \left[\frac{\left(\sigma_1 - \sigma_2\right)^2 + \left(\sigma_2 - \sigma_3\right)^2 + \left(\sigma_1 - \sigma_3\right)^2}{2} \right]^{1/2}. \tag{5.24}$$

Failure occurs when $\sigma_{eq} > \sigma_y$.

In the case of a plane stress-state, i.e. one of the principal stress components equals zero, we rewrite Eq. 5.24 as

$$\sigma_{eq} = \left(\sigma_A^2 - \sigma_A\sigma_B + \sigma_B^2\right)^{1/2}. \tag{5.25}$$

Here, we use σ_A and σ_B instead of σ_1, σ_2, and σ_3, because we do not know which principal stress equals zero, until we calculate them. However, we always need to rank $\sigma_1 > \sigma_2 > \sigma_3$ and, therefore, can only do so after we determine the sign of both σ_A and σ_B.

We can find several other expressions of the maximum distortion energy criterion, for instance, expressed as a function of *xyz* coordinate directions instead of principal directions. The simplest form of the maximum distortion energy criterion is when both σ_y and σ_z equal zero and the equivalent stress σ_{eq} reduces to

$$\sigma_{eq} = \left(\sigma_x^2 + 3\tau_{xy}^2\right)^{1/2}. \tag{5.26}$$

5.4.1.4 Comparison Between Different Failure Criteria

It is interesting to compare the different failure criteria to each other, and understand under which conditions the failure criteria predict different results. Figure 5.17 shows a graphical comparison of the three failure criteria for ductile materials in a plane stress-state, i.e. one of the principal stresses equals zero. We identify the maximum normal stress criterion with a dashed line, the maximum shear stress criterion with a solid line, and the maximum distortion energy criterion with a dash-dot line. We show the two nonzero principal stress components, σ_A and σ_B, on the coordinate axes. We draw a contour along which $\sigma_{eq} = \sigma_y$ for each failure criterion. Thus, each of these contours defines the envelope within which the safety factor, $n = \sigma_y/\sigma_{eq} \geq 1$, according to each respective failure criterion. If a stress-state falls within the contour of a failure criterion, then it predicts that

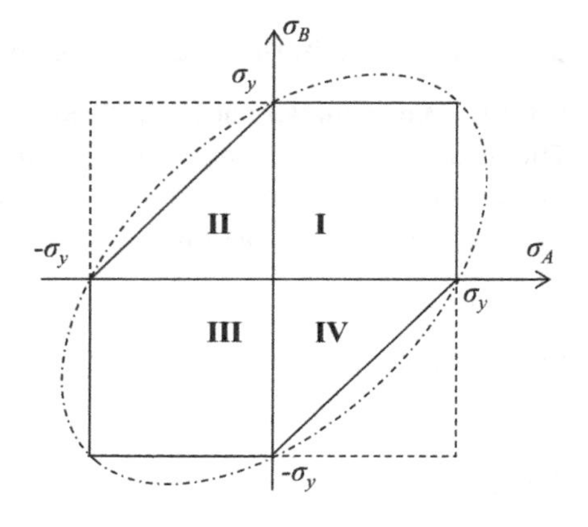

Figure 5.17 Comparison of the three failure criteria for ductile materials in plane stress-state; maximum normal stress criterion (dashed line), maximum shear stress criterion (solid line), and maximum distortion energy criterion (dash-dot line).

the mechanical element will not fail and $n \geq 1$. Conversely, if a stress-state falls outside the contour of a failure criterion, then it predicts that the mechanical element will fail because $n \leq 1$.

Examining the different quadrants of Figure 5.17 allows understanding the relationship between the graphical representation and the mathematical equations discussed earlier. First, constructing the square contour (dashed line) for the maximum normal stress criterion from Eqs. (5.19) and (5.20) is intuitive.

Next, we relate the maximum shear stress criterion of Eq. (5.23) to the contour shown in Figure 5.17 quadrant-by-quadrant. In the first quadrant, $\sigma_A, \sigma_B > 0$. Thus, $\sigma_1 = \sigma_A$, $\sigma_2 = \sigma_B$, and $\sigma_3 = 0$. Applying Eq. (5.23), $\sigma_{eq} = \sigma_A$. Thus, failure (inception of yielding) occurs when $\sigma_{eq} = \sigma_y$ or, when $\sigma_A = \sigma_y$. Similarly, in the second quadrant, $\sigma_A < 0, \sigma_B > 0$. Thus, $\sigma_1 = \sigma_B, \sigma_2 = 0$, and $\sigma_3 = \sigma_A$. Applying Eq. (5.23), $\sigma_{eq} = \sigma_B - \sigma_A$. Thus, failure (inception of yielding) occurs when $\sigma_{eq} = \sigma_y$ or, when $\sigma_B - \sigma_A = \sigma_y$, which represents the equation of a line in the coordinate system of Figure 5.17. The analysis of quadrant 3 is similar to that of quadrant 1, but now both nonzero principal stress components are negative, and the analysis of quadrant 4 is similar to that of quadrant 2.

Finally, we recognize that for the distortion energy criterion the shape (dash-dot line in Figure 5.17) of the contour represents an inclined ellipse. Using the equation for this geometry, the contour is given as $\sigma_{eq} = \sqrt{\sigma_A^2 - \sigma_A\sigma_B + \sigma_B^2}$. Thus, failure (yielding) occurs when $\sigma_{eq} = \sigma_y$ or, when $\sigma_y = \sqrt{\sigma_A^2 - \sigma_A\sigma_B + \sigma_B^2}$.

The graphical representation of the different failure criteria in tandem with the physical interpretation provided above, helps to understand how each failure criterion evaluates a stress-state resulting from combined loading, relative to each other. Indeed, the smallest contour defines the strictest or most conservative failure criterion. Figure 5.17 shows that the maximum shear stress criterion results in the smallest contour, whereas the maximum distortion energy criterion is slightly larger. Hence, for a specific stress-state, the maximum shear stress criterion always predicts an equivalent stress σ_{eq} that is equal or higher than what the distortion energy criterion predicts. As a result, using the maximum shear stress criterion will result in designing mechanical elements with larger dimensions compared to using the maximum distortion energy criterion. In contrast, the maximum normal stress criterion is a good criterion to use when the stress-state falls into the first or third quadrant; however, it significantly underestimates σ_{eq} in the second and fourth quadrant. Thus, it should not be used for a stress-state that falls in either one of these two quadrants. In general, engineers consider the maximum distortion energy criterion to offer a good balance between accuracy and conservative interpretation of a stress-state resulting from combined loading. For that reason, it is perhaps the most commonly used failure criterion for ductile materials.

5.4.2 Failure Criteria for Brittle Materials

5.4.2.1 Maximum Normal Stress Theory (Rankine)

The maximum normal stress theory predicts that failure occurs when the largest principal stress in the mechanical element equals the strength. Note that we always rank the principal stresses σ_1, σ_2, and σ_3 in descending order, i.e. $\sigma_1 > \sigma_2 > \sigma_3$. Thus, failure occurs when

$$\sigma_1 > \sigma_{ut} \tag{5.27}$$

for tension, and

$$\sigma_3 < -\sigma_{ut} \tag{5.28}$$

for compression.

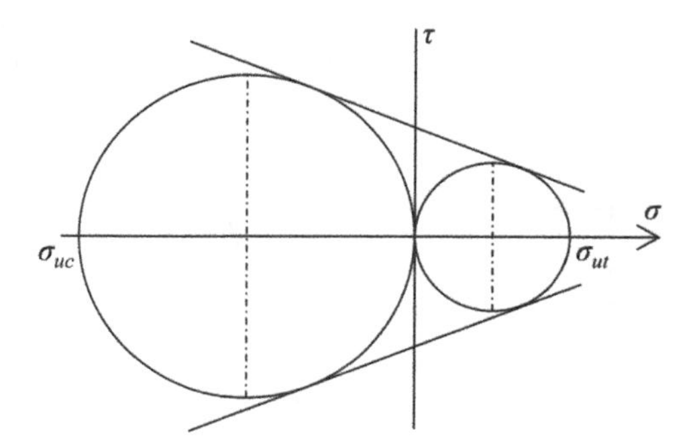

Figure 5.18 Coulomb–Mohr failure criterion for a plane stress-state.

We point out that the maximum normal stress theory is similar for ductile and brittle materials, with the exception that we use σ_y for ductile materials and σ_{ut} for brittle materials, as the strength of the material.

5.4.2.2 Coulomb–Mohr Theory

Failure occurs for any stress-state that produces a circle tangent to the envelope of Mohr circles for the stress-states of uniaxial compression and uniaxial tension (for cases of plane stress only).

Figure 5.18 shows the shear stress τ versus normal stress σ coordinate system, together with the Mohr circles for uniaxial tension (right) and compression (left), and two tangent lines to those circles that define the envelope mentioned above. Here, we refer to the ultimate tensile stress as σ_{ut} and the ultimate compressive stress as σ_{uc}. Brittle materials, such as ceramics, typically display much larger compressive than tensile strength. Hence, the Mohr circle for uniaxial compression is significantly larger than the circle for uniaxial tension.

Three different situations can occur:

First, $\sigma_A \geq 0$ and $\sigma_B \leq 0$. Thus,

$$\frac{\sigma_A}{\sigma_{ut}} - \frac{\sigma_B}{\sigma_{uc}} = \frac{1}{n}. \tag{5.29}$$

Second, $\sigma_A \geq \sigma_B \geq 0$. Thus,

$$\frac{\sigma_A}{\sigma_{ut}} = \frac{1}{n}. \tag{5.30}$$

Third, $0 \geq \sigma_A \geq \sigma_B$. Thus,

$$\frac{\sigma_B}{\sigma_{uc}} = -\frac{1}{n}. \tag{5.31}$$

Here, n is the safety factor. Also, note that both σ_{ut} and σ_{uc} are positive, i.e. we account for the negative compressive stress in the nomenclature by using the subscript *uc*.

5.4.2.3 Comparison Between Different Failure Criteria

Similar to failure criteria for ductile materials, it is interesting to compare the different failure criteria for brittle materials to each other, and understand under which conditions each failure criterion predicts different results. Figure 5.19 shows a graphical comparison of the two failure criteria for brittle materials, in a plane stress-state. We identify the maximum normal stress criterion with a dashed line and the Coulomb–Mohr theory with a solid line. We show the two non-zero principal stress components, σ_A and σ_B, on the coordinate axes. Each of these contours defines the envelope within which the safety factor, $n = \sigma_{ut}/\sigma_{eq} \geq 1$, according to each respective failure criterion. If a stress-state falls within the contour of a failure criterion, then it predicts that the mechanical

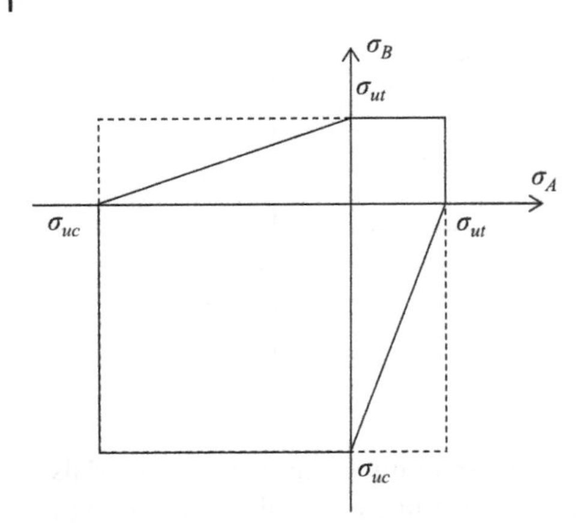

Figure 5.19 Comparison of the two failure criteria for brittle materials in plane stress-state; maximum normal stress criterion (dashed line) and Coulomb–Mohr theory (solid line).

element will not fail and $n \geq 1$. Conversely, if a stress-state falls outside the contour of a failure criterion, then it predicts that the mechanical element will fail because $n \leq 1$.

Example problem 5.2
We consider a brittle material with $\sigma_{ut} = 300\,\text{MPa}$ and $\sigma_{uc} = 600\,\text{MPa}$. Calculate the safety factor against fracture using the Coulomb–Mohr failure theory for the following cases of plane stress (note that the values are not principal stresses):

(a) $\sigma_x = 150\,\text{MPa}$, $\sigma_y = 150\,\text{MPa}$
(b) $\sigma_x = 80\,\text{MPa}$, $\tau_{xy} = 40\,\text{MPa}$
(c) $\sigma_x = 150\,\text{MPa}$, $\sigma_y = -50\,\text{MPa}$, $\tau_{xy} = 50\,\text{MPa}$

Plot the contour of the Mohr–Coulomb theory in the σ_A versus σ_B plane and identify the cases (a) through (c) with black square markers.

Solution
(a) We first calculate the principal stresses (see Appendix C):

$$\sigma_{A,B} = \frac{\sigma_x+\sigma_y}{2} \pm \sqrt{\left(\frac{\sigma_x-\sigma_y}{2}\right)^2 + \tau_{xy}^2},$$

$$\sigma_{A,B} = \frac{150+150}{2} \pm \sqrt{\left(\frac{150-150}{2}\right)^2 + 0^2},$$

$\sigma_A = 150$ MPa, $\sigma_B = 150$ MPa.

Thus, $\sigma_1 = 150$ MPa, $\sigma_2 = 150$ MPa, $\sigma_3 = 0$

$\sigma_A/\sigma_{ut} = 1/n \Rightarrow n = \sigma_{ut}/\sigma_A = 300/150 = 2.$

(b) We first calculate the principal stresses:

$$\sigma_{A,B} = \frac{80+0}{2} \pm \sqrt{\left(\frac{80-0}{2}\right)^2 + 40^2},$$

$\sigma_A = 96.56$ MPa, $\sigma_B = -16.56$ MPa.

Thus, $\sigma_1 = 96.56$ MPa, $\sigma_2 = 0$, $\sigma_3 = -16.56$ MPa

$$\frac{\sigma_A}{\sigma_{ut}} - \frac{\sigma_B}{\sigma_{uc}} = \frac{1}{n} \Rightarrow n = \left[\frac{\sigma_A}{\sigma_{ut}} - \frac{\sigma_B}{\sigma_{uc}}\right]^{-1},$$

$$n = \left[\frac{96.56}{300} - \frac{(-16.56)}{600}\right]^{-1} = 2.86.$$

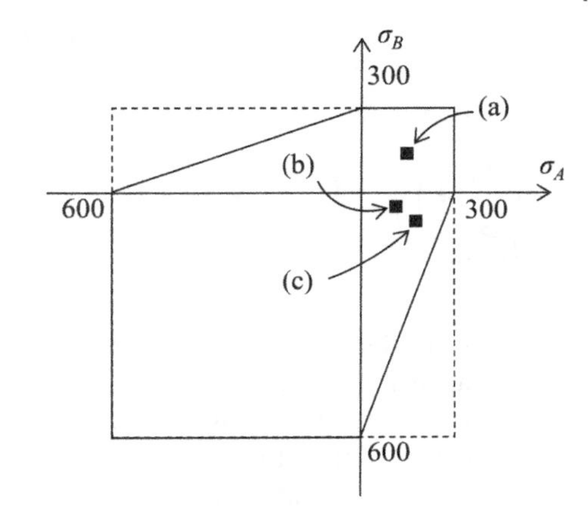

Figure 5.20 Graphical representation of the three stress states (a), (b), and (c), with respect to the contours of the maximum normal stress and Coulomb–Mohr failure criteria.

(c) We first calculate the principal stresses:

$$\sigma_{A,B} = \frac{\sigma_x+\sigma_y}{2} \pm \sqrt{\left(\frac{\sigma_x-\sigma_y}{2}\right)^2 + \tau_{xy}^2},$$

$$\sigma_{A,B} = \frac{150-50}{2} \pm \sqrt{\left(\frac{150-(-50)}{2}\right)^2 + 50^2},$$

$\sigma_A = 161.8$ MPa, $\sigma_B = -61.8$ MPa.

Thus, $\sigma_1 = 161.8$ MPa, $\sigma_2 = 0$, $\sigma_3 = -61.8$ MPa

$$\frac{\sigma_A}{\sigma_{ut}} - \frac{\sigma_B}{\sigma_{uc}} = \frac{1}{n} \Rightarrow n = \left[\frac{\sigma_A}{\sigma_{ut}} - \frac{\sigma_B}{\sigma_{uc}}\right]^{-1},$$

$$n = \left[\frac{161.8}{300} - \frac{(-61.8)}{600}\right]^{-1} = 1.56.$$

Figure 5.20 shows the different stress states with respect to both failure criteria.

5.5 Key Takeaways

1. We use the basic equations for different types of simple loading to calculate the internal stress resulting from axial loading, bending, and torsion.
2. A geometric discontinuity or irregularity locally increases the internal stress in a mechanical element. We use the theoretical and shear stress concentration factors to relate the local stress to the nominal stress in the mechanical element. Stress concentration factors depend on the type of loading and the geometry, but are independent of the material type.
3. We use failure criteria for both ductile materials (maximum normal stress, maximum shear stress, and distortion energy criteria) and brittle materials (maximum normal stress and Coulomb–Mohr criteria) to calculate the internal stress in a mechanical element subject to combined loading, where different types of simple loading occur simultaneously.

5.6 Problems

5.1 Figure 5.21 shows a circular aluminum tube with outer radius $R_o = 0.1$ m. A torque moment $M_t = 10\,000$ Nm twists the tube. Calculate the maximum inner radius R_i of the tube to limit the angle of twist ψ to 0.15°/ m.

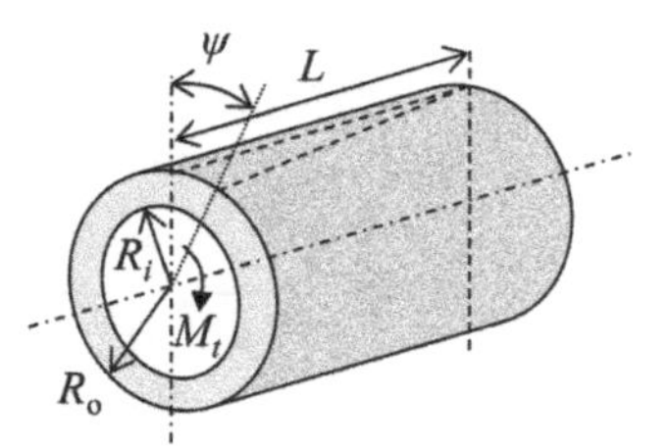

Figure 5.21 Circular tube.

5.2 Consider a ductile material with yield stress $\sigma_y = 150$ MPa. Calculate the safety factor against yielding using the maximum shear stress and distortion energy failure criteria, for each of the following cases of plane stress:
(a) $\sigma_1 = 100$ MPa, $\sigma_2 = 100$ MPa, $\sigma_3 = 0$ MPa
(b) $\sigma_1 = 100$ MPa, $\sigma_2 = 20$ MPa, $\sigma_3 = 0$ MPa
(c) $\sigma_1 = 80$ MPa, $\sigma_2 = 0$ MPa, $\sigma_3 = -60$ MPa
(d) $\sigma_1 = 0$ MPa, $\sigma_2 = -20$ MPa, $\sigma_3 = -60$ MPa
(Note: σ_1, σ_2, σ_3 are principal stresses.)

5.3 Consider a brittle material with ultimate tensile stress $\sigma_{ut} = 300$ MPa and ultimate compressive stress $\sigma_{uc} = 600$ MPa. Calculate the safety factor against fracture using the Coulomb–Mohr failure theory, for each of the following cases of plane stress:
(a) $\sigma_x = 150$ MPa, $\sigma_y = 150$ MPa
(b) $\sigma_x = 80$ MPa, $\tau_{xy} = 40$ MPa
(c) $\sigma_x = 150$ MPa, $\sigma_y = -50$ MPa, $\tau_{xy} = 50$ MPa
(Note: σ_x, σ_y, are not principal stresses.)

5.4 Figure 5.22 shows a cantilever beam with circular cross-section made from 1035 HR steel, subject to the following static loading: a transverse force $F = 2000$ N, an axial force $P = 15\,000$ N, and a torque moment $M_t = 150$ Nm. $L = 0.1$ m. Calculate the diameter d using the Von Mises (distortion energy) failure criterion and consider a design factor $n_D = 4$.

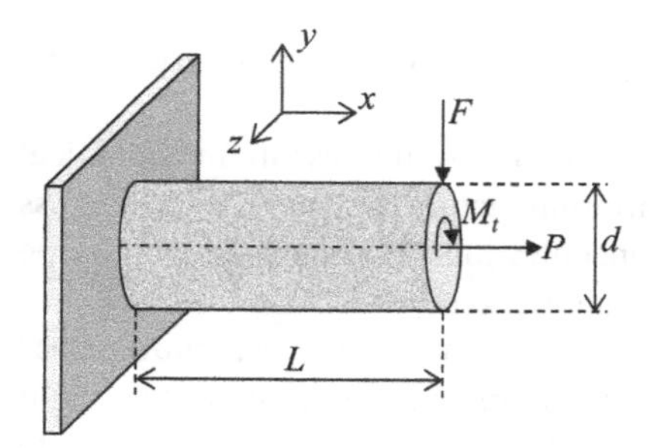

Figure 5.22 Cantilever beam with circular cross-section.

5.5 Figure 5.23 shows the end of a stationary shaft with circular cross-section. Determine the theoretical stress concentration factor to accurately calculate the stress at the shoulder fillet of the shaft. $D = 60$ mm, $d = 45$ mm, $r = 6$ mm, $F = 1000$ N.

5.6 Figure 5.24 shows a stationary shaft with circular cross-section and with an oil groove, supported in two bearings, and made from AISI 1020 HR steel. $D = 50$ mm, $d_r = 49$ mm, $r = 2.5$ mm, $L_1 = 0.4$ m, $L_2 = 0.2$ m.

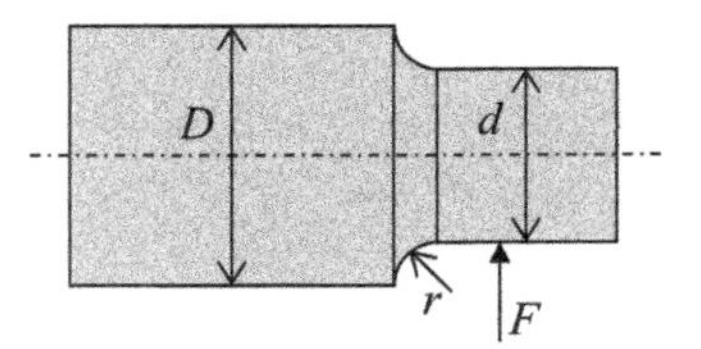

Figure 5.23 Stationary shaft with circular cross-section and with shoulder fillet.

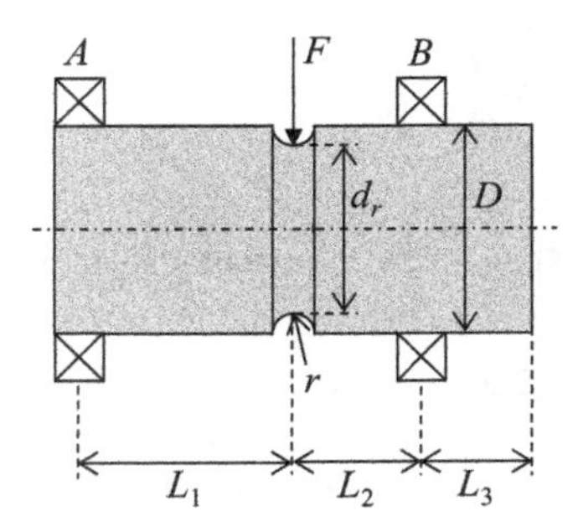

Figure 5.24 Stationary shaft with circular cross-section and with an oil groove.

(a) Determine the theoretical stress concentration factor to accurately calculate the stress at the oil groove of the shaft.
(b) Calculate the maximum force F to maintain a safety factor against yielding $n = 2$.

5.7 Figure 5.25 shows a stationary shaft with circular cross-section and with a shoulder fillet, supported in two bearings, and made from AISI 1010 HR steel. $D = 30$ mm, $d_r = 20$ mm, $r = 5$ mm, $L_1 = L_2 = 0.4$ m.
(a) Determine the theoretical stress concentration factor to accurately calculate the stress at the shoulder fillet of the shaft.
(b) Calculate the maximum force F to maintain a safety factor against yielding $n = 2$.

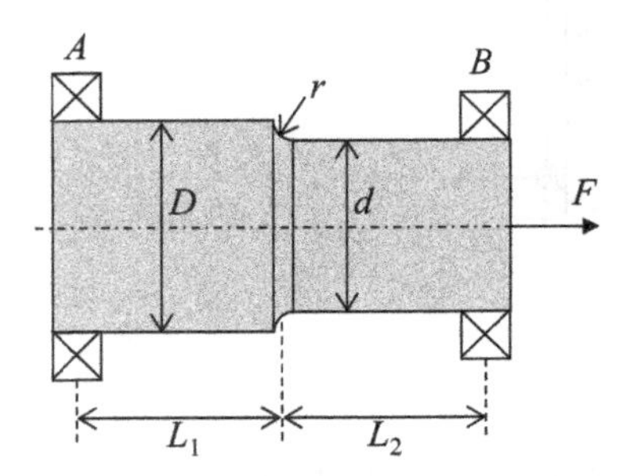

Figure 5.25 Stationary shaft with circular cross-section and with a shoulder fillet.

5.8 Figure 5.26 shows a horizontal beam supported by a vertical tension link. $L_1 = 2.5$ m, $L_2 = 1.5$ m, $F = 12.5$ kN. The cross-sections of both the beam and the link are square with $b = 100$ mm. All connections use 8 mm diameter cylindrical pins in "double shear" (meaning that there are two shear planes, one on each side of the connector).
(a) Calculate the tensile stress in the vertical tension link $A - B$.
(b) Calculate the maximum normal stress due to bending in the horizontal beam $C - D$.
(c) Graph the maximum tensile stress in the horizontal beam as a function of the location between C and D.
(d) Calculate the shear stress in pins A and C.

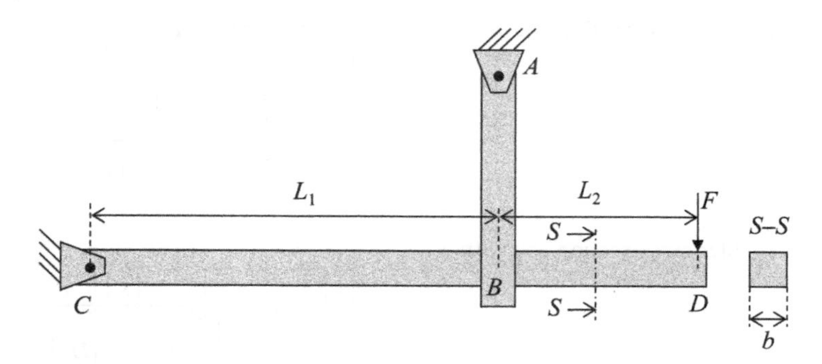

Figure 5.26 Horizontal beam with vertical tension link.

5.9 Figure 5.27 shows a structure to hang a decorative store sign to the outside wall of a store. Both trusses are square tubes with size d and thickness t. $\sigma_y = 215$ MPa. $L = 2$ m and $\alpha = 15°$. The decorative sign weighs 250 kg, indicated by force F. All connections use 10 mm diameter cylindrical pins in "double shear" (meaning that there are two shear planes, one on each side of the connector).

(a) Calculate the size of the square tubes required to obtain a design factor $n_D = 6$ to account for wind, snow, and other unanticipated loading. Table 5.1 shows an excerpt from the supplier catalog.

(b) Calculate the shear stress in pins A and B.

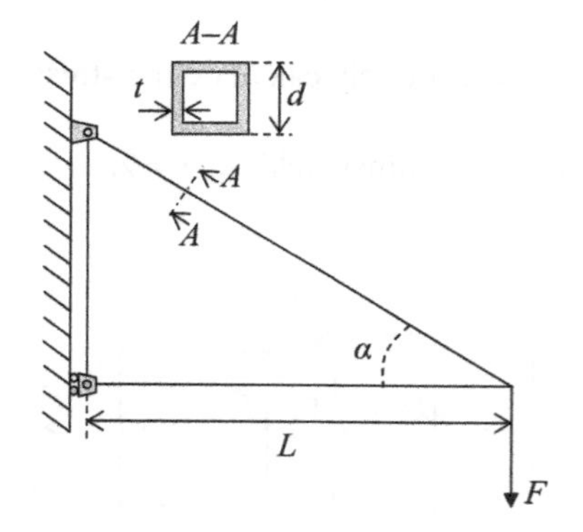

Figure 5.27 Structure to attach a sign to a store front.

Table 5.1 Excerpt from supplier catalog.

Size d (m)	Thickness t (m)
0.012	0.001
0.025	0.002
0.040	0.002
0.055	0.002
0.100	0.003

5.10 Figure 5.28 shows a T-beam, made from AISI 1040 HR steel, and subject to a bending moment $M = 15$ Nm. Calculate the parameter d, and consider a design factor $n_d = 2$.

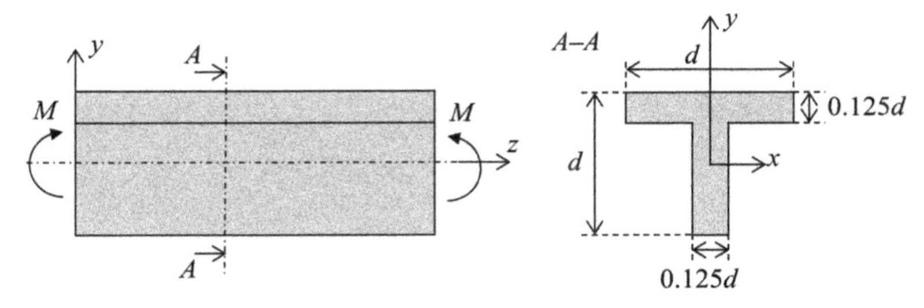

Figure 5.28 Horizontal T-beam.

6

Design for Fatigue Strength

6.1 Introduction

We define a dynamic load as an external force or moment that changes as a function of time. This contrasts a static load, which remains constant as a function of time.

When designing or verifying a mechanical element for dynamic strength, similar to static strength, we first analyze the physical connection between the mechanical element and its environment, and replace it with boundary conditions and reaction forces. Together, the external forces and the reaction forces describe the free-body diagram of the mechanical element. From the free-body diagram, we determine the internal forces in the mechanical element as an axial force, a shear force, a bending moment, and a torque moment diagram. These diagrams show each internal force component as a function of the location in the mechanical element and, thus, we can determine which section of the mechanical element experiences the highest load. Subsequently, when considering the internal force diagrams together with the (local) geometry of the mechanical element, we convert internal forces to internal stresses.

The key difference between designing a mechanical element for static and dynamic strength is that in the former case, all stress components remain constant as a function of time, whereas in the latter case, we must account for stress components that vary with time.

6.1.1 Types of Dynamic Loads

Two types of dynamic loads exist: an alternating and a fluctuating load, which result in an alternating and fluctuating stress, respectively. Figure 6.1 schematically shows the magnitude σ as a function of time t of (a) an alternating stress, and (b) a fluctuating stress. We observe that the alternating stress fully reverses around the horizontal axis, i.e. each loading cycle the element is subject to both tension (positive) and compression (negative). To understand this concept, we consider the following experiment. Figure 6.1a shows a shaft that rotates with angular velocity ω and is subject to a transverse load F that creates a bending moment in the shaft. If an observer is glued to the surface of the shaft, then the observer experiences compression when on top of the shaft. However, after we rotate the shaft over π radians, the observer is now underneath the shaft and experiences tension. If we rotate the shaft over another π radians, the observer is on top of the shaft again and experiences compression. Thus, each rotation of the shaft, the observer experiences compression and tension, resulting in a perfectly reversing stress magnitude as a function of time.

In contrast, a fluctuating stress does not perfectly reverse around the horizontal axis. Rather, it is a combination of an alternating stress and a steady stress. The steady stress applies a vertical shift to

Design of Mechanical Elements: A Concise Introduction to Mechanical Design Considerations and Calculations,
First Edition. Bart Raeymaekers.

Companion website: www.wiley.com/go/raeymaekers/designofmechanicalelements

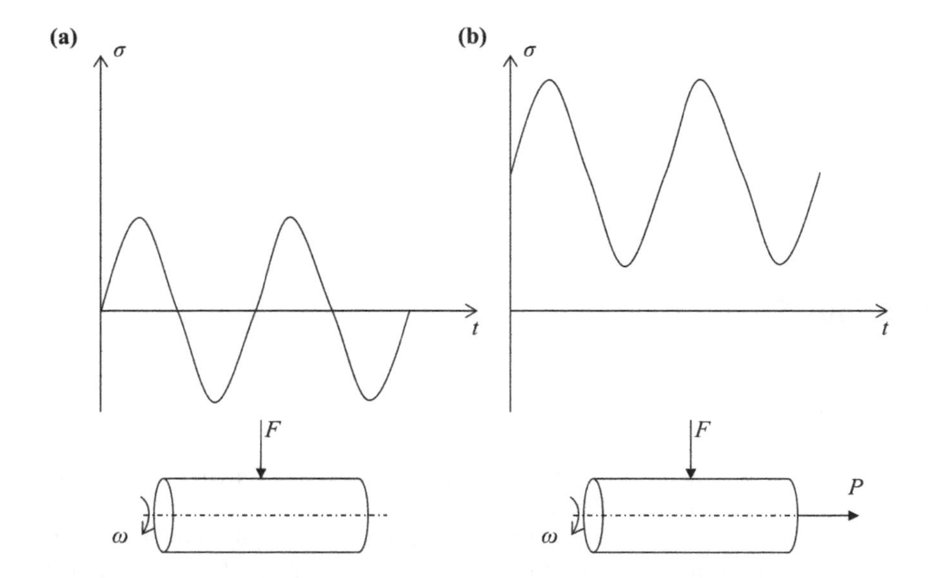

Figure 6.1 (a) Alternating stress and (b) fluctuating stress, showing magnitude as a function of time.

the alternating stress, shifting it upward when the steady stress is positive, and shifting it downward when the steady stress is negative. Figure 6.1b shows a shaft that rotates with angular velocity ω and is subject to a transverse load F that creates an alternating stress, in addition to an axial load P that superimposes a steady stress on the alternating stress, thus resulting in a fluctuating stress.

6.1.2 Fatigue Failure Mechanism

Mechanical elements routinely fail when subject to an alternating or fluctuating stress that is well below the yield stress or ultimate tensile stress of the material. For instance, if we consider a thin metal wire, say when we unfold a paper clip, then we intuitively know that it is difficult to fracture the metal wire under a static axial load, i.e. when we axially pull on the metal wire. In contrast, if we bend the metal wire back-and-forth, applying an alternating stress, then the metal wire fractures in just a few loading cycles. The number of cycles to failure depends on how much we bend the metal wire, i.e. the stress amplitude. This simple example highlights the difference between a static and dynamic load and their effect on failure of a mechanical element.

A static and dynamic load causes material failure under different conditions because the physics that govern material failure under static and dynamic loading is fundamentally different. Specifically, a dynamic load causes fatigue failure. Figure 6.2 shows a cross-sectional view of a mechanical element, illustrating the different stages that lead to fatigue failure.

- **Step 1**: Fatigue failure always originates at an initial (sub)surface crack. This crack can be the result of a material defect, surface topography, or a manufacturing process.
- **Step 2**: The initial crack represents a geometric discontinuity on the surface of the mechanical element, which when subject to an external load creates a stress concentration. Locally, the maximum stress around the discontinuity might exceed the yield stress of the material and, thus, cause the material to deform around the initial crack, and the crack to grow each loading cycle.
- **Step 3**: The mechanical element is subject to a dynamic load, i.e. it experiences an alternating or fluctuating stress that loads and unloads the mechanical element. During these loading and

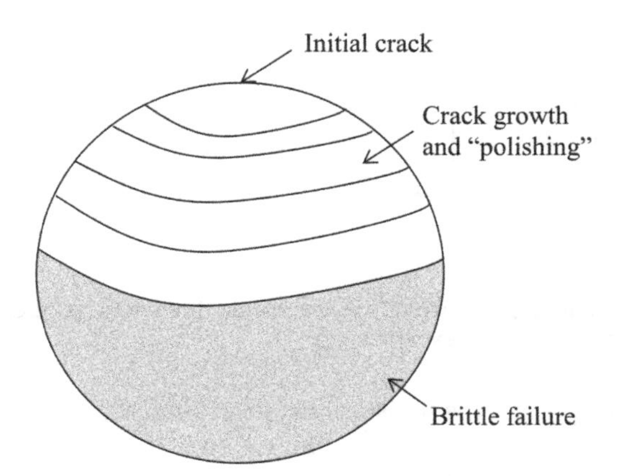

Figure 6.2 Cross-sectional view of a mechanical element, showing the different stages of fatigue failure, including crack initiation, crack growth and polishing, and brittle failure.

unloading cycles, the crack grows continuously as a result of the stress concentration around the crack geometry. Additionally, the two crack surfaces rub against each other because of the loading and unloading cycles, which causes both crack surfaces to polish each other. Polished fracture surfaces are a characteristic sign of fatigue failure, when performing a *postmortem* analysis of a fractured mechanical element.

- **Step 4**: The cross-sectional area of the mechanical element decreases with increasing number of loading and unloading cycles and, thus, increasing crack size. Thus, the nominal stress in the remaining material of the mechanical element increases, ultimately reaching the yield stress or ultimate tensile stress, which results in brittle failure of the mechanical element. The combination of a polished and grain-like fracture surface is a telltale sign of fatigue failure. The polished portion of the fracture surface evidences crack growth under repeated loading and unloading cycles, whereas the grain-like fracture surface results from the nominal stress in the mechanical element (the portion that has not cracked) exceeding the strength of the material.

6.2 Fatigue-life Methods

We use fatigue-life methods to predict the life of a mechanical element, subject to a dynamic load, as a function of the number of loading cycles to failure N. Three commonly used methods exist.

1. **Stress-life method:** Typically used for high-cycle fatigue, $N > 10^3$ loading cycles.
2. **Strain-life method:** Typically used for low-cycle fatigue, $N < 10^3$ loading cycles.
3. **Fracture mechanics:** This is the most accurate, yet most difficult method (and represents a research field of its own). Fracture mechanics methods assume a crack exists and then attempt to mathematically predict crack growth as a function of local stress around the crack geometry.

In this textbook, we only discuss the *stress-life method* because it covers cases of high-cycle fatigue, which are relevant to and very common in the design of mechanical elements.

In the stress-life method, we subject specimens to an alternating or fluctuating stress of specific magnitude, while we count the number of loading cycles until the specimen fails. The most widely used stress-life method experiment is the R.R. Moore experiment, or high-speed rotating beam experiment. This is a standardized experiment, in which the specimen geometry, surface quality, and external load are defined by a standard. Standardizing experiments is useful because it allows comparing experimental results obtained by different laboratories with each other. We also

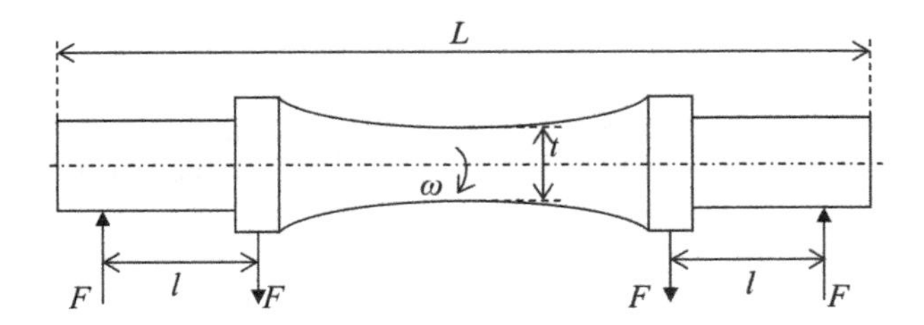

Figure 6.3 Schematic of the standardized material specimen typically used in an R.R. Moore high-speed rotating beam experiment.

emphasize the statistical nature of fatigue failure. Therefore, experiments must be repeated several times to obtain meaningful results.

Figure 6.3 schematically shows the geometry and loading of a typical R.R. Moore high-speed rotating beam experiment specimen.[1] ω indicates the angular velocity of the specimen around its symmetry axis, whereas the force components F indicate that the specimen is subject to simple bending only. The geometry is further defined by L, l, and t. The surface of the specimen is polished to minimize the effect of surface topography on surface cracks, where fatigue failure originates.

Performing the R.R. Moore high-speed rotating beam experiment involves applying different stress amplitudes to the specimens (by changing the magnitude of F) and counting the number of loading cycles to failure for each different stress amplitude. These results allow determining the so-called S–N curve, which shows the stress amplitude versus the number of loading cycles to failure. The stress amplitude is also called the *fatigue strength*. Thus, the fatigue strength σ_f, which we always define in conjunction with a number of loading cycles N, expresses the stress amplitude that we can impose on a mechanical element (in this case the R.R. Moore specimen) to achieve a specific fatigue life expressed in number of loading cycles N.

Figure 6.4 schematically shows a typical S–N curve of a carbon steel. We schematically show a cloud of data points obtained from experiments and fit a piecewise continuous line through the data points. The intercept with the vertical axis is the ultimate tensile stress σ_{ut} because if there is only one loading cycle, then the specimen experiences static rather than dynamic loading. Furthermore, we distinguish between low- and high-cycle fatigue at 10^3 loading cycles. The S–N curve shows a *knee* at 10^6 loading cycles. The corresponding intercept with the vertical axis is the endurance limit σ'_e, indicating that a stress amplitude smaller than σ'_e leads to infinite fatigue life of the mechanical element. Thus, we also indicate finite and infinite fatigue life in Figure 6.4. Here, the $'$ in σ'_e specifically refers to the R.R. Moore high-speed rotating beam experiment. In contrast, σ_e (without $'$) refers to the endurance limit of an actual mechanical element, different from the R.R. Moore specimen.

Determining σ'_e is routine yet time-consuming because we must repeat each experiment several times to obtain an S–N curve like the one of Figure 6.4, with statistically significant data. The literature offers data for commonly used steels and other materials. In addition, methods exist to approximate the endurance limit and bypass lengthy experiments. For instance, experiments show that the endurance limit of steel relates to its ultimate tensile stress and that we can use the following approximation:

$$\sigma'_e = \begin{cases} 0.5\,\sigma_{ut} & \text{if } \sigma_{ut} \leq 1400\,\text{MPa}, \\ 700\ \text{MPa} & \text{if } \sigma_{ut} \geq 1400\,\text{MPa}. \end{cases} \tag{6.1}$$

1 See, e.g. ISO 1143:2010. *Metallic Materials – Rotating Bar Bending Fatigue Testing*. International Organization for Standardization.

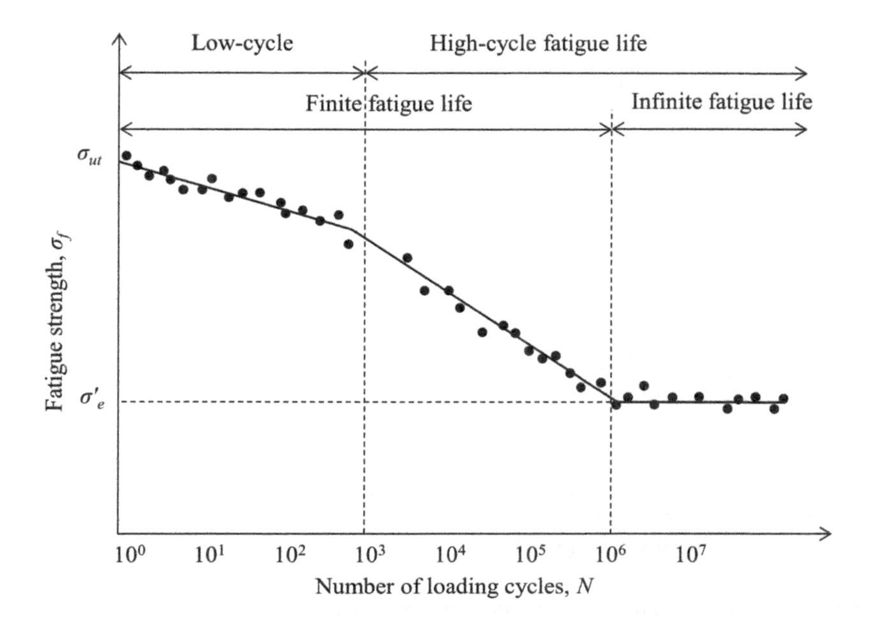

Figure 6.4 Typical *S–N* curve for carbon steel, showing the fatigue strength versus the number of (fully reversed) loading cycles until failure.

We point out that nonferrous materials and alloys (e.g. titanium, aluminum, and magnesium) typically do not display a *knee* at 10^6 loading cycles in their *S–N* curve and, thus, do not have an endurance limit. Rather, for these types of materials, the *S–N* curve monotonically decreases with increasing number of loading cycles N, and obtaining infinite fatigue life is impossible.

6.3 Fatigue Strength

When designing a mechanical element for infinite fatigue life, we determine the endurance limit and calculate the dimensions of the mechanical element such that the stress amplitude imposed on the mechanical element does not exceed the endurance limit. However, when designing a mechanical element for finite fatigue life, we determine the stress amplitude or fatigue strength σ_f that corresponds to a specific number of loading cycles N for which we design that element.

When designing a mechanical element for low-cycle fatigue, i.e. fewer than 10^3 loading cycles, the fatigue strength is only slightly smaller than the ultimate tensile stress, and strain-life methods exist to determine the fatigue strength. However, when designing a mechanical element for high-cycle fatigue, we must describe the relationship between σ_f and N between $10^3 \leq N \leq 10^6$ loading cycles. We approximate the *S–N* curve in this region as

$$\sigma_f = aN^b, \tag{6.2}$$

where a and b represent two parameters that we calculate as

$$a = \frac{(f\sigma_{ut})^2}{\sigma_e}, \tag{6.3}$$

$$b = -\frac{1}{3}\log\left(\frac{f\sigma_{ut}}{\sigma_e}\right). \tag{6.4}$$

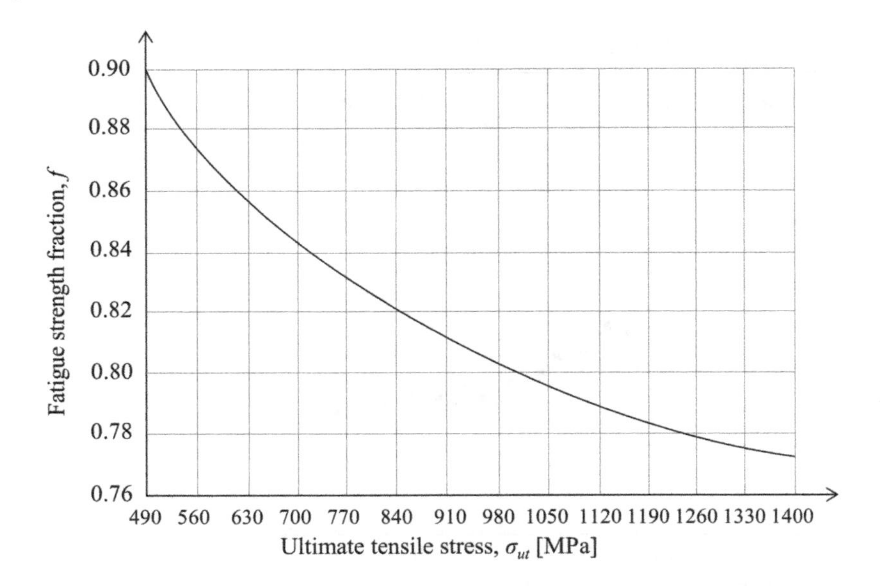

Figure 6.5 Fatigue strength fraction as a function of ultimate tensile stress.

The parameter f in Eqs. (6.3) and (6.4) is the *fatigue strength fraction*, which we derive mathematically[2] as a function of the ultimate tensile stress of the material. Physically, the fatigue strength fraction represents the maximum stress, expressed as a fraction of the ultimate tensile stress, which leads to a fatigue life of 10^3 loading cycles. We determine f graphically from Figure 6.5.

Additionally, note that Eqs. (6.3) and (6.4) use σ_e instead of σ_e'. We can use these equations to calculate the fatigue life of actual mechanical elements (using σ_e) or to calculate the theoretical case of an R.R. Moore high-speed rotating beam specimen (using $\sigma_e = \sigma_e'$). In Section 6.4, we describe how to determine σ_e for any mechanical element, starting from the ideal R.R. Moore high-speed rotating beam experiment σ_e'.

6.4 Endurance-limit Modifying Factors

We determine the endurance limit of any mechanical element σ_e, based on the endurance limit of an R.R. Moore high-speed rotating beam specimen σ_e' of the same material. We accomplish this by *modifying* σ_e' to account for the physical differences between the R.R. Moore specimen and the actual mechanical element. Practically, we consider the R.R. Moore specimen as the *ideal* specimen, and any modification of the specimen will reduce its endurance limit. We implement this concept using *endurance-limit modifying factors*. Thus,

$$\sigma_e = k_a k_b k_c k_d k_e k_f \sigma_e', \tag{6.5}$$

with k_a, k_b, k_c, k_d, k_e, and k_f the endurance-limit modifying factors. Each endurance-limit modifying factor modifies the endurance limit σ_e' to account for a specific difference between the R.R. Moore specimen and the actual mechanical element.

2 See, e.g. Budynas, R.G. and Nisbett, J.K. (2015). *Shigley's Mechanical Engineering Design*, 10e. McGraw-Hill.

Table 6.1 Surface factor k_a.

Surface quality	Parameter a	Parameter b
Ground	1.58	−0.085
Machined	4.51	−0.265
Hot-rolled	57.7	−0.718
As-forged	272	−0.995

Source: Noll, G. and Lipson, C. (1946). Allowable working stresses. Proc. Soc. Exp. Anal. 3 (2): 89–109.

6.4.1 k_a: Surface Factor

Fatigue failure starts with an initial crack on the surface of the mechanical element. The surface topography significantly affects the presence of cracks or geometric discontinuities on the surface of the mechanical element that could initiate fatigue failure. Smooth, polished surfaces are less likely to develop stress concentrations that would cause crack growth. The surface of the R.R. Moore specimen is polished. We account for the surface quality of the actual mechanical element, using $0 \leq k_a \leq 1$, where $k_a = 1$ indicates that the surface quality of the mechanical element is identical to that of the R.R. Moore specimen. Otherwise, we reduce k_a to recognize that the endurance limit of the mechanical element is lower than that of the R.R. Moore specimen of the same material, for the reasons mentioned above. We calculate k_a as

$$k_a = a\sigma_{ut}^b, \tag{6.6}$$

where a and b are parameters we determine from Table 6.1 (note these are different parameters than those used in Eqs. (6.3) and (6.4)), and σ_{ut} is expressed in MPa. Parameters a and b derive from curve fitting experimental data and are tabulated in reference works.

Example problem 6.1

Calculate k_a for a machined steel specimen with $\sigma_{ut} = 700$ MPa.

Solution

We find $a = 4.51$ and $b = -0.265$ from table 6.1 for surface quality *machined.*
Thus, using Eq. (6.6), $k_a = a\sigma_{ut}^b = 4.51(700)^{-0.265} = 0.79$.
Note that the endurance limit is reduced by more than 20% by using a *machined* instead of a *polished* surface. Thus, k_a often has a substantial impact on σ_e, and we should give it serious consideration.

Finally, we re-emphasize that σ_{ut} requires input in MPa in Eq. (6.6). The equations to determine endurance limit modifying factors are largely *empirical*, i.e. they derive from curve-fitting large data sets, instead of *first principles*. For that reason, there does not always exist a logical relationship between the units in those equations, and we need to pay close attention to using the tables and equations correctly.

6.4.2 k_b: Size Factor

The R.R. Moore specimen has a well-defined geometry, as depicted in Figure 6.3. A mechanical element has a geometry that deviates from this standardized specimen. Thus, we must use k_b to account for the different geometry and size of the mechanical element compared to the R.R. Moore specimen.

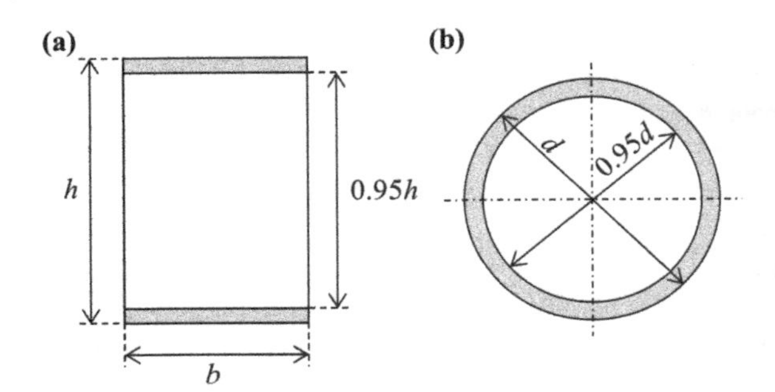

Figure 6.6 Equivalent diameter d_{eq}, showing (a) a nonrotating rectangular cross-section, and (b) a rotating circular cross-section, indicating the area at or above 95% of the maximum stress.

First, k_b depends on the type of dynamic load:

$$k_b = \begin{cases} 1 & \text{for a dynamic axial load,} \\ f(d) & \text{for a dynamic bending and torsion load.} \end{cases} \tag{6.7}$$

In the case of a dynamic bending or torsion load, k_b depends on the diameter d of the mechanical element.[3]

$$k_b = \begin{cases} 1.24d^{-0.107} & \text{when } 2.79 \leq d \leq 51 \text{ mm} \\ 1.51d^{-0.157} & \text{when } 51 \leq d \leq 254 \text{ mm.} \end{cases} \tag{6.8}$$

If the mechanical element does not rotate or does not have a circular cross-section, we first determine an equivalent diameter d_{eq}, which we then use in Eq. (6.8) to calculate k_b.

We determine d_{eq} by equating the volume of the mechanical element stressed at or above 95% of the maximum stress to the same volume of a rotating circular cross-section with d_{eq}. Figure 6.6 illustrates an example of a rectangular, nonrotating cross-section of a beam subject to simple alternating bending. The area of the rectangle stressed at or above 95% of the maximum stress is $0.05hb$. Correspondingly, the area of a circular cross-section of a rotating beam at or above 95% of the maximum stress is $\pi(d_{eq}^2 - (0.95d_{eq})^2)/4$. Thus, by equating both areas (or volumes, when we assume identical length of both elements), we find that $d_{eq} = 0.808\sqrt{bh}$. Another commonly used d_{eq} is that for a nonrotating beam with circular cross-section of diameter d, for which a similar analysis shows $d_{eq} = 0.370d$. Note that this discussion also implies that determining k_b is identical for a solid and a hollow circular cross-section because the maximum stress occurs at the surface of the circular cross-section.

6.4.3 k_c: Load Factor

The R.R. Moore specimen is subject to simple rotating bending. A mechanical element might experience a different type of dynamic loading. Hence, we must use k_c to account for a different type of load imposed on the mechanical element compared to the R.R. Moore specimen.[4] However, because mechanical elements are most often subject to combined loading, k_c is almost always equal to one. Note that the choice of k_c only considers dynamic loads.

$$k_c = \begin{cases} 1 & \text{Simple bending,} \\ 0.85 & \text{Simple axial loading,} \\ 0.59 & \text{Simple torsion,} \\ 1 & \text{Combined loading.} \end{cases} \tag{6.9}$$

3 See, e.g. Mischke, C.R. (1987). Prediction of stochastic endurance strength. J. Vib. Acoust. Stress Reliab. Design Trans. ASME 109 (1): 113–122.

4 See, e.g. Grover, H.J., Gordon, S.A., and Jackson, L.R. (1960). *Fatigue of Metals and Structures*. Bureau of Naval Weapons, Department of the Navy; and Mischke, C.R. (1987). Prediction of stochastic endurance strength. Journal of Vibration Acoustics Stress Reliability in Design Transactions of the ASME 109 (1): 113–122.

6.4.4 k_d: Temperature Factor

We perform the R.R. Moore high-speed rotating beam experiment at room temperature. However, a mechanical element might operate in a different temperature environment. It is well-known that the strength of materials (and specifically steel) decreases with increasing temperature. For instance, the strength of steel decreases by half when its temperature reaches 600 °C. For that reason, we must use k_d to account for any temperature difference between the operating environment of a mechanical element and the R.R. Moore experiment.

A complex relationship exists between the fatigue strength and the temperature; if critical, lab testing is advisable. Furthermore, when the operating temperature is substantially lower than room temperature, the material is more susceptible to brittle fracture, whereas when the temperature is higher than room temperature, yielding as a result of reduced strength is the critical problem.

We determine k_d as

$$k_d = \begin{cases} \frac{(\sigma_y)_T}{(\sigma_y)_{RT}} & \text{for ductile materials,} \\ \frac{(\sigma_{ut})_T}{(\sigma_{ut})_{RT}} & \text{for brittle materials.} \end{cases} \tag{6.10}$$

Here, the subscripts RT and T refer to *room temperature* and *operating temperature*, respectively. We determine the ratio of the yield or ultimate tensile stress at the operating and room temperature in Table 6.2, based on the operating temperature. $(\sigma)_T/(\sigma)_{RT}$ refers to either the yield stress or ultimate tensile stress, depending on whether we evaluate a ductile or brittle material.

6.4.5 k_e: Reliability Factor

The R.R. Moore high-speed rotating beam experiment implicitly assumes that half of the specimens have an endurance limit larger than σ'_e, i.e. the reliability is 50%. If we assume that the results from fatigue experiments follow a normal distribution, and transform that normal distribution into a standard normal distribution, then a reliability of 50% corresponds to $z = (\sigma_f - \mu_{\sigma_f})/s_{\sigma_f} = 0$. Here,

Table 6.2 Temperature factor k_d.

Temperature	$(\sigma)_T/(\sigma)_{RT}$
20	1.000
50	1.010
100	1.020
150	1.025
200	1.020
250	1.000
300	0.975
350	0.943
400	0.900
450	0.843
500	0.768
550	0.672
600	0.549

Brandes, E.A. (1983). *Smithells Metals Reference Book*, 6e. Butterworths.

Table 6.3 Reliability factor k_e.

Reliability	z	Reliability factor k_e
50	0	1.000
90	1.288	0.897
95	1.645	0.868
99	2.326	0.814
99.9	3.091	0.753
99.99	3.719	0.702
99.999	4.265	0.659
99.9999	4.753	0.620

we define the fatigue strength σ_f as a random variable that follows a normal distribution with mean μ_{σ_f} and standard deviation s_{σ_f}. z is the standard normal distribution random variable. We use z to determine the reliability factor k_e as

$$k_e = 1 - 0.08z, \tag{6.11}$$

where z directly relates to the standard normal distribution. Increasing the reliability increases z and, thus, decreases k_e (see Table 6.3).

6.4.6 k_f: Miscellaneous Effects Factor

The R.R. Moore high-speed rotating beam experiment requires specific operating parameters. The previous sections show that deviating from these specific operating parameters affects (almost always reduces) the endurance limit of the specimen. However, parameters we did not explicitly consider, such as corrosion or residual stresses, may also affect the endurance limit. We can define a new endurance limit modifying factor k_f that accounts for any parameter or effect that we have not yet included in the other endurance limit modifying factors. Hence, we define the k_f factor based on experiments or simulations to measure the effect of any additional parameter on the endurance limit.

Example problem 6.2
Estimate the endurance limit of a ground shaft with diameter 10 mm, machined from hot rolled steel with $\sigma_{ut} = 1020$ MPa. The shaft is subject to rotating bending and operates in a 300 °C environment.

Solution
We first estimate the endurance limit of the R.R. Moore specimen made of the same material as the shaft using Eq. (6.1), i.e. $\sigma'_e = 0.5\sigma_{ut} = 0.5 \times 1020 = 510$ MPa.
Next, we determine the endurance limit modifying factors to calculate the endurance limit of the shaft, using Eq. (6.5).

Surface factor: We find $a = 1.58$ and $b = -0.085$ from Table 6.1 considering the surface finish of the shaft is *ground*. Even though the problem statement indicates, hot rolled, machined, and ground, we always consider the final surface operation, as this will determine the surface quality of the mechanical element when in use and subject to a dynamic load. Thus, using Eq. (6.6), $k_a = a\sigma_{ut}^b = 1.58(1020)^{-0.085} = 0.877$.

Size factor: We find $k_b = 1.24d^{-0.107} = 1.24(10)^{-0.107} = 0.969$. Note that Eq. (6.8) requires using d in mm.
Load factor: The shaft is subject to rotating bending and, thus, $k_c = 1$.
Temperature factor: Table 6.2 shows that $(\sigma_y)_T/(\sigma_y)_{RT} = 0.975$ in a 300 °C environment. Thus, $k_d = 0.975$.

Because we have no information about k_e and k_f, we set them equal to 1.
We calculate the endurance limit of the shaft σ_e using Eq. (6.5) as
$\sigma_e = 0.877 \times 0.969 \times 1 \times 0.975 \times 510 = 497.25$ MPa.

6.5 Fluctuating Stresses

Thus far, we have only considered alternating stresses. However, a mechanical element is often subject to combined static and dynamic loading that creates a fluctuating stress, i.e. the combination of a steady and an alternating stress, as illustrated in Figure 6.1b. Figure 6.7 schematically shows the amplitude σ versus time t of a fluctuating stress. We indicate the minimum stress σ_{min} and maximum stress σ_{max}, the stress amplitude σ_a, the steady stress σ_m, and the stress range $\sigma_{max} - \sigma_{min}$. Furthermore, we define the stress ratio as $\sigma_{min}/\sigma_{max}$, and the amplitude ratio as σ_a/σ_m.

We first consider two limiting cases of a fluctuating stress.

1. **Alternating stress:** $\sigma_a \neq 0$ and $\sigma_m = 0$. Failure occurs when $\sigma_a \geq \sigma_e$ when considering infinite fatigue life, or when $\sigma_a \geq \sigma_f$ when considering finite fatigue life. We re-emphasize that σ_f corresponds to a specific number of loading cycles N.
2. **Steady stress:** $\sigma_a = 0$ and $\sigma_m \neq 0$. Failure occurs when $\sigma \geq \sigma_y$ (for a ductile material such as steel).

When both $\sigma_a \neq 0$ and $\sigma_m \neq 0$, we use a failure criterion for fluctuating stresses. Several failure criteria exist, including the Goodman and Soderberg line, and the Gerber parabola, among other examples. Figure 6.8 graphically illustrates the Goodman and Soderberg line. Note that the failure criteria describe an envelope, which predicts failure if a stress-state falls outside this envelope. Thus, the Goodman and Soderberg line effectively represent lines along which the safety factor n is constant and equal to 1. We can also describe the Goodman and Soderberg line analytically, i.e.

$$\frac{\sigma_m}{\sigma_{ut}} + \frac{\sigma_a}{\sigma_e} = \frac{1}{n} \text{ (Goodman line)}, \tag{6.12}$$

$$\frac{\sigma_m}{\sigma_y} + \frac{\sigma_a}{\sigma_e} = \frac{1}{n} \text{ (Soderberg line)}. \tag{6.13}$$

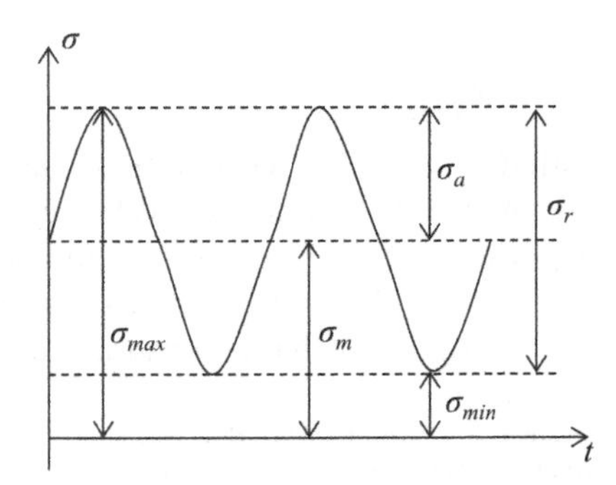

Figure 6.7 Fluctuating stress, illustrating different characteristic parameters.

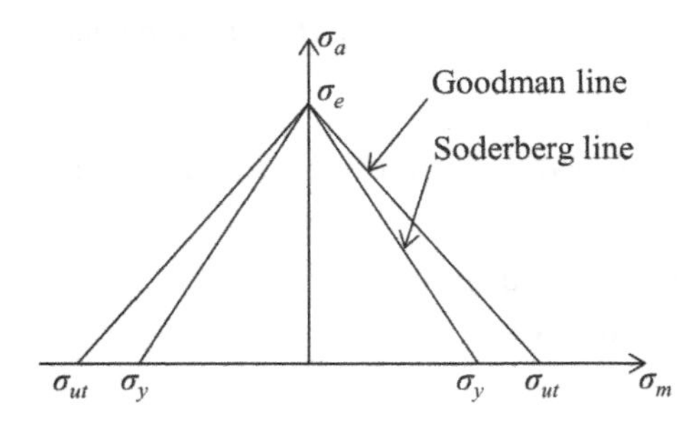

Figure 6.8 Failure criteria for fluctuating stresses, showing the Goodman and Soderberg line.

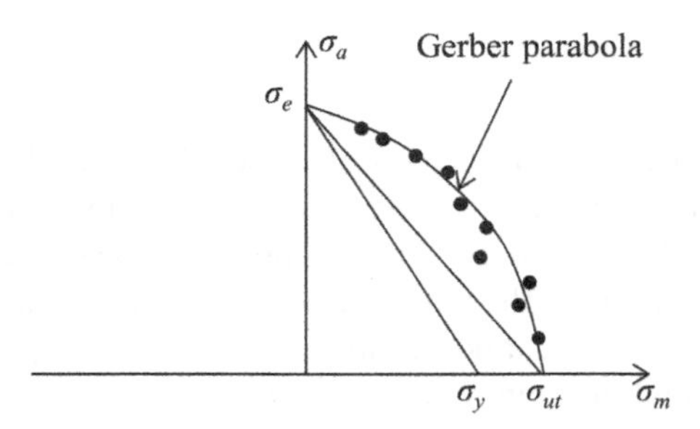

Figure 6.9 Failure criteria for fluctuating stresses, showing the Gerber parabola, relative to the Goodman and Soderberg line.

Experimental evidence shows that the Goodman and Soderberg failure criteria provide conservative predictions of the fatigue life of a mechanical element. Nonlinear failure criteria fit the experimental data better than the linear fits provided by the Goodman and Soderberg line. For instance, Figure 6.9 shows the Gerber parabola relative to the Goodman and Soderberg line. We also schematically show a cloud of experimental data points to illustrate that the Gerber parabola provides a better fit of the experimental data than the Goodman and Soderberg lines.

We describe the Gerber parabola analytically as

$$\left(\frac{n\sigma_m}{\sigma_{ut}}\right)^2 + \frac{n\sigma_a}{\sigma_e} = 1. \tag{6.14}$$

These failure criteria assume a constant stress amplitude throughout the fatigue life of the mechanical element. To consider varying stress amplitudes throughout the life of the mechanical element, we can use the *Palmgren–Miner* theory, or *cumulative fatigue damage* theory. Essentially, the Palmgren–Miner theory explains that the fatigue life of a mechanical element can be *consumed* in an infinite number of ways. One can expose the mechanical element to a few loading cycles of large stress amplitude or, equivalently, to a large number of loading cycles of small stress amplitude. We present the Palmgren–Miner theory as[5]

$$\sum_{i=1}^{k} \frac{n_i}{N_i} = C, \tag{6.15}$$

with ($0.8 \leq C \leq 2.2$) a calibration constant.

Here, n_i is the number of loading cycles that a mechanical element is subject to stress amplitude $\sigma_{a,i}$, whereas N_i is the maximum number of loading cycles to which the mechanical element can be exposed at $\sigma_{a,i}$ before failure occurs (see the S–N curve). Thus, the ratio n_i/N_i represents the fraction of the fatigue life that is consumed by subjecting the mechanical element to n_i loading cycles of

5 See, e.g. Grover, H.J., Gordon, S.A., and Jackson, L.R. (1960). *Fatigue of Metals and Structures*. Bureau of Naval Weapons, Department of the Navy.

stress amplitude $\sigma_{a,i}$. Summing the fractions of the fatigue life that is consumed by subjecting the mechanical element to different stress amplitudes $\sigma_{a,i}$ throughout its lifetime ultimately allows predicting the fatigue life according to the Palmgren–Miner theory.

Example problem 6.3

Figure 6.10 shows a rotating shaft with length $L = 0.5$ m and diameter D, subject to a transverse load $F = 1$ kN. The shaft is machined from AISI 1035 HR steel ($\sigma_y = 270$ MPa, $\sigma_{ut} = 500$ MPa).

(a) Calculate diameter D for infinite fatigue life and consider a design factor $n_D = 2$.
(b) Calculate diameter D for a finite fatigue life of $N = 3 \cdot 10^5$ loading cycles and consider a design factor $n_D = 2$.

Solution

We quantify the external load and internal stress in the shaft to solve both (a) and (b). The reaction force in the bearings is $F/2$ because of symmetry. The maximum bending moment in the shaft occurs in the middle of the shaft where the external load is applied, and is equal to $M_{max} = FL/4 =$ 125 Nm. Figure 6.11 shows the free-body diagram and the internal shear force V and internal bending moment M diagrams, registered with the free-body diagram.

We estimate the endurance limit of the R.R. Moore specimen made of the same material than the rotating shaft using Eq. (6.1), i.e.

$$\sigma_e' = 0.5\sigma_{ut} = 0.5 \times 500 = 250 \text{ MPa}.$$

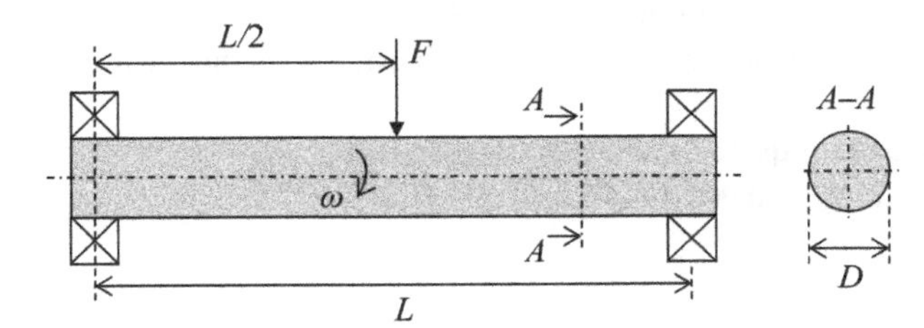

Figure 6.10 Rotating shaft with diameter D.

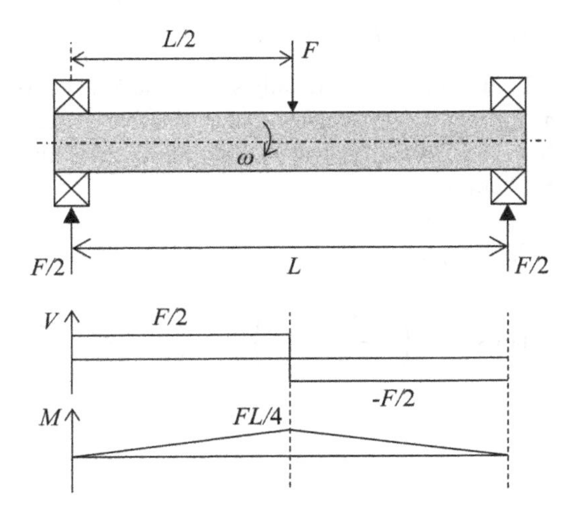

Figure 6.11 Free-body diagram and internal force diagrams.

Next, we determine the endurance-limit modifying factors to calculate the endurance limit of the shaft, using Eq. (6.5).

Surface factor: We find $a = 4.51$ and $b = -0.265$ from Table 6.1, considering that the shaft is *machined*. Using Eq. (6.6), $k_a = a\sigma_{ut}^b = 4.51(500)^{-0.265} = 0.868$.

Size factor: We find $k_b = 1.24D^{-0.107}$. However, we note that D is the shaft diameter we attempt to calculate in both (a) and (b), i.e. it is unknown. Hence, we estimate $k_b = 0.85$, and will verify this assumption after calculating D.

Load factor: The shaft is subject to rotating bending and, thus, $k_c = 1$.

Because we have no information about k_d, k_e, and k_f, we set them equal to 1.

We calculate the endurance limit of the shaft σ_e using Eq. (6.5) as

$$\sigma_e = 0.868 \times 0.850 \times 1 \times 500 = 368.90 \text{ MPa}.$$

(a) Design for infinite fatigue life: $n_D = \sigma_e/\sigma_a$, where σ_a is the maximum stress amplitude in the shaft.

$$\sigma_a = \frac{M_{max}D/2}{\pi D^4/64} = \frac{32M_{max}}{\pi D^3},$$

$$n_D = \frac{\pi D^3 \sigma_e}{32M_{max}}.$$

$$\text{Thus, } D = \left(\frac{32M_{max}n_D}{\pi\sigma_e}\right)^{1/3} = \left(\frac{32 \times 125 \times 2}{\pi \times 368.90 \cdot 10^6}\right)^{1/3} = 0.019 \text{ m}.$$

We now check the assumption we made for k_b. $k_b = 1.24(19)^{-0.107} = 0.90$. Since $0.90 > 0.85$, the assumption we made was appropriate, as we slightly underestimated k_b.

(b) Design for finite fatigue life of $N = 3 \cdot 10^5$ loading cycles: $n_D = \sigma_f/\sigma_a$, and σ_a is the maximum stress amplitude in the shaft.

$\sigma_f = aN^b$, where we calculate a and b using Eqs. (6.3) and (6.4), respectively.

We determine the fatigue strength fraction $f = 0.90$ from Figure 6.5. Thus,

$$a = \frac{(f\sigma_{ut})^2}{\sigma_e} = \frac{(0.90 \times 500 \cdot 10^6)^2}{368.90 \cdot 10^6} = 548.93 \text{ MPa},$$

$$b = -\frac{1}{3}\log\left(\frac{f\sigma_{ut}}{\sigma_e}\right) = -\frac{1}{3}\log\left(\frac{0.90 \times 500 \cdot 10^6}{368.90 \cdot 10^6}\right) = -0.0287.$$

Thus, $\sigma_f = aN^b = 548.93 \cdot 10^6(3 \cdot 10^5)^{-0.0287} = 382.25$ MPa.

Figure 6.12 schematically shows the *S–N* curve for this material, indicating the fatigue strength σ_f for $N = 3 \cdot 10^5$ loading cycles, relative to the endurance limit σ_e. Note that $\sigma_f = 382.25$ MPa for $N = 3 \cdot 10^5$ loading cycles, which is slightly higher than $\sigma_e = 368.90$ MPa, as we expect from the *S–N* curve.

$$n_D = \frac{\pi D^3 \sigma_f}{32M_{max}}.$$

$$\text{Thus, } D = \left(\frac{32M_{max}n_D}{\pi\sigma_f}\right)^{1/3} = \left(\frac{32\times125\times2}{\pi\times382.25\cdot10^6}\right)^{1/3} = 0.0188 \text{ m}.$$

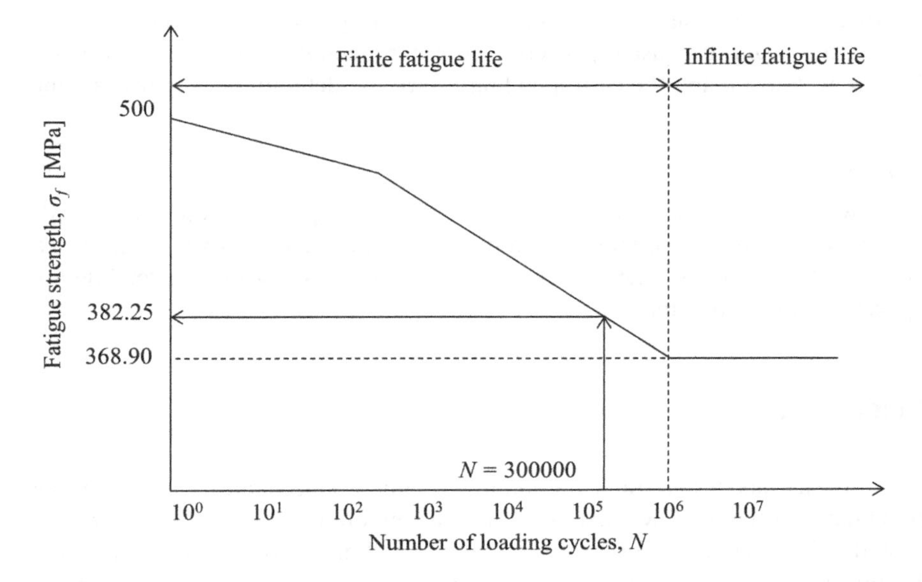

Figure 6.12 *S*–*N* curve, indicating the fatigue life corresponding to $N = 3 \cdot 10^5$ loading cycles.

6.6 Stress Concentrations

Stress concentrations relate the maximum local stress around irregularities or discontinuities in the geometry of a mechanical element to the nominal stress if that irregularity or discontinuity would not exist. We have previously defined the theoretical stress concentration factor K_t (see Eq. (5.15)) and the theoretical shear stress concentration factor K_{ts} (see Eq. (5.16)). These stress concentration factors are valid in the case of a static load. Here, we briefly discuss how to account for stress concentrations in the context of a dynamic load.

Specifically, we define the fatigue stress concentration factor, or sometimes called fatigue notch factor, as

$$K_f = \frac{\text{Fatigue strength of notch-free specimen}}{\text{Fatigue strength of notched specimen}}. \tag{6.16}$$

K_f represents the reduced effect of stress concentrations under dynamic compared to static load. However, the reason for the reduced sensitivity is not well-understood. To quantify the sensitivity of a material to notches (characterized by a notch radius), we define the so-called *notch sensitivity* q, i.e.,

$$q = \frac{K_f - 1}{K_t - 1}, \tag{6.17}$$

or the equivalent relationship for shear stresses using K_{ts} instead of K_t. Note that $0 \leq q \leq 1$. We observe that if $q = 0$, then $K_f = 1$, i.e. there is no effect of the stress concentration on the alternating stress component. Alternatively, if $q = 1$ ("full notch sensitivity"), then $K_f = K_t$, and the effect of the stress concentration is identical under dynamic and static loading.

The notch sensitivity of a specific material depends on the material properties, the type of loading, and the geometry of the notch. It is usually determined experimentally and tabulated in reference works. Thus, if we know q and the corresponding K_t, we can determine K_f by rearranging Eq. (6.17) as

$$K_f = q\left(K_t - 1\right) + 1. \tag{6.18}$$

In this textbook, we will always assume full notch sensitivity, i.e. $q = 1$ because it represents the worst-case. However, in reality, we must determine q from a diagram that shows the notch sensitivity versus the notch radius for different values of the ultimate tensile stress, and for different types of loading (e.g. bending versus torsion).[6]

6.7 Key Takeaways

1. We use the stress-life method to determine the endurance limit of a mechanical element subject to high-cycle fatigue. To that end, we first determine the endurance limit of the R.R. Moore high-speed rotating beam specimen of the same material and, then, we use endurance-limit modifying factors to account for the differences between the mechanical element we design and the R.R. Moore specimen.
2. Two types of dynamic loading exist: alternating and fluctuating stresses.
3. We calculate a mechanical element subject to an alternating stress for infinite fatigue life by comparing the maximum stress amplitude to the endurance limit. Similarly, we calculate a mechanical element subject to an alternating stress for finite fatigue life by comparing the maximum stress amplitude to the fatigue strength for any finite number of loading cycles.
4. We use failure criteria such as the Goodman and Soderberg line, and the Gerber parabola, to calculate a mechanical element for infinite fatigue life, when subject to a fluctuating stress.
5. An alternative way to think about fatigue life is the Palmgren–Miner or cumulative damage theory. This theory allows considering mechanical elements that are subject to varying stress amplitude throughout their lifetime.

6.8 Problems

6.1 Figure 6.13 shows a solid, rotating shaft with diameter $D = 50$ mm, machined from AISI 1080 HR steel. The shaft is subject to a force F, with $\alpha = 45°$, and it operates in an environment of 400 °C. 99% reliability is required. Calculate the endurance limit of the shaft.

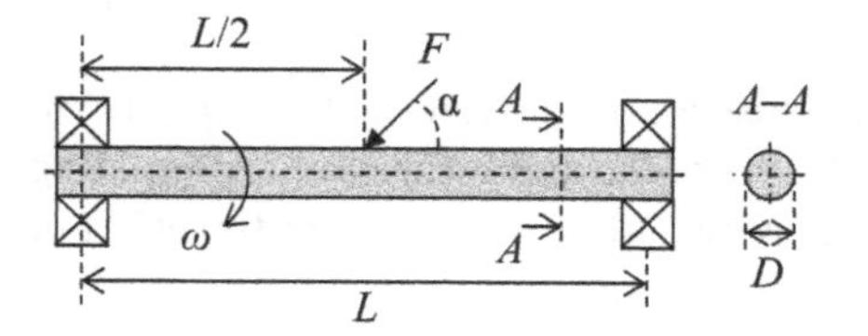

Figure 6.13 Solid rotating shaft.

6 See, e.g. Sines, G., Waisman, J.L., and Dolan, T.J. (1959). *Metal Fatigue*. McGraw-Hill.

6.2 Figure 6.14 shows a cantilever beam with square cross-section and $b = 40$ mm, machined from AISI 1080 HR steel. The cantilever beam is subject to an alternating transverse load $F = F_a \sin \omega t$. Calculate the endurance limit of the cantilever beam.

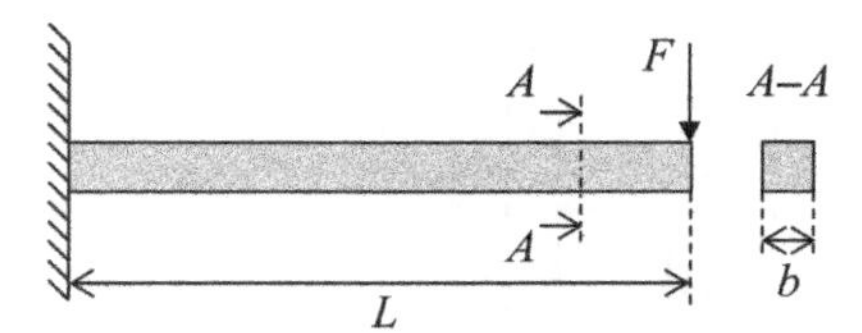

Figure 6.14 Cantilever beam with square cross-section.

6.3 Figure 6.15 shows a solid, rotating shaft, machined from AISI 1035 HR steel and ground after machining. $L = 0.5$ m. $F = 1$ kN.

(a) Calculate the fatigue strength that corresponds to $2 \cdot 10^5$ loading cycles.
(b) Calculate the diameter D required to obtain a fatigue life of $2 \cdot 10^5$ loading cycles and consider a design factor $n_D = 2$.
(c) Calculate the safety factor against static failure for the value of D calculated in (b).

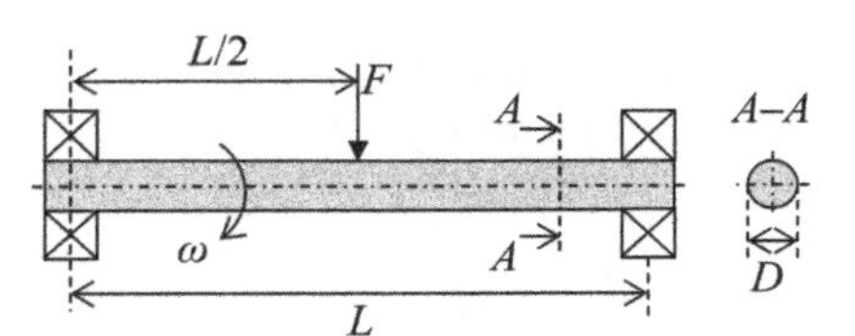

Figure 6.15 Solid, rotating shaft with transverse load.

6.4 Figure 6.16 shows a solid, rotating shaft, machined from AISI 1050 HR steel. $L = 0.5$ m. $F = 1$ kN.

(a) Calculate the fatigue strength that corresponds to $3 \cdot 10^5$ loading cycles.
(b) Calculate the diameter D required to obtain a fatigue life of $3 \cdot 10^5$ loading cycles and consider a design factor $n_D = 3$.
(c) Calculate the safety factor against static failure for the value of D calculated in (b).

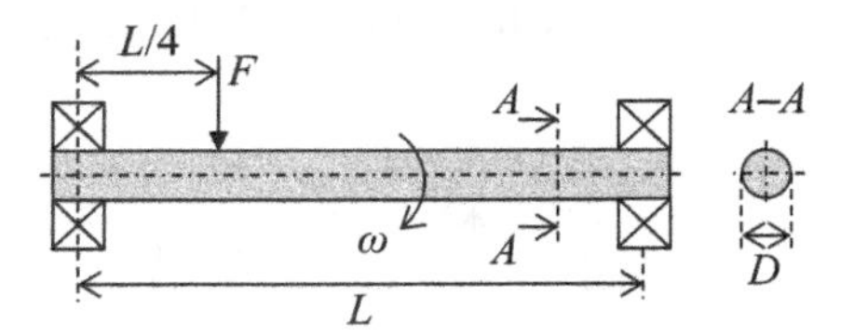

Figure 6.16 Solid rotating shaft with transverse load.

6.5 Re-consider Figure 6.16 but consider a shaft diameter $D = 30$ mm. Calculate the maximum alternating torque moment that can be superimposed on the existing loading of Figure 6.16, to achieve a fatigue life of $1.5 \cdot 10^5$ cycles, and consider a design factor $n_D = 2$.

6.6 Figure 6.17 shows a mixer, mounted on a rotating shaft, machined from AISI 1020 HR steel, and ground after machining. The shaft is subject to a transverse force $F = 2$ kN. $L = 0.5$ m. Calculate the diameter D of the shaft for infinite fatigue life and consider a design factor $n_D = 2$.

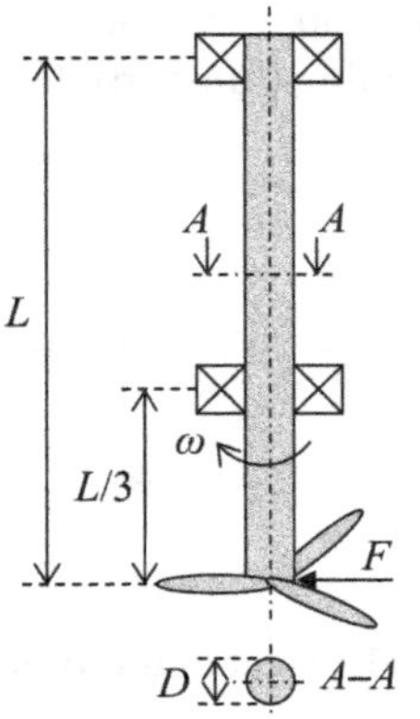

Figure 6.17 Mixer mounted on a rotating shaft.

6.7 Figure 6.18 shows a rotating shaft with $D = 50$ mm and $d = 20$ mm, machined from AISI 1040 HR steel. $L = 0.5$ m.

(a) Calculate the maximum force F that can be imposed on the shaft to achieve a fatigue life of 50 000 loading cycles and consider a design factor $n_D = 4$.

(b) Calculate the safety factor against static failure.

(c) How does the answer to (a) change when considering infinite fatigue life and a design factor $n_D = 4$?

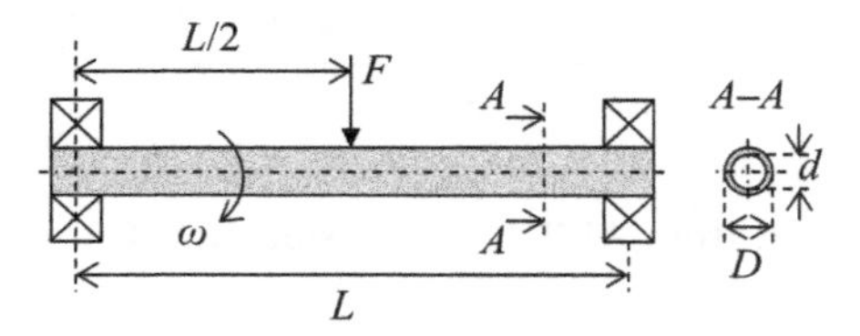

Figure 6.18 Rotating shaft.

6.8 Figure 6.19 shows a cantilever beam with circular cross-section and diameter D, machined from AISI 1055 HR steel. A handle is mounted on the end of the cantilever beam and is subject to a force $F = 200\sin(\omega t)$. $L_1 = 0.2$ m, $L_2 = 0.125$ m. Calculate D to achieve infinite fatigue life using the Soderberg fatigue failure criterion, and consider a design factor $n_D = 4$.

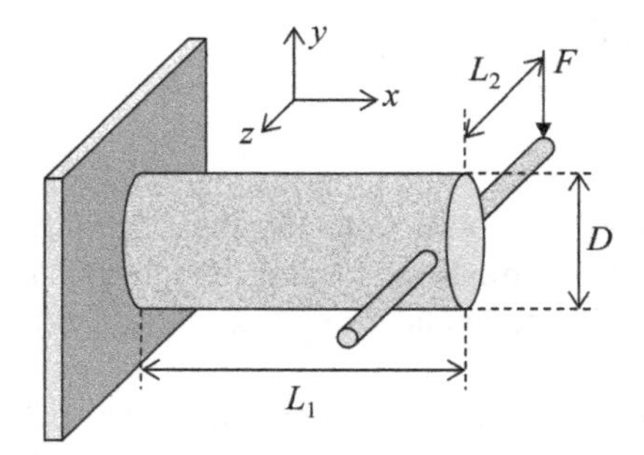

Figure 6.19 Cantilever beam with circular cross-section.

6.9 A mechanical element is made from AISI 1018 HR steel and is subject to a fluctuating stress. The alternating component $\sigma_a = 25$ MPa results from rotating bending and the steady component $\sigma_m = 40$ MPa results from an axial load. The mechanical element has a circular cross-section with diameter $D = 40$ mm.

(a) Calculate the endurance limit.
(b) Calculate the safety factor for infinite fatigue life using the Gerber theory.
(c) Calculate the safety factor for infinite fatigue life using the Goodman theory.
(d) Calculate the safety factor for infinite fatigue life using the Soderberg theory.
(e) Calculate the safety factor against yielding (static failure).

6.10 A mechanical element is made from AISI 1018 HR steel and is subject to a fluctuating stress. The alternating component $\sigma_a = 25$ MPa results from rotating bending and the steady component $\sigma_m = 40$ MPa results from an axial load. Furthermore, it is subject to a steady shear stress $\tau_m = 20$ MPa that results from a steady torque moment. The mechanical element has a circular cross-section with diameter $D = 40$ mm.

(a) Calculate the endurance limit.
(b) Calculate the safety factor for infinite fatigue life using the Gerber theory.
(c) Calculate the safety factor for infinite fatigue life using the Goodman theory.
(d) Calculate the safety factor for infinite fatigue life using the Soderberg theory.
(e) Calculate the safety factor against yielding (static failure).

6.11 A mechanical element is made from AISI 1080 HR steel and is subject to a fluctuating stress. The alternating component $\sigma_a = 50$ MPa results from rotating bending and the steady component $\sigma_m = 100$ MPa results from an axial load. Furthermore, it is subject to a shear stress $\tau = 20 + 40\sin(20\pi t)$ MPa that results from a torque moment. The mechanical element has a circular cross-section with diameter $D = 50$ mm.

(a) Calculate the endurance limit.
(b) Calculate the safety factor for infinite fatigue life using the Gerber theory.
(c) Calculate the safety factor for infinite fatigue life using the Goodman theory.
(d) Calculate the safety factor for infinite fatigue life using the Soderberg theory.
(e) Calculate the safety factor against yielding (static failure).

7

Shafts

7.1 Introduction

In this chapter, we will use the knowledge of Chapters 5 and 6, and apply it directly to the design of shafts. We define a shaft as a rotating mechanical element that transmits power or motion. This is different than an axle, which is a nonrotating member that does not transmit torque and only supports, e.g. rotating wheels or pulleys. Nothing is unique about the design of shafts. However, it is perhaps the most common machine element and, therefore, we pay extra attention to it.

7.1.1 Practical Considerations Related to Shaft Design

Shaft design involves both local stress and deflection calculations. Local stress in the shaft depends on the local geometry and dimensions of the shaft. In contrast, deflection of the shaft depends on the entire (global) geometry of the shaft. Hence, we first calculate the local stress in the shaft to determine the dimensions of the shaft geometry and, afterwards, we calculate the deflection of the shaft. We design the shaft dimensions to achieve the required design factor and account for the material we select through its yield stress. Alternatively, we select a material with yield stress that provides the required safety (design) factor for a given shaft geometry. Thus, the material choice primarily depends on the local stress in the shaft.

Deflection depends on the stiffness of the shaft, which in turn depends on the length of the shaft L, the area moment of inertia of the cross-section of the shaft I (which might vary along the shaft if the shaft dimensions change), and the Young's modulus of the shaft material E. For instance, if we consider a shaft supported on both ends and subject to a transverse load F in the middle between both ends, then the maximum deflection occurs in the middle of the shaft and is $\delta = FL^3/48EI$ (considering small elastic deformations)(see Appendix B). Hence, the (bending) stiffness of the shaft $k = F/\delta$ or, $k = 48EI/L^3$. The stiffness of the shaft is proportional to the Young's modulus of the shaft material and the area moment of inertia of the shaft cross-section and inverse proportional to the third power of the length of the shaft. Therefore, the material choice of the shaft, which enters the stiffness equation through the Young's modulus E, has a limited effect on k and, thus, on the deflection of the shaft. Indeed, the Young's modulus is almost constant for most steels (approximately $E = 210\,\text{GPa}$), independent of its alloying elements. On the other hand, changing the length of the shaft has a significant influence on the stiffness of the shaft, i.e. shortening the shaft increases its stiffness by a cubic factor. However, the length of the shaft might not necessarily be a design variable because it is likely defined by the dimensions of a machine, or an application in which it must transmit power or motion. Finally, the geometry

Design of Mechanical Elements: A Concise Introduction to Mechanical Design Considerations and Calculations, First Edition. Bart Raeymaekers.

Companion website: www.wiley.com/go/raeymaekers/designofmechanicalelements

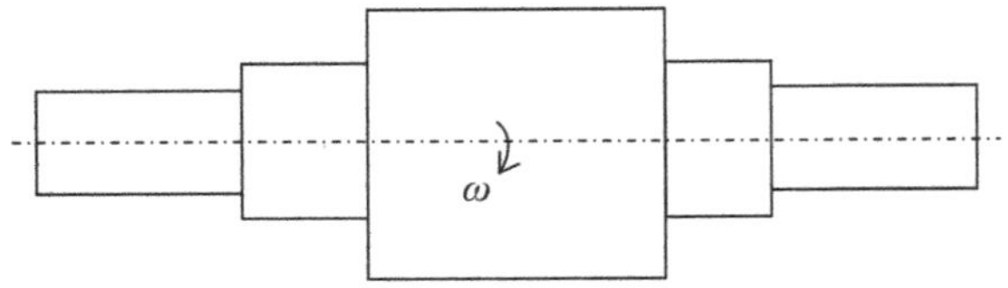

Figure 7.1 Schematic of a shaft layout, showing a typical stepped-cylinder design.

and dimensions of the cross-section of the shaft have a substantial influence on the stiffness on the shaft. For instance, the area moment of inertia of a shaft with a circular cross-section changes with the fourth power of the diameter (see Eq. (5.12)) and, increasing the diameter of the shaft (even a circular tube, see Eq. (5.11)) substantially increases its stiffness. Thus, it is more desirable to control the deflection of a shaft by changing the geometry and dimensions of its cross-section (area moment of inertia) than by changing the material.

The question arises how to weigh strength and deflection when designing a shaft. Practically, we start by selecting an inexpensive low or medium-carbon steel (e.g. AISI 1020-1050) as an economical choice. We then perform local stress calculations. If strength requirements dominate over deflection considerations, i.e. the material could locally yield as a result of the external load, we select a steel with a higher yield stress (and often higher price tag!). Conversely, if deflection requirements dominate over strength considerations, we change the dimensions of the shaft. In general, when performing this analysis, we weigh the cost of the material against the need for reducing the shaft diameter.

The shaft layout is also an important consideration, and it is usually specified early in the design process because we account for it in the stress and deflection calculations. Figure 7.1 schematically illustrates a typical stepped-cylinder shaft layout. This layout allows designing the shaft with different diameter sections to ensure that the local stress does not exceed the yield stress anywhere in the shaft. Furthermore, it creates a *shoulder* to position different mechanical elements such as a bearing, a pulley, or a gear, needed to transmit motion or power. As such, different sections of the shaft might also require a different surface finish and/or material properties such as hardness, prescribed by the manufacturer of the mechanical element. For instance, bearing manufacturers usually require a specific surface hardness and roughness to maximize bearing durability and longevity. Finally, when deciding on a shaft layout, we must also consider assembly and disassembly of the shaft and the different mechanical elements that we will mount on it.

7.1.2 Torque Transmission

A shaft transfers torque or motion by means of, e.g. gears, couplings, clutches, and belt/pulley combinations. Here, we will limit the discussion to two ways of torque transmission because the focus of this chapter is on the design and verification calculations of the shaft itself. Indeed, the type of torque transmission only determines how we convert the external load to boundary conditions and reaction forces in the free-body diagram. It does not alter the methodology we use to calculate local stress and/or deflection of a shaft.

7.1.2.1 Relationship Between Torque, Power, and RPM

The first method of torque transfer we consider in this textbook requires understanding the relationship between torque, power, and rotational velocity (RPM). We recognize that the power P relates to the torque M_t and the angular frequency of the shaft ω as

$$P = M_t\omega. \tag{7.1}$$

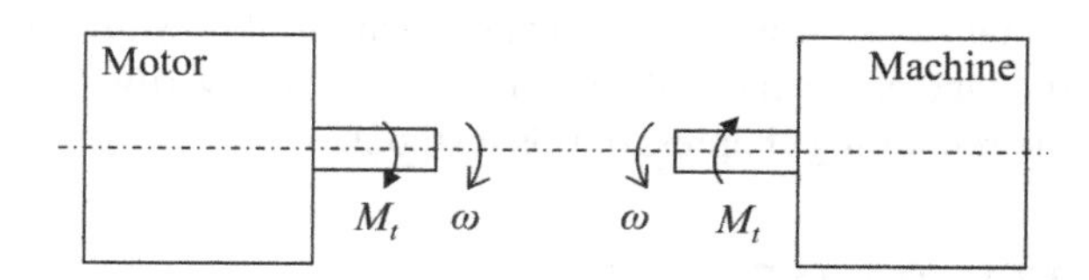

Figure 7.2 Motor-machine torque transmission.

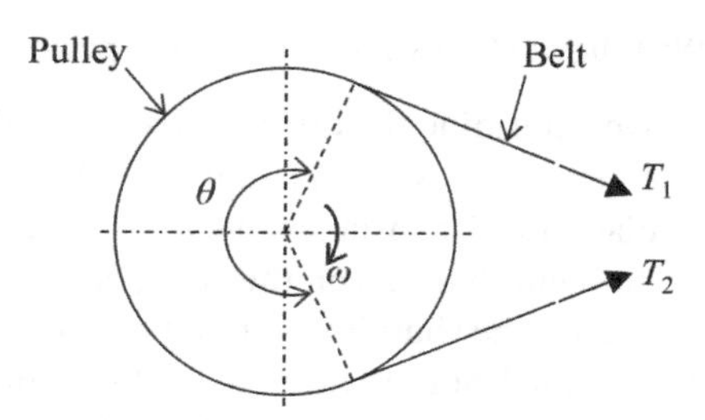

Figure 7.3 Belt–pulley torque transmission.

We relate the angular frequency ω to the shaft RPM N as

$$\omega = \frac{2\pi N}{60}. \tag{7.2}$$

Combining Eqs. (7.1) and (7.2) yields

$$P = \frac{2\pi N M_t}{60}. \tag{7.3}$$

This relationship is useful to understand *motor-machine* torque and motion transmission combinations. Figure 7.2 shows a motor and a machine. We characterize a motor with a torque moment M_t and an angular frequency ω oriented in the same direction and sense, i.e. the motor drives the shaft and it rotates as a result. Conversely, we characterize a machine with its torque moment and angular frequency oriented in the same direction and opposite sense. The machine acts as *inertia* driven by a motor and, thus, resists the torque of the motor that makes the machine shaft rotate in the same direction and sense as that of the motor.

7.1.2.2 Belt–Pulley Torque Transmission

The second method of torque transfer we consider in this textbook requires understanding the belt–pulley relationship. Figure 7.3 shows a belt and pulley combination. The *tight-side* tension in the belt is T_1 and the *slack-side* tension is T_2, and we identify the tight- from the slack-side of the belt by considering the direction in which the pulley rotates, as indicated by its angular frequency ω. Additionally, we indicate the *wrap angle* θ, which describes the angle over which the belt makes contact with the pulley. As such, the belt–pulley equation, or sometimes also called the capstan equation, relates the tight-side to the slack-side tension as

$$\frac{T_1}{T_2} = e^{\mu\theta}, \tag{7.4}$$

with μ the friction coefficient between the belt and the pulley.

7.2 Recipe for Shaft Calculations

A *design calculation* involves determining an unknown dimension, e.g. the diameter of a shaft, considering specific external loading and boundary conditions. In contrast, a *verification calculation* requires calculating the safety factor of an already finished design, i.e. the geometry and dimensions of the mechanical element are known *a priori*. These definitions are independent of the type of mechanical element we design or verify, and we will use this terminology throughout the remaining chapters of this textbook.

7.2.1 Design Calculation

We follow a five-step procedure to perform a design calculation of a shaft.

1. We determine the external loading and boundary conditions. The purpose of this first step is to convert any external loading and boundary conditions into a format that fits into the free-body diagram. For instance, we may need to convert the power and RPM of a motor into a torque moment that acts on the shaft, or we must convert the force acting on a pulley into a force that acts on the shaft in combination with a torque moment.
2. Starting from the external loading, boundary conditions, and reaction forces, we determine the free-body diagram of the shaft. Then, we derive the internal force diagrams, including axial and shear force diagrams, and the bending and torque moment diagrams.
3. We determine the cross-section of the shaft that is subject to the heaviest load. Unless we design a specific portion of the shaft, we always design or verify the shaft at the cross-section that is subject to the heaviest load. If the shaft does not fail at that cross-section, it will not fail anywhere else. Determining the cross-section that is subject to the heaviest load is often straightforward, but can also require calculation.
4. We determine all steady and alternating stress components present in the shaft, i.e. steady normal stress σ_m, steady shear stress τ_m, alternating normal stress σ_a, and alternating shear stress τ_a. We first determine which stress components are zero and nonzero. Then, we calculate the nonzero stress components, which we express as a function of the unknown dimension that we design, e.g. the diameter of a shaft. If multiple stress components of the same type exist, then we superimpose them. For instance, several axial loading components each cause a steady normal stress $\sigma_{m,1}, \sigma_{m,2}, \ldots$. Then, superposition of these stress components yields $\sigma_m = \sigma_{m,1} + \sigma_{m,2}$, and we account for the sense of the stress vectors by either adding or subtracting the different stress components.
   After determining each nonzero stress component, we combine the normal and shear stress components into an equivalent normal stress. Because of a static and dynamic load result in fundamentally different failure mechanisms, we keep them separate, but use the Von Mises failure criterion for ductile materials to calculate the equivalent steady normal stress $\sigma_{m,eq} = \sqrt{\sigma_m^2 + 3\tau_m^2}$ and the equivalent alternating normal stress $\sigma_{a,eq} = \sqrt{\sigma_a^2 + 3\tau_a^2}$.
5. Finally, we determine the unknown dimension of the mechanical element by using the definition of the design factor, which is a design specification. For a static load, the design factor relates the yield stress to the maximum stress occurring in the design, i.e. $n_D = \sigma_y/(\sigma_{m,eq} + \sigma_{a,eq})$. For a dynamic load, we use a fatigue failure criterion such as the Soderberg or Goodman line to calculate the unknown dimension of the mechanical element.

7.2.2 Verification Calculation

The process of performing a verification calculation is almost identical to that of a design calculation and, thus, we use the same five-step methodology as described above. However, the critical difference between the verification and design calculation is that in the verification calculation, we know the entire geometry of the mechanical element *a priori*, and we solve for the unknown safety factor. In contrast, in a design calculation, we know the design factor and solve for an unknown dimension of the mechanical element.

7.3 Example Calculations

Example problem 7.1

Figure 7.4 shows a package delivery drone. A 3 kW motor powers the drone. The shaft is machined from AISI 1060 HR steel and rotates at $\omega = 104.7$ RAD/s. Design the diameter D of the motor shaft for infinite fatigue life using the Soderberg failure criterion and consider a design factor $n_D = 2$. Assume the loading shown in Figure 7.4 remains constant throughout the life of the shaft and that the drone and package are weightless. When the drone moves it creates a drag force $F = 500$ N, as indicated in Figure 7.4. $L_1 = 0.1$ m and $L_2 = 0.3$ m. Calculate the safety factor against yielding.

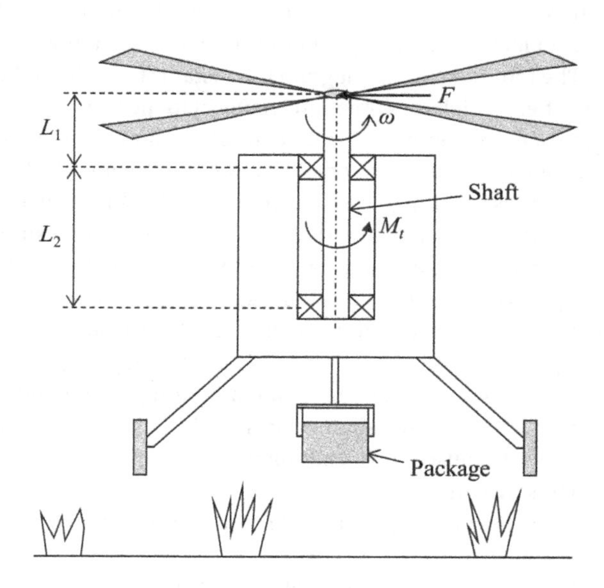

Figure 7.4 Package delivery drone.

Solution

The material properties of AISI 1060 HR steel are $\sigma_y = 370$ MPa, $\sigma_{ut} = 680$ MPa.

Step 1: We convert the power and RPM of the motor to a torque moment that applies to the motor shaft using Eq. (7.3), i.e.

$P = M_t 2\pi N/60 \Rightarrow M_t = 60P/2\pi N = 60 \times 3000/(2\pi \times 1000) = 28.64$ Nm.

Step 2: We determine the free-body diagram (Figure 7.5) and calculate the reaction forces $V_A = 166.67$ N and $V_C = -666.67$ N, using equilibrium of forces and the moment equation around a fixed point. We note that the torque moment caused by the motor remains constant throughout the entire length of the shaft because there is no other torque moment reinforcing or counteracting it.

Step 3: We determine the cross-section of the shaft that is subject to the heaviest load because the problem statement does not specify a location where to design the shaft. Thus, we determine

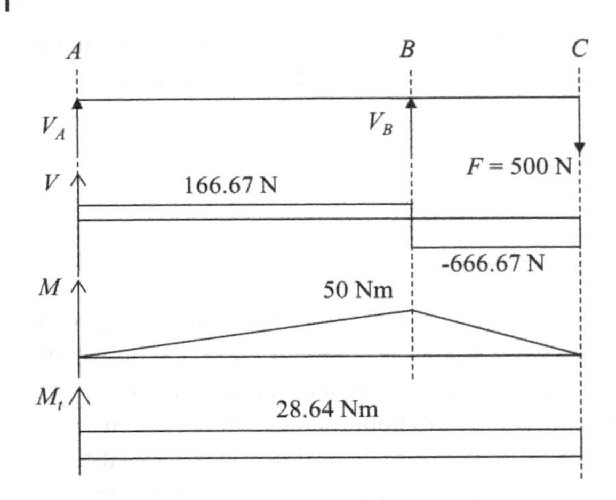

Figure 7.5 Shaft design, free-body diagram.

where the shaft will fail first. If we design the shaft at that location such that it satisfies the specified design factor, then every other location of the shaft will also at least satisfy that design factor.

The torque moment diagram is constant throughout the entire length of the shaft and, thus, does not affect which cross-section is subject to the heaviest load. The bending moment diagram shows a zero bending moment at both ends of the shaft (A and C). Thus, the shaft is unlikely to fail at these locations. Because we consider long, thin beams (Euler–Bernoulli beam theory), we neglect the shear stress resulting from any transverse loading (but account for the bending moment it creates!). Thus, the location where the bending moment is maximum, B, is subject to the heaviest load.

Step 4: This is a design calculation, and we determine that

$\sigma_m = 0$ because the shaft is not subject to an axial load. In addition, the shaft rotates and is subject to a transverse load. Thus, the normal stress resulting from bending is alternating, not steady.

$\tau_m \neq 0$ because the shaft is subject to a steady torque moment. A steady torque moment results in a steady shear stress, whereas an alternating torque moment results in an alternating shear stress.

$\sigma_a \neq 0$ because the shaft rotates while subject to a transverse load.

$\tau_a = 0$ because no alternating torque moment exists.

Thus, we must calculate σ_a and τ_m.

Using the flexure formula (Eq. (5.7)), we find that $\sigma_a = \frac{M_{max}D/2}{\pi/64\ D^4}$, with M_{max} the maximum bending moment.

Rearranging the terms yields $\sigma_a = 32M_{max}/\pi D^3 = 509.30/D^3$.

Using the torque formula (Eq. (5.10)), we find that $\tau_m = \frac{M_t D/2}{\pi/32\ D^4}$, with M_t the torque moment.

Rearranging the terms yields $\tau_m = 16M_t/\pi D^3 = 145.86/D^3$.

After determining the individual stress components, we calculate the equivalent normal stress using the Von Mises criterion to combine the respective static and dynamic stress components.

$$\sigma_{a,eq} = \sqrt{\sigma_a^2 + 3\tau_a^2} = 509.30/D^3,$$

$$\sigma_{m,eq} = \sqrt{\sigma_m^2 + 3\tau_m^2} = \sqrt{3}\tau_m = 252.64/D^3.$$

To use the Soderberg fatigue failure criterion, we determine the endurance limit of the shaft. We first calculate the endurance limit of the R.R. Moore specimen of the same material.

$$\sigma_e' = 0.5 \times 680 = 340 \text{ MPa}.$$

Then, we use the endurance limit modifying factors to modify the endurance limit of the R.R. Moore specimen into that of the shaft we design.

$k_a = 4.51 \times 680^{-0.265} = 0.80$ (see Table 6.1, and Eq. (6.6)).
$k_b =$ unknown. We assume $k_b = 0.85$ and will verify after the design calculation.
$k_c = 1$ (combined loading).

$$\sigma_e = k_a \times k_b \times k_c \times \sigma_e' = 0.80 \times 0.85 \times 1 \times 340 = 231.2 \text{ MPa}.$$

Step 5: We apply the Soderberg fatigue failure criterion, i.e.

$$\frac{1}{2} = \frac{509.30}{231.2 \cdot 10^6 D^3} + \frac{252.64}{370 \cdot 10^6 D^3}.$$

Thus, $D = 0.018$ m, or $D = 18$ mm.
Check $k_b = 1.24 \times 18^{-0.107} = 0.91$ (see Eq. (6.8)).
Thus, the assumption of $k_b = 0.85$ was conservative.
Finally, we calculate the safety factor against static yielding.

$$\sigma_{a,eq} = 509.30/D^3 = 509.30/0.018^3 = 87.33 \text{ MPa},$$
$$\sigma_{m,eq} = 252.64/D^3 = 252.64/0.018^3 = 43.32 \text{ MPa},$$
$$n = \sigma_y/(\sigma_{a,eq} + \sigma_{m,eq}) = 370/(87.33 + 43.32) = 2.83.$$

Example problem 7.2
Figure 7.6 shows a rotating shaft supported in two bearings in A and C, respectively. A belt and pulley combination drives the shaft. The diameter of the pulley $D_{pulley} = 0.1$ m, and its weight is 4 kg. The tight-side tension $T_1 = 500$ N, the friction coefficient between the belt and pulley $\mu = 0.6$, and the wrap angle of the belt around the pulley $\theta = 180°$. A 50 kg flywheel is mounted in B. The shaft is a circular tube made from AISI 1050 HR steel, with length $L = 0.9$ m, outer diameter $D = 0.040$ m, and inner diameter d.

(a) Calculate the slack-side tension T_2 in the belt.
(b) Calculate the inner diameter d of the circular tube to obtain infinite fatigue life using the Soderberg criterion and consider a design factor $n_D = 4$.
(c) Calculate the safety factor against yielding.

Solution
(a) We use Eq. (7.4) to calculate the slack-side tension in the belt.
Thus, $T_2 = T_1 \exp(-\mu\theta) = 500 \exp(-0.6\pi) = 75.92$ N.
(b) The material properties of AISI 1050 HR steel are $\sigma_y = 340$ MPa, $\sigma_{ut} = 620$ MPa.

Step 1: We determine the external loading and boundary conditions. We note that external forces and moments act in two distinct planes; the *xy*-plane and the *xz*-plane, referring to the Cartesian coordinate system of Figure 7.6. The weight of the pulley and flywheel act in the negative *y*-direction, and reaction forces exist in the bearings at A and C. Furthermore, the shaft is subject to the tension in both sides of the belt, oriented in the positive *z*-direction, which also causes reaction forces in the bearings at A and C.

Step 2: We determine the free-body diagram of the shaft and note that we must use two free-body diagrams; one in the *xy*-plane and one in the *xz*-plane because Step 1 reveals that the external loading acts in two different coordinate planes. Thus, Figure 7.7 shows the two free-body diagrams and the corresponding shear force, bending moment, and torque moment diagrams.

With respect to the *xy*-plane, we determine reaction forces $V_A = 230$ N and $V_C = 310$ N, using equilibrium of forces and the moment equation around a fixed point. Similarly, the reaction force in the *xz*-plane are $V_A = -287.96$ N and $V_C = 863.88$ N. We note that in the *xz*-plane, we account for

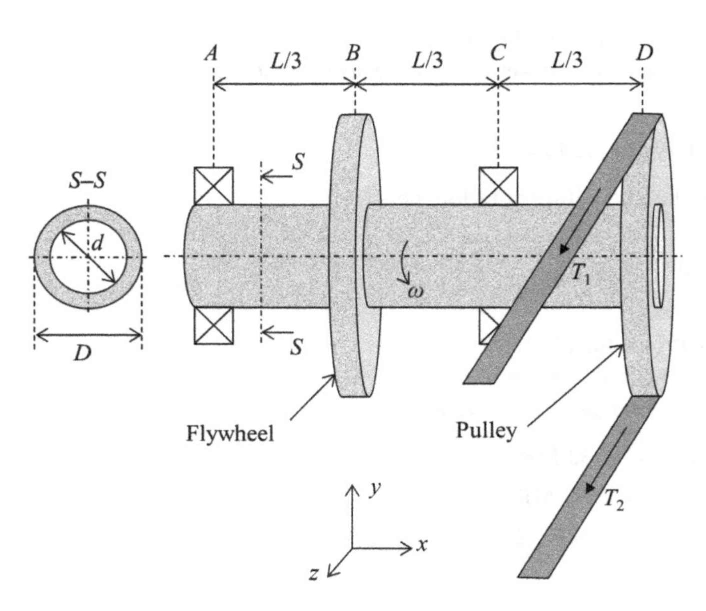

Figure 7.6 Shaft design.

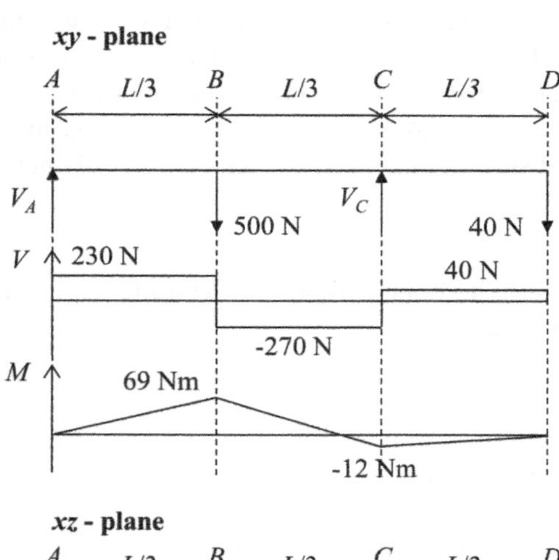

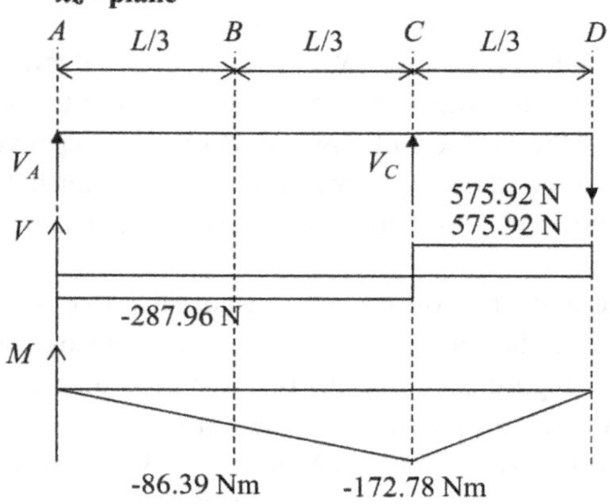

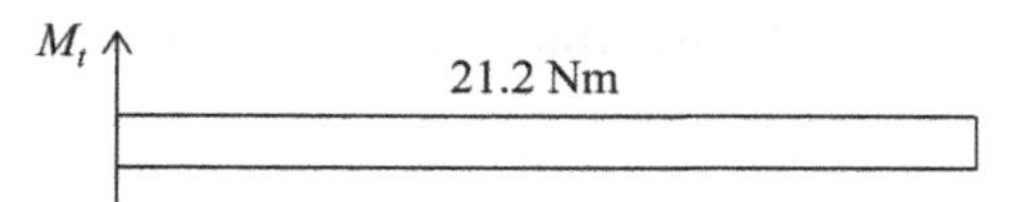

Figure 7.7 Free-body diagrams in the *xy*-plane and *xz*-plane.

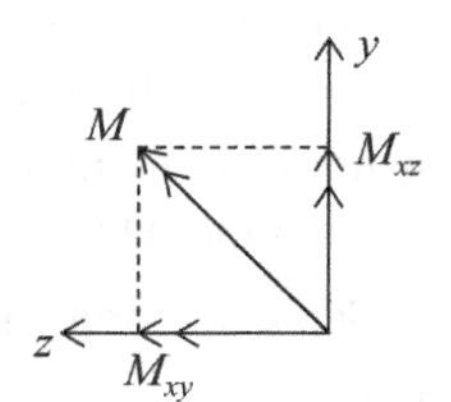

Figure 7.8 Vector sum of two moment vectors.

the tension in both sides of the belt by the sum of the tight-side and slack-side tension, i.e. $T_1 + T_2 =$ 575.92 N. However, in doing so, we move the forces to the centerline of the shaft, whereas in reality they act tangentially to the pulley. Thus, we must compensate with the torque moment caused by the difference between the tight-side and slack-side tension, i.e. the friction force between the belt and pulley (assuming no slip between the pulley and the belt). This friction force acts tangentially to the pulley. Hence, the torque moment is $M_t = (500 - 75.92) \times 0.05 = 21.2$ Nm, with $D_{pulley} = 0.1$ m. We show this torque moment diagram at the bottom of Figure 7.7, registered with the other internal force diagrams. We note that the torque moment caused by the belt–pulley transmission remains constant throughout the entire length of the shaft because there is no other torque moment to reinforce or counteract it.

Step 3: We determine the cross-section of the shaft that is subject to the heaviest load because the problem statement does not specify a location where to design the shaft. Thus, we determine where the shaft will fail first. If we design the shaft at that location such that it satisfies the specified design factor, then every other location of the shaft will also at least satisfy that design factor.

The torque moment diagram is constant throughout the entire length of the shaft and, thus, does not determine which cross-section experiences the heaviest load. The bending moment diagram shows a zero bending moment at both ends of the shaft (A and D). Thus, the shaft is unlikely to fail at these locations. Because we consider long, thin beams (Euler–Bernoulli beam theory), we neglect the shear stress resulting from any transverse loading (but account for the bending moment it creates!). Thus, two candidate locations, B and C, remain. In each of these locations, the shaft experiences a bending moment in two orthogonally oriented planes. We represent each moment as a vector and compute the vector sum of both moments to find the resulting bending moment in B and C. Figure 7.8 graphically shows the vector sum of two moment vectors.

Therefore, we find the resulting bending moments in

$$B : \sqrt{86.39^2 + 69^2} = 110\,\text{Nm},$$

$$C : \sqrt{12^2 + 172.78^2} = 173.2\,\text{Nm}.$$

Hence, cross-section C is subject to the heaviest load, which is where we will design the shaft.

Step 4: We determine which stress components are zero and nonzero:

$\sigma_m = 0$ because the shaft is not subject to an axial load. In addition, the shaft rotates and is subject to a transverse load. Thus, the normal stress resulting from bending is alternating, not steady.

$\tau_m \neq 0$ because the shaft is subject to a steady torque moment. Since we neglect the shear stress resulting from transverse loading, a torque moment is the only remaining source of shear stress in the shaft. A steady torque moment results in a steady shear stress, whereas an alternating torque moment results in an alternating shear stress.

$\sigma_a \neq 0$ because the shaft rotates while subject to a transverse load. Thus, the shaft experiences alternating cycles of tension and compression, as shown in Figure 6.1.

$\tau_a = 0$ because no alternating torque moment exists.

Thus, we must calculate σ_a and τ_m.

Using the flexure formula (Eq. (5.7)), we find that $\sigma_a = \frac{M_{max}D/2}{\pi/64(D^4-d^4)}$, with M_{max} the maximum bending moment in C.

Rearranging the terms yields $\sigma_a = \frac{173.4\times0.040/2}{\pi/64(D^4-d^4)} = \frac{70.65}{(D^4-d^4)}$.

Using the torque formula (Eq. (5.10)), we find that $\tau_m = \frac{M_tD/2}{\pi/32(D^4-d^4)}$, with M_t the torque moment in C.

Rearranging the terms yields $\tau_m = \frac{21.2\times0.040/2}{\pi/32(D^4-d^4)} = \frac{4.32}{(D^4-d^4)}$.

After determining the individual stress components, we calculate the equivalent normal stress, using the Von Mises failure criterion, to combine the respective static and dynamic stress components.

$$\sigma_{a,eq} = \sqrt{\sigma_a^2 + 3\tau_a^2} = 70.65/\left(D^4 - d^4\right),$$

$$\sigma_{m,eq} = \sqrt{\sigma_m^2 + 3\tau_m^2} = \sqrt{3}\tau_m = 7.48/\left(D^4 - d^4\right).$$

To use the Soderberg fatigue failure criterion, we determine the endurance limit of the shaft. Thus, we first calculate the endurance limit of the R.R. Moore specimen of the same material.

$$\sigma_e' = 0.5 \times 620 = 310 \text{ MPa}.$$

Then, we use the endurance limit modifying factors to modify the endurance limit of the R.R. Moore specimen into that of the shaft we design.

$$k_a = 57.7 \times 620^{-0.718} = 0.57 \text{ (see Table 6.1, and Eq. (6.6))},$$

$$k_b = 1.24 \times 40^{-0.107} = 0.84 \text{ (see Eq. (6.8))},$$

$$k_c = 1 \text{ (bending)},$$

$$\sigma_e = k_a \times k_b \times k_c \times \sigma_e' = 0.57 \times 0.84 \times 1 \times 310 = 148 \text{ MPa}.$$

Step 5: We use the Soderberg fatigue failure criterion to design the unknown dimension, i.e.

$$\frac{1}{4} = \frac{7.48/340 \cdot 10^6}{0.040^4 - d^4} + \frac{70.65/148 \cdot 10^6}{0.040^4 - d^4}.$$

Thus, $d = 0.0274$ m, or 27.4 mm.

(c) We calculate the safety factor against static yielding:

$n = \sigma_y/\left(\sigma_{a,eq} + \sigma_{m,eq}\right)$, with

$$\sigma_{a,eq} = 70.65/\left(0.040^4 - 0.0274^4\right) = 37.00 \text{ MPa},$$

$$\sigma_{m,eq} = 7.48/\left(0.040^4 - 0.0274^4\right) = 3.92 \text{ MPa}.$$

Thus, $n = 340/(37.00 + 3.92) = 8.32$.

Example problem 7.3

Figure 7.9 shows a portion of a rotating shaft, machined from SAE 2340 steel. We are specifically interested in designing the shaft at the oil groove, which creates a stress concentration. At that location, the shaft experiences an alternating bending moment of 70 Nm and a steady torque moment of 45 Nm. Note that $D/D_r = 1.15$ and that $r/d_r = 0.077$. Assume full notch sensitivity. Calculate the shaft for infinite fatigue life and consider a design factor $n_D = 2.5$.

Solution

The material properties of SAE 2340 steel are $\sigma_y = 1120$ MPa, $\sigma_{ut} = 1225$ MPa.

Step 1: The external loading and boundary conditions are specified in the problem description.

Figure 7.9 Parametric shaft design.

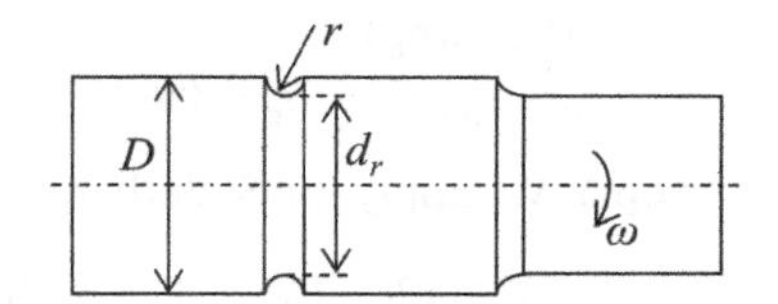

Step 2: We do not need to determine the free-body diagram because the internal bending moment and torque moment in the oil groove are specified in the problem description.

Step 3: We do not need to determine the cross-section that is subject to the heaviest load because the problem description specifies where to calculate the shaft.

Step 4: This is a design calculation and based on the problem description, we determine that

$\sigma_m = 0$ because the shaft is not subject to an axial load.

$\tau_m \neq 0$ because a steady torque moment exists. A steady torque moment results in a steady shear stress, whereas an alternating torque moment results in an alternating shear stress.

$\sigma_a \neq 0$ because the shaft is subject to an alternating bending moment.

$\tau_a = 0$ because no alternating torque moment exists.

Thus, we calculate σ_a and τ_m.

Using the flexure formula (Eq. (5.7)), we find that $\sigma_a = K_f \frac{M_{max} d_r/2}{\pi/64\ d_r^4}$, with M_{max} the maximum bending moment.

Rearranging the terms yields $\sigma_a = K_f \frac{70 \times d_r/2}{\pi/64\ d_r^4} = K_f \frac{713}{d_r^3}$.

Using the torque formula (Eq. (5.10)), we find that $\tau_m = K_{ts} \frac{M_t d_r/2}{\pi/32\ d_r^4}$, with M_t the torque moment.

Rearranging the terms yields $\tau_m = K_{ts} \frac{45 \times d_r/2}{\pi/32\ d_r^4} = K_{ts} \frac{229}{d_r^3}$.

Note that we account for the stress concentration at the oil groove by including the fatigue stress concentration factor K_f with the alternating stress, and the theoretical shear stress concentration factor K_{ts} with the static shear stress. We find K_t and K_{ts} based on $D/D_r = 1.15$ and $r/d_r = 0.077$ using Figures 5.9 and 5.10, respectively. We estimate that $K_t \cong 1.9$, $K_{ts} \cong 1.5$, and because $q = 1$, we find that $K_f \cong 1.9$.

We update the stress calculation and account for the stress concentration factors, i.e.

$$\sigma_a = 1.9 \times 713/d_r^3 = 1355/d_r^3,$$

$$\tau_m = 1.5 \times 229/d_r^3 = 344/d_r^3.$$

After determining the individual stress components, we calculate the equivalent normal stress, using the Von Mises criterion to combine the respective static and dynamic stress components.

$$\sigma_{a,eq} = \sqrt{\sigma_a^2 + 3\tau_a^2} = 1355/d_r^3,$$

$$\sigma_{m,eq} = \sqrt{\sigma_m^2 + 3\tau_m^2} = \sqrt{3}\tau_m = 595/d_r^3.$$

To use the Soderberg fatigue failure criterion, we determine the endurance limit of the shaft. We first calculate the endurance limit of the R.R. Moore specimen of the same material.

$$\sigma_e' = 0.5 \times 1225 = 612.5 \text{ MPa}.$$

Then, we use the endurance limit modifying factors to modify the endurance limit of the R.R. Moore specimen into that of the shaft we design.

$k_a = 4.51 \times 1225^{-0.265} = 0.685$ (see Table 6.1, and Eq. (6.6)).

k_b = unknown. We assume $k_b = 0.85$ and will verify after the design calculation.

$k_c = 1$ (bending).
$\sigma_e = k_a \times k_b \times k_c \times \sigma_e' = 0.685 \times 0.85 \times 1 \times 612.5 = 356.63$ MPa.

Step 5: We apply the Soderberg fatigue failure criterion, i.e.

$$\frac{1}{2.5} = \frac{595/1120 \cdot 10^6}{d_r^3} + \frac{1355/356.63 \cdot 10^6}{d_r^3}.$$

Solving for d_r we find that $d_r = 0.022$ m, or $d_r = 22$ mm.
Check $k_b = 1.24 \times 22^{-0.107} = 0.89$ (see Eq. (6.8)).
The assumption of $k_b = 0.85$ was conservative.

7.4 Critical Rotation Frequency of a Shaft

A mechanical element vibrates when an external force that acts on the mechanical element creates elastic deformation and is suddenly removed. Intuitively, we know that the frequency of vibration increases with decreasing mass and increasing stiffness. However, a *critical* vibration frequency exists for which the vibration amplitude may increase continuously and ultimately lead to failure of the mechanical element. Therefore, it is imperative to verify that the operating frequency of a mechanical element is substantially lower than its *critical* frequency. We note that several critical vibration frequencies exist for each mechanical element, which coincide with the natural vibration modes of the mechanical element and its harmonics, i.e. resonance frequencies. Specifically, we are interested in the critical bending RPM or critical torsion RPM (we note that we switch from rotational frequency in RAD/sec to RPM). Here, we focus on the critical bending RPM of a shaft, as it often is the lowest resonance frequency of the shaft. Since we must keep the operating frequency or operating RPM of the shaft substantially lower than its critical RPM, the critical bending RPM is typically the resonance frequency closest to the operating frequency.

Figure 7.10 shows a schematic of a shaft supported in two bearings that rotates with angular frequency ω. The center of mass of the shaft does not coincide with the rotation axis of the shaft, which may occur as a result of manufacturing or material imperfections. We indicate the distance between the center of mass of the shaft and its rotation axis as the eccentricity e.

Furthermore, we note that the center of mass m is subject to

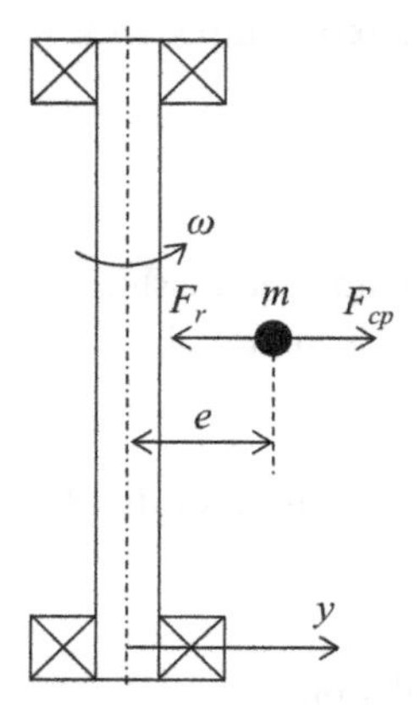

Figure 7.10 Rotating shaft, showing the eccentricity between its center of mass and its rotation axis.

- A centripetal force, $F_{cp} = m\omega^2(e + y)$, where y is the deflection of the shaft as a result of bending.
- A restoring force, $F_r = ky$, which depends on the bending stiffness k that resists bending, and the deflection of the shaft y. For small, elastic deformations, the relationship between the restoring force and the deflection is linear.

In equilibrium, we write that

$$F_{cp} = F_r, \tag{7.5}$$

and, thus,

$$m\omega^2(e + y) = ky. \tag{7.6}$$

Rearranging the terms yields

$$m\omega^2 e + (m\omega^2 - k)y = 0, \tag{7.7}$$

or

$$y = \frac{m\omega^2 e}{k - m\omega^2} = \frac{e}{\frac{k}{m\omega^2} - 1}. \tag{7.8}$$

If the operating frequency approaches the critical frequency, then $\omega = \omega_c$ and, thus, $y \to \infty$. This implies that the denominator of Eq. (7.8) must approach 0. Hence,

$$\frac{k}{m\omega_c^2} - 1 = 0, \tag{7.9}$$

and,

$$\omega_c = \sqrt{\frac{k}{m}}. \tag{7.10}$$

We already know the relationship between the angular frequency and the RPM from Eq. (7.2). Thus,

$$N_c = \frac{60}{2\pi}\omega_c = \frac{30}{\pi}\sqrt{\frac{k}{m}}. \tag{7.11}$$

Multiple masses on a shaft, such as a flywheel and a pulley, or accounting for the shaft weight itself, result in multiple critical bending RPM values, one for each mass. We approximate the lowest critical bending RPM as

$$\frac{1}{N_c} = \frac{1}{N_{c,1}} + \frac{1}{N_{c,2}} + \cdots + \frac{1}{N_{c,n}}, \tag{7.12}$$

with $N_{c,i}$ the critical bending RPM when only considering mass i.

Finally, if an analysis of the critical bending RPM reveals that it is too close to the operating frequency, then we may decrease the operating frequency if that is possible (often it might not be because the output frequency of a motor or gearbox is fixed), or we might increase the stiffness of the shaft by, e.g. changing its cross-section, which changes its moment of inertia.

Example problem 7.4

Figure 7.11 shows a rotating shaft with diameter $D = 15$ mm, machined from AISI 1010 HR steel. $L_1 = 55$ mm, $L_2 = 120$ mm, and $L_3 = 65$ mm. Two flywheels mount on the shaft in B and C, and their weight $m_B = 47.6$ kg and $m_C = 15.3$ kg, respectively. We assume the shaft itself is massless. We also assume the shaft is spinning but no external torque moment is applied to the shaft.

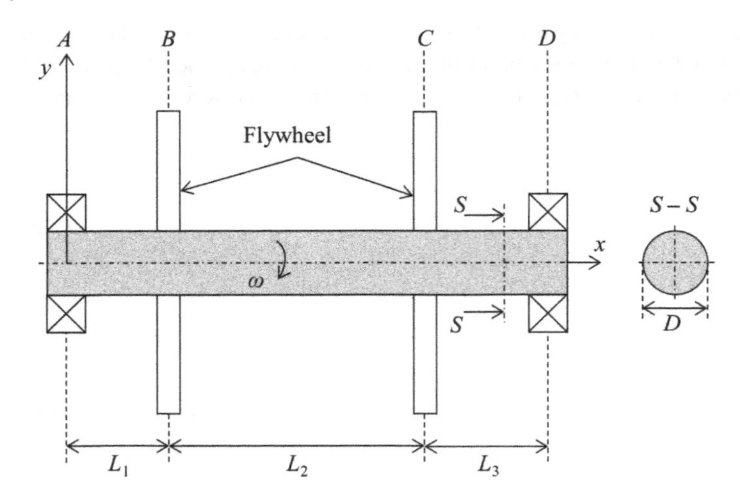

Figure 7.11 Shaft design.

(a) Calculate the safety factor against yielding.
(b) Calculate the safety factor against fatigue failure for infinite fatigue life.
(c) Calculate the lowest critical bending RPM.

Solution

(a) The material properties of AISI 1010 HR steel are $\sigma_y = 180$ MPa, $\sigma_{ut} = 320$ MPa.

Step 1: The boundary conditions and the external loading are given in the problem description. The weight of the flywheels can be converted to a transverse load.

Step 2: Figure 7.12 shows the free-body diagram of the shaft. Using equilibrium of forces and the moment equation around a fixed point, the reaction forces are $V_A = 408.35$ N and $V_D = 220.65$ N.

Step 3: We determine that cross-section B is subject to the heaviest load because the bending moment is maximum in B.

Step 4: This is a verification calculation and based on the problem description, we determine that

$\sigma_m = 0$ because the shaft is not subject to an axial load. In addition, the shaft rotates and is subject to a transverse load. Thus, the normal stress resulting from bending is alternating, not steady.

$\tau_m = 0$ because no steady torque moment exists. A steady torque moment results in a steady shear stress, whereas an alternating torque moment results in an alternating shear stress.

$\sigma_a \neq 0$ because the shaft rotates while subject to a transverse load.

$\tau_a = 0$ because no alternating torque moment exists.

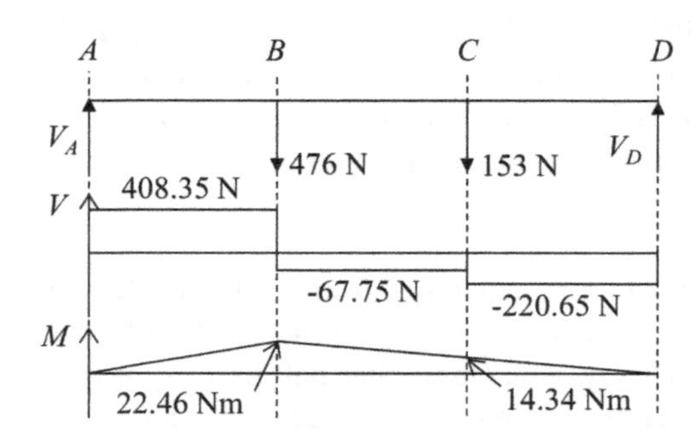

Figure 7.12 Free-body diagram.

Thus, we calculate σ_a.

Using the flexure formula (Eq. (5.7)), we find that $\sigma_a = \frac{M_{max}D/2}{\pi/64\ D^4}$, with M_{max} the maximum bending moment.

Rearranging the terms yields $\sigma_a = 32 \times 22.46/\left(\pi \times 0.015^3\right) = 67.79$ MPa.

After calculating the individual stress components, we calculate the equivalent normal stress using the Von Mises criterion to combine the respective dynamic stress components.

$$\sigma_{a,eq} = \sqrt{\sigma_a^2 + 3\tau_a^2} = 67.79 \text{ MPa}.$$

To use a fatigue failure criterion, we must determine the endurance limit of the shaft.

We first calculate the endurance limit of the R.R. Moore specimen of the same material.

$$\sigma_e' = 0.5 \times 320 = 160 \text{ MPa}.$$

Then, we use the endurance limit modifying factors to modify the endurance limit of the R.R. Moore specimen into that of the shaft we design.

$k_a = 4.51 \times 320^{-0.265} = 0.978$ (see Table 6.1, and Eq. (6.6)).
$k_b = 1.24 \times 15^{-0.107} = 0.928$ (see Eq. (6.8)).
$k_c = 1$ (bending).
$\sigma_e = k_a \times k_b \times k_c \times \sigma_e' = 0.978 \times 0.928 \times 1 \times 160 = 145.21$ MPa.

Step 5: We calculate the safety factor against yielding:
$n = \sigma_y/\sigma_{a,eq} = 180/67.79 = 2.66$.

(b) We also calculate the safety factor against fatigue failure for infinite fatigue life:
$n = \sigma_e/\sigma_{a,eq} = 145.21/67.79 = 2.14$.

(c) We calculate the lowest critical bending RPM of the shaft. Since we assume that the shaft itself is massless, there are only two masses mounted on the shaft, i.e. the two flywheels. We first calculate the critical bending RPM of each mass individually, and we do so by considering that only one mass is mounted on the shaft at a time. Then we use Eq. (7.12) to combine the critical bending RPM of each individual mass on the shaft, into the lowest critical bending RPM of the entire assembly.

We first consider that only flywheel B is mounted on the shaft. The shaft with flywheel is a mass-spring system, i.e. the mass deflects the spring, which is the bending stiffness of the shaft. For small, elastic deformations, we relate the force to the deflection as $F = k\delta$, with δ the deflection of the shaft as a result of bending and k its bending stiffness.

The deflection of the shaft between A and B is given as $\delta_{AB} = \frac{Fbx}{6EIL}(x^2 + b^2 - L^2)$, where F is the load applied in B, x is the coordinate between A and B where we evaluate the deflection, E is the Young's modulus of the shaft material, I is the area moment of inertia of the cross-section of the shaft, L is the length of the shaft, and a and b represent the distance between A and B, and B and D, respectively (see Appendix B).

Since $L = a + b$, and evaluating the maximum deflection at B $(x = a)$, we write that $\delta_{AB} = \frac{Fba}{6EIL}(a^2 + b^2 - (a + b)^2)$.

Thus, $\delta_{AB} = \frac{Fba}{6EIL}(a^2 + b^2 - a^2 - 2ab - b^2)$,

or $|\delta_{AB}| = Fa^2b^2/3EIL$, since we know that the deflection is in the gravitational direction.

Because $F = k\delta$ and, thus, $k_B = F/\delta_{AB}$, we find that the bending stiffness of the shaft with only flywheel B present is $k_B = 3EIL/a^2b^2$.

Noting that $I = \pi D^4/64 = \pi \times 0.015^4/64 = 2.485 \cdot 10^{-9}$ m^4, we calculate

$$k_B = 3 \times 200 \cdot 10^9 \times 2.485 \cdot 10^{-9} \times 0.24/(0.055^2 \times 0.185^2) = 3.456 \cdot 10^6 \text{ N/m}.$$

Thus, the critical bending RPM with only flywheel B on the shaft is

$$N_{c,B} = \frac{30}{\pi}\sqrt{\frac{3.456 \cdot 10^6}{47.6}} = 2573 \text{ RPM.}$$

We then consider that only flywheel C is mounted on the shaft and repeat the calculation. We determine that

$$k_C = 3 \times 200 \cdot 10^9 \times 2.485 \cdot 10^{-9} \times 0.24/(0.175^2 \times 0.065^2) = 2.765 \cdot 10^6 \text{ N/m.}$$

Thus, the critical bending RPM with only flywheel C on the shaft is

$$N_{c,C} = \frac{30}{\pi}\sqrt{\frac{2.765 \cdot 10^6}{15.3}} = 4060 \text{ RPM.}$$

Using Eq. (7.12), we calculate the lowest critical bending RPM as

$1/N_c = \sqrt{1/N_{c,B}^2 + 1/N_{c,C}^2}$, or

$1/N_c = \sqrt{1/2573^2 + 1/4060^2}$ and, thus,

$N_c = 2173$ RPM.

7.5 Key Takeaways

1. Shafts are probably the most common mechanical element. We use a methodical approach to perform both design and verification calculations of shafts. We also emphasize important practical aspects that we consider during design of shafts.
2. We include stress concentration factors in shaft design calculations and note that stress concentrations are less harmful in dynamic than in static loading situations.
3. We must ensure that the critical bending RPM of a shaft design is substantially higher than the operating RPM of the shaft, to avoid damage and, ultimately, failure. The critical bending RPM is mostly affected by the bending stiffness of the shaft, which depends on its area moment of inertia.

7.6 Problems

7.1 Figure 7.13 shows a shaft supported in two bearings in A and C, and driven by a belt and pulley transmission. The weight of the pulley $m_{pulley} = 6$ kg. The tight-side tension in the belt $T_1 = 900$ N, the friction coefficient between the belt and the pulley $\mu = 0.7$, and the wrap angle of the belt around the pulley $\theta = 180°$. The diameter of the pulley $D_{pulley} = 0.06$ m. The shaft is made from an AISI 1050 HR circular tube with $D = 0.025$ m. A flywheel that weighs $m_{flywheel} = 50$ kg attaches to the end of the shaft. $L_1 = L_2 = L_3 = 0.2$ m.

(a) Calculate the slack-side tension T_2 in the belt.
(b) Calculate the inner diameter d of the shaft to obtain infinite fatigue life using the Soderberg failure criterion and consider a design factor $n_D = 1.5$.
(c) Calculate safety factor against yielding.

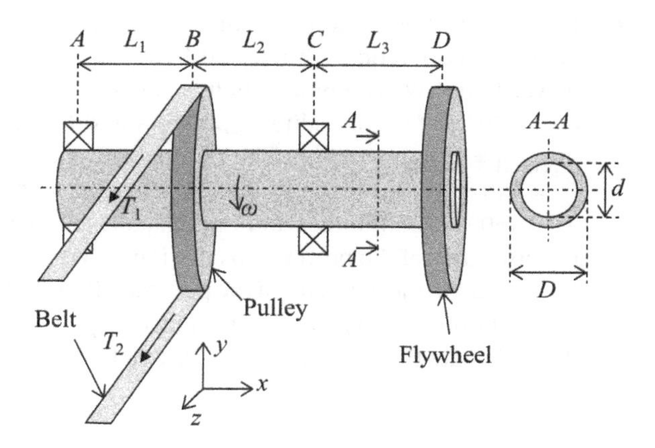

Figure 7.13 Shaft with belt and pulley transmission.

7.2 Figure 7.14 shows a rotating shaft machined from AISI 1040 HR steel and supported in two bearings. $F = 1$ kN. $L_1 = 0.1$ m, $L = 0.5$ m.

(a) Calculate the diameter D of the shaft to obtain infinite fatigue life and consider a design factor $n_D = 1.5$.

(b) Write a computer program to iteratively solve for the correct diameter D. Determine the number of iterations to find the correct solution such that the change in k_b value between the last iteration and the previous one does not exceed 0.1%, if the initial guess $k_b = 1$.

(c) How does the answer to (b) change if the initial guess $k_b = 0.1$?

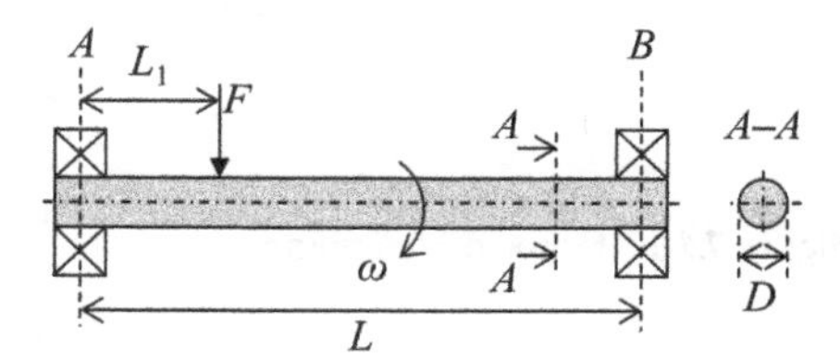

Figure 7.14 Rotating shaft.

7.3 Figure 7.15 shows a rotating shaft machined from AISI 1035 HR steel and supported in two bearings. $F = 2$ kN, $r = 0.05D$, $L_1 = L_2 = 0.25$ m, the notch sensitivity $q = 1$, $\omega = 1000$ RPM. Calculate the diameter D of the shaft to obtain infinite fatigue life and consider a design factor $n_D = 1.2$.

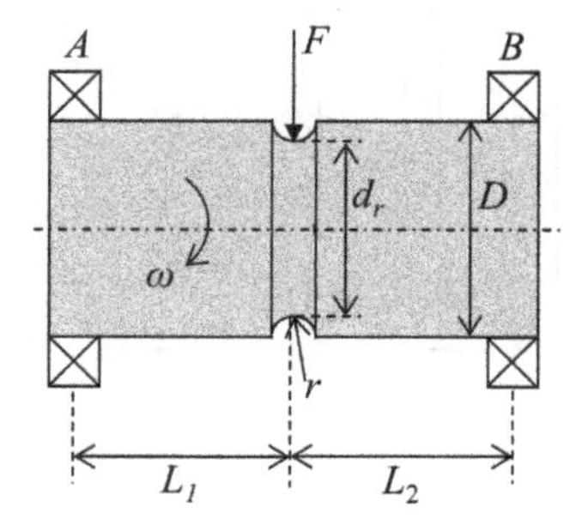

Figure 7.15 Rotating shaft with oil groove.

7.4 Figure 7.16 shows a top view of a motor-driven pump assembly. The direction of the gravitational acceleration is indicated by g and is oriented orthogonal to the page. The motor power is 10 kW. The motor shaft is supported by two bearings in A and B and rotates with $\omega_1 = 1500$ RPM. The radius of pulley A and B, $r_{pulley,A} = r_{pulley,B} = 50$ mm, and their weight $m_{pulley,A} = m_{pulley,B} = 5$ kg. The pump shaft is supported by two bearings in E and F. $L_1 = L_2 = L_3 = L_4 = 0.1$ m, and $L = 0.2$ m. A belt connects both pulleys and the tight-side tension $T_1 = 1500$ N. Both motor and pump shafts are machined from AISI 1010 HR steel. Assume the efficiency of the power transmission is 100%.

(a) Calculate the diameter of the pump shaft to obtain infinite fatigue life using the Soderberg criterion and consider a design factor $n_D = 2$.

(b) Calculate the safety factor against yielding.

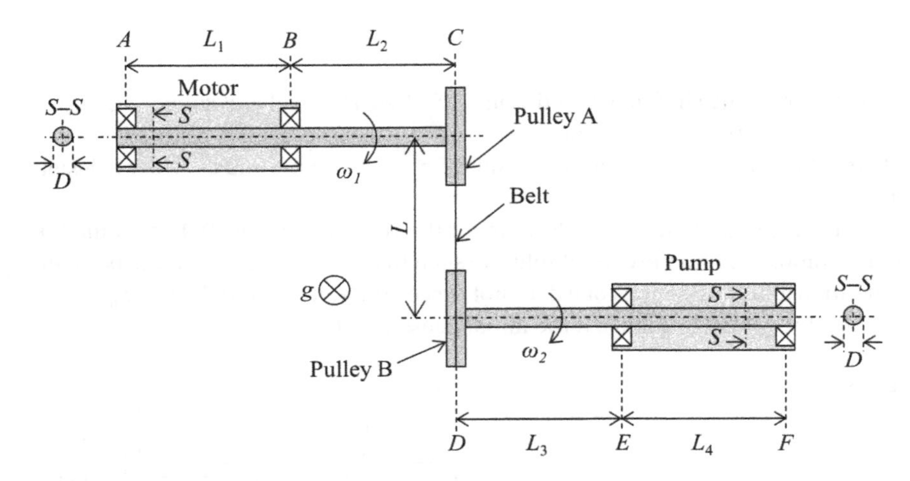

Figure 7.16 Motor-driven pump.

7.5 Figure 7.17 shows a rotating shaft machined from AISI 1040 HR steel and supported in two bearings. $F = 4$ kN, $r = 0.05d$, $L_1 = L_2 = 0.25$ m, the notch sensitivity $q = 1$, $\omega = 1000$ RPM. Calculate the diameter D of the shaft to obtain infinite fatigue life and consider a design factor $n_D = 1.2$.

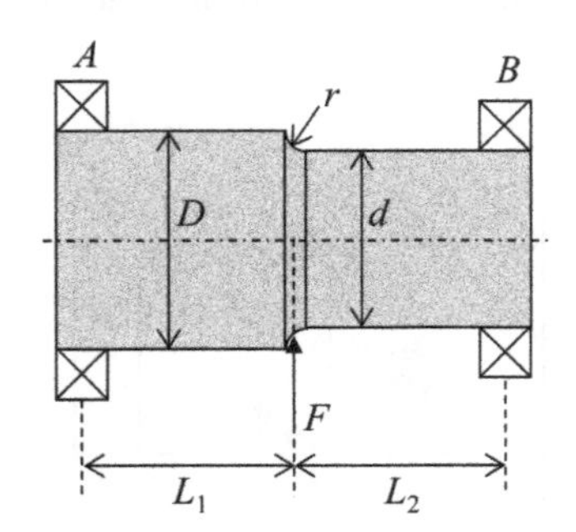

Figure 7.17 Rotating shaft with shoulder fillet.

7.6 Figure 7.18 shows a rotating shaft machined from AISI 1040 HR steel that drives a fan, and is supported in two bearings. $F = 1.5$ kN. $L_1 = 0.08$ m, $L_2 = 0.10$ m, $L = 0.5$ m. The shaft is driven by a belt and pulley transmission. The friction coefficient between the belt and the pulley $\mu = 0.7$ and the wrap angle $\theta = 180°$. The tight-side tension in the belt $T_1 = 500$ N. The diameter of the pulley $D_{pulley} = 0.06$ m.

a) Calculate the slack-side tension in the belt T_2.
b) Calculate the diameter of the shaft D to obtain infinite fatigue life using the Soderberg failure criterion and consider a design factor $n_D = 1.5$.

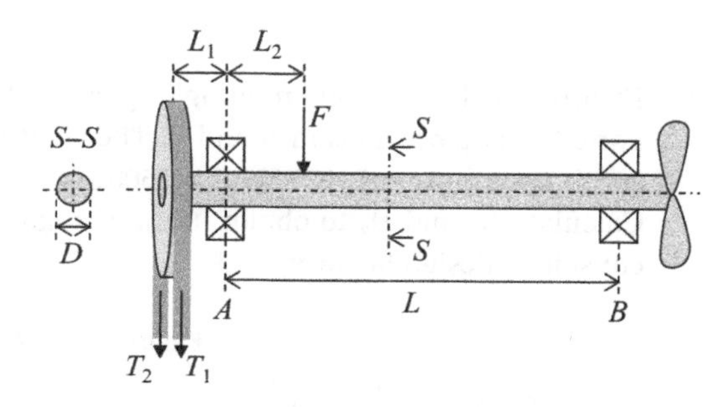

Figure 7.18 Shaft that drives a fan with a belt and pulley transmission.

7.7 Figure 7.19 shows a rotating shaft machined from AISI 1060 HR steel that is part of a paper printing machine. The shaft carries a cylindrical paper roll (indicated as a distributed load F) and is supported in two bearings in A and B. The paper roll has a density of $\rho = 984$ kg/m^3, an outside diameter (OD) of 1.5 m and an inside diameter (ID) of 22 cm. $L_1 = 2$ m, $L_2 = 5$ m, $L = 6$ m. Assume the dimensions of the paper roll remain unchanged as the shaft rotates. Calculate the inner diameter d of the shaft to obtain infinite fatigue life using the Goodman criterion and consider a design factor $n_D = 2$. The outer diameter of the shaft $D = 22$ cm and the shaft rotates at $\omega = 50$ RPM driven by a continuous power $P = 900$ W.

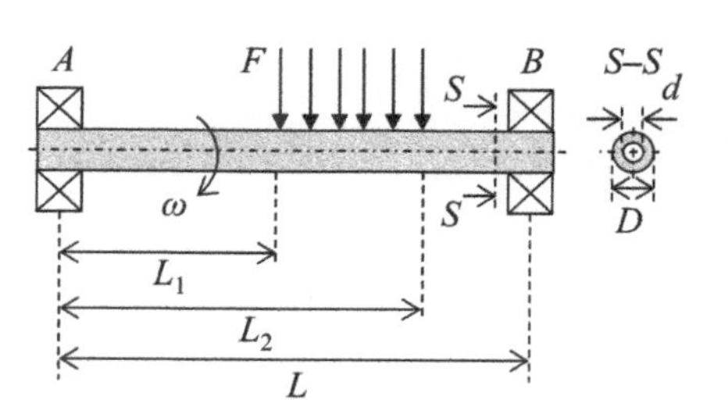

Figure 7.19 Shaft of a paper-printing machine that carries a paper roll.

7.8 Figure 7.20 shows a solid rotating shaft machined from AISI 1030 HR steel and supported in two bearings in A and C, and with a flywheel mounted in D that weighs $m_{flywheel} = 20$ kg. Diameter $D = 40$ mm, $L_1 = L_3 = 0.2$ m, $L_2 = 0.8$ m. A pulley with diameter $D_{pulley} = 100$ mm and a belt drive the shaft. The tight-side tension in the belt $T_1 = 1500$ N, the friction coefficient between belt and pulley $\mu = 0.8$, and the wrap angle $\theta = 180°$.

(a) Calculate the safety factor for infinite fatigue life using the Soderberg failure criterion.
(b) Calculate the safety factor against yielding.

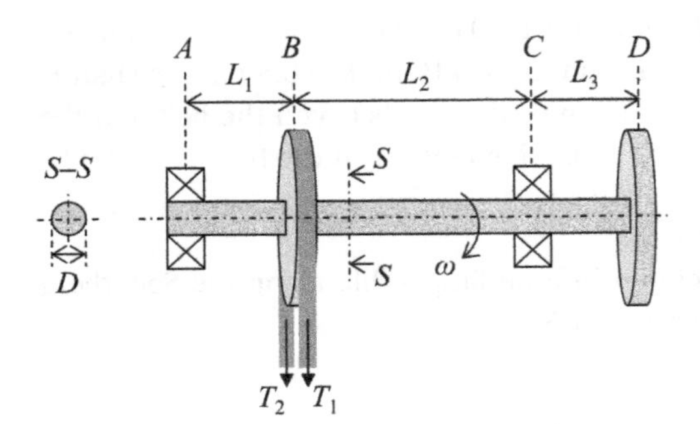

Figure 7.20 Rotating shaft with a flywheel and a belt and pulley transmission.

7.9 Figure 7.21 shows a solid rotating stepped shaft, machined from AISI 1040 HR steel, and supported in two bearings in A and B. The shaft rotates with $\omega = 1000$ RPM and is driven by a continuous power of 2000 W. $F = 1500$ N. $L_1 = 0.35$ m, $L_2 = 0.55$ m, $L_3 = 0.95$ m, $L = 1.5$ m. Calculate D_1 and D_2 to obtain infinite fatigue life using the Soderberg failure criterion and consider a design factor $n_D = 1.5$.

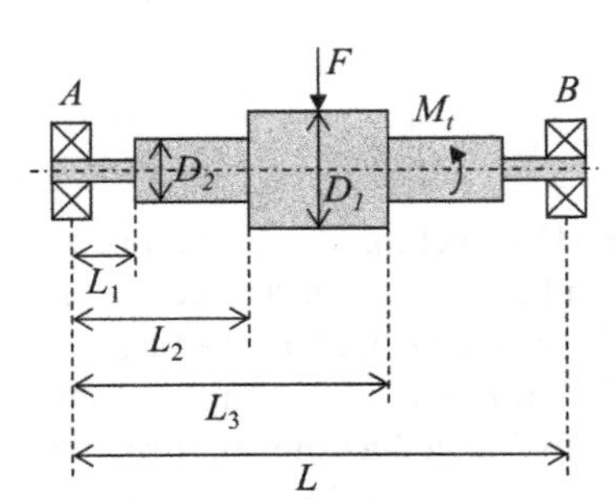

Figure 7.21 Rotating stepped shaft.

8

Bolted Joints

8.1 Introduction

In this chapter, we will use the knowledge of the chapters on design for static and dynamic strength and apply it to the design of bolted joints. Bolted joints are connections between mechanical elements that we can assemble and disassemble in a nondestructive manner, using screw thread. This contrasts, e.g. welded joints, which once assembled, require destructive methods to disassemble.

We first recognize that to understand bolted joints, we must understand screw thread. Two categories of screw thread exist; screw thread that is intended to create motion between two mechanical elements and screw thread that is meant to fasten mechanical elements together. We refer to the former category as *power screws* (sometimes also referred to as *lead screws* or *translation screws*), and to the latter category as *fasteners*.

8.2 Power Screws

Power screws are a category of screw thread specifically designed to facilitate motion between mechanical elements. It is most often used to convert a rotary into a linear motion to displace a load. For instance, a car jack, a vise, or a lathe are examples of devices that use a power screw to convert rotary into linear motion and to displace a load.

8.2.1 Screw Thread Nomenclature and Geometry

Figure 8.1 illustrates the geometry of two common types of screw thread used in power screws. Figure 8.1a shows square screw thread, whereas Figure 8.1b shows trapezoidal screw thread, sometimes also referred to as *ACME thread*. Square screw thread offers the lowest friction force between the threaded rod and nut and, thus, is the most energy efficient to convert rotary to linear motion and displace a load. However, it is difficult and expensive to manufacture because of its 90° angle. In contrast, trapezoidal screw thread is characterized by a 29° angle, which simplifies manufacturing, structurally strengthens the screw thread and, thus, allows displacing a larger load compared to square screw thread. However, it is also less energy efficient because the screw thread creates a larger friction force than the square screw thread.

Several parameters define the geometry of the screw thread. d is the *major diameter* and d_r is the *root diameter* (sometimes also referred to as the *minor diameter*), which are the largest and smallest

Design of Mechanical Elements: A Concise Introduction to Mechanical Design Considerations and Calculations, First Edition. Bart Raeymaekers.

Companion website: www.wiley.com/go/raeymaekers/designofmechanicalelements

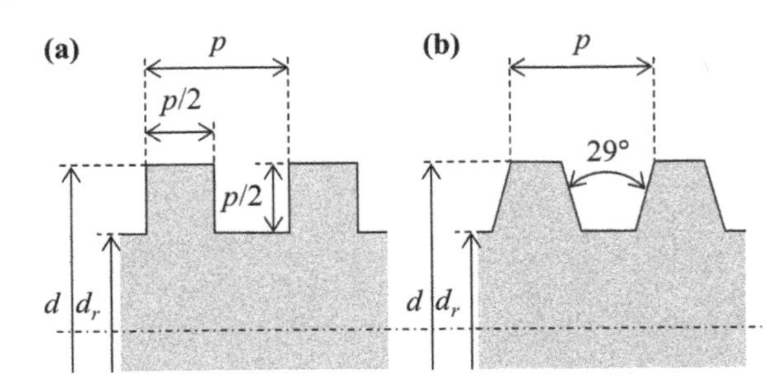

Figure 8.1 Power screw thread types, showing (a) square screw thread and (b) trapezoidal (ACME) screw thread. Source: Modified from ASME B1.5:1997 (R2014). *Acme Screw Threads*. American Society of Mechanical Engineers.

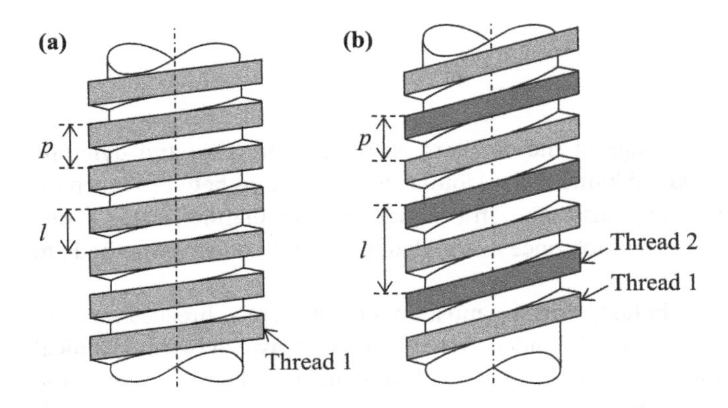

Figure 8.2 Power screw thread made of a square screw thread wrapped around a cylinder, showing (a) simplex and (b) duplex thread.

diameter of the screw thread, respectively. p is the *pitch* of the screw thread, which is the distance between adjacent crests.

Note that the pitch p is closely related to the *lead* l. The lead is the axial distance over which the screw translates with one full turn over 2π radians. Most screws have *simplex* thread, which means that a single thread is wrapped around the cylinder that constitutes the power screw. Thus, the pitch and lead are identical ($l = p$) because one full turn of the screw linearly advances it over exactly the distance of the pitch p. However, sometimes two or even three threads are wrapped in parallel around the cylinder that constitutes the power screw. In these cases, the pitch and lead are different, but related as $l = n \times p$, where n is the number of threads wrapped in parallel around the cylinder. For simplex thread $n = 1$, for duplex thread $n = 2$, and for triplex thread $n = 3$. Figure 8.2 illustrates this concept, showing (a) *simplex* and (b) *duplex* thread, which comprises a single and double thread wrapped around a cylinder, respectively. We observe that in the case of duplex thread, the pitch remains equal to the distance between adjacent crests, even though those crests now belong to different screw threads. We note that the lead is double the pitch.

8.2.2 Power Screw Torque

One of the most important questions when designing power screws is to determine the torque required to displace a specified load. Or conversely, for a specific power screw setup, determine the maximum load it can displace. To answer those questions, we examine the power screw more closely.

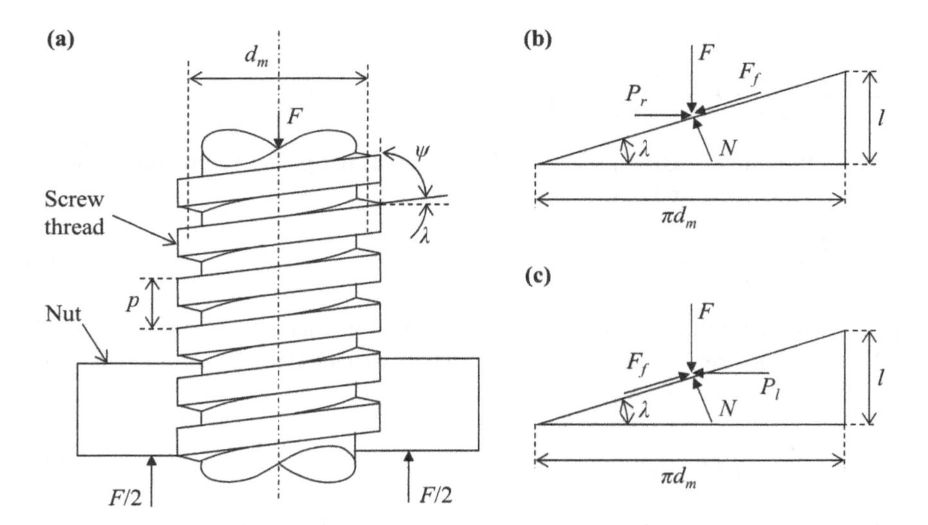

Figure 8.3 (a) Power screw setup with nut, showing a single developed screw thread turn, and identifying the different force components for (b) raising and (c) lowering the load F.

Figure 8.3a shows a frontal view of a section of a power screw with square screw thread and a corresponding nut. The power screw is subject to an axial load F, which transfers to the nut. We indicate the mean diameter $d_m = (d + d_r)/2$, and the screw thread helix angle λ. Figure 8.3b,c shows a single turn of the screw thread, developed in a 2D plane, i.e. if we axially cut through a single turn of the screw thread and *unroll* ("develop") that single turn in the 2D plane of this page. We observe a right triangle, where the smallest angle of the triangle equals the helix angle λ of the screw thread. Furthermore, Figure 8.3b,c shows the forces acting between the power screw and the nut when raising and lowering the load F, respectively.

We calculate the torque required to raise the axial load F. First, we express equilibrium of forces in both horizontal and vertical directions. The forces we consider are as follows: F is the external load we displace with the power screw. We apply a torque $M_{t,r}$ to raise the load F with the power screw, which is equal to the product of a force P_r tangential to the power screw and its leverage arm equal to the radius of the power screw. We use the mean diameter d_m or mean radius $d_m/2$ for all power screw calculations. Thus, $M_{t,r} = P_r \times d_m/2$. F loads the nut onto the screw thread, creating a normal force N (normal to the screw thread), and a friction force F_f resulting from the motion between screw thread and nut, while we rotate the power screw with respect to the nut to raise the load F. Thus, we write

$$\Sigma F_H = 0 \Rightarrow P_r - N \sin \lambda - F_f \cos \lambda = 0, \tag{8.1}$$

$$\Sigma F_V = 0 \Rightarrow F + F_f \sin \lambda - N \cos \lambda = 0. \tag{8.2}$$

If we consider Coulomb friction, i.e. $F_f = \mu N$, with μ the friction coefficient between the screw thread of the power screw and nut, then we rewrite Eqs. (8.1) and (8.2) as

$$P_r - N \sin \lambda - \mu N \cos \lambda = 0 \Rightarrow P_r = N(\sin \lambda + \mu \cos \lambda), \tag{8.3}$$

$$F - N \cos \lambda + \mu N \sin \lambda = 0 \Rightarrow F = N(\cos \lambda - \mu \sin \lambda). \tag{8.4}$$

Dividing Eq. (8.3) by (8.4) yields

$$P_r = F\frac{\sin\lambda + \mu\cos\lambda}{\cos\lambda - \mu\sin\lambda} = F\left(\frac{\tan\lambda + \mu}{1 - \mu\tan\lambda}\right). \tag{8.5}$$

Figure 8.3b,c shows that $\tan\lambda = l/(\pi d_m)$, where l is the lead of the power screw thread. Substituting this term into Eq. (8.5), we find that

$$P_r = F\left(\frac{l/(\pi d_m) + \mu}{1 - \mu l/(\pi d_m)}\right). \tag{8.6}$$

We previously defined the torque to raise the load F as the product of a force P_r tangential to the power screw and its leverage arm, i.e. $M_{t,r} = P_r \times d_m/2$. Thus, we calculate the torque required to raise load F as

$$M_{t,r} = \frac{Fd_m}{2}\left(\frac{l + \mu\pi d_m}{\pi d_m - \mu l}\right). \tag{8.7}$$

We perform a similar analysis to determine the torque required to lower load F,

$$M_{t,l} = \frac{Fd_m}{2}\left(\frac{\mu\pi d_m - l}{\pi d_m + \mu l}\right). \tag{8.8}$$

$M_{t,r}$ and $M_{t,l}$ describe the torque required to displace the load F and to overcome the friction force between the screw thread of the power screw and the nut. We note that Eqs. (8.7) and (8.8) account for the thread geometry (d_m, l, and $\lambda = \arctan(l/\pi d_m)$), the material properties (μ), and the operating parameters of the power screw (F).

When we consider trapezoidal as opposed to square screw thread, we account for the angle $2\alpha = 29°$. α increases the friction force by wedging the threads of the power screw and the nut. Figure 8.4 illustrates this concept. Hence, we divide the friction terms in Eqs. (8.7) and (8.8) by $\cos\alpha$. Noting that $\sec\alpha = 1/\cos\alpha$, we write the torque required to raise load F with trapezoidal screw thread as

$$M_{t,r} = \frac{Fd_m}{2}\left(\frac{l + \mu\pi d_m\sec\alpha}{\pi d_m - \mu l\sec\alpha}\right) \tag{8.9}$$

and, similarly, the torque required to lower load F with trapezoidal screw thread as

$$M_{t,l} = \frac{Fd_m}{2}\left(\frac{\mu\pi d_m\sec\alpha - l}{\pi d_m + \mu l\sec\alpha}\right). \tag{8.10}$$

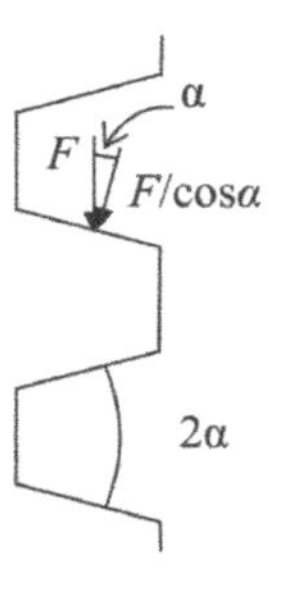

Figure 8.4 Geometry of trapezoidal power screw thread, showing the projection of load F over angle α.

8.2.3 Self-locking

A power screw can be self-locking, i.e. it does not lower itself when loaded, but instead requires a positive torque to lower the load F. Thus, a power screw is self-locking if $M_{t,l} > 0$, because the friction force F_f between the power screw thread and nut is sufficiently large to prevent motion under load F. We express the condition for self-locking as

$$M_{t,l} = \frac{Fd_m}{2}\left(\frac{\mu\pi d_m - l}{\pi d_m + \mu l}\right) > 0 \tag{8.11}$$

The denominator of Eq. (8.11) is always positive because d_m, l, and μ are always positive. Hence, the condition for self-locking requires a positive numerator of Eq. (8.11). Therefore,

$$\mu\pi d_m - l > 0, \tag{8.12}$$

or,

$$\mu > \frac{l}{\pi d_m}, \tag{8.13}$$

which reduces to

$$\mu > \tan\lambda. \tag{8.14}$$

Equation (8.14) shows that we obtain a self-locking power screw whenever the friction coefficient μ between the power screw thread and nut exceeds the tangent of the helix angle λ of the power screw thread.

Whether or not we design a power screw to be self-locking depends on its function and application. For instance, the power screw of a car jack is self-locking because it provides an important safety feature. Indeed, it would be dangerous if a car supported by the jack could lower the power screw by its own weight. However, if the power screw drives a freight elevator, it might be desirable if the weight of the elevator load could lower the power screw in a controlled fashion, thus not requiring energy to do so. Hence, an analysis of the application in which we use the power screw reveals whether a self-locking design is desirable.

8.2.4 Efficiency of a Power Screw

We quantify the efficiency with which a power screw raises a load F. To do so, we first define the *ideal* power screw with no friction force between the power screw thread and nut, i.e. $\mu = 0$. Thus, from Eq. (8.7), we determine that the torque to raise a load F with the ideal power screw is

$$M_{t,r} = \frac{Fl}{2\pi}. \tag{8.15}$$

Then, we define the efficiency of the power screw as

$$\varepsilon = \frac{\text{Torque required for an ideal power screw}}{\text{Torque required for a realistic power screw}} \tag{8.16}$$

or,

$$\varepsilon = \frac{Fl}{2\pi M_{t,r}}. \tag{8.17}$$

A power screw with high efficiency is desirable because it reduces energy consumption. However, one may need to balance power screw efficiency with the need for self-locking, if it exists.

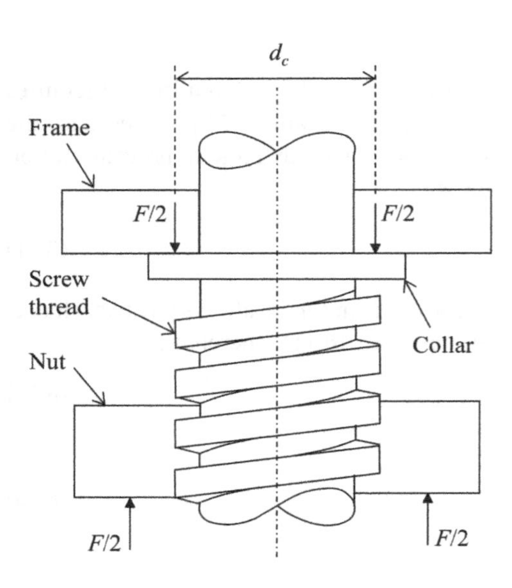

Figure 8.5 Power screw setup with nut, and a thrust collar against the frame.

8.2.5 Collar Friction

An axially loaded power screw uses a thrust collar or bearing between the rotary (power screw) and stationary (frame) parts of the power screw assembly. This thrust collar also creates a friction force that must be overcome when rotating the power screw to displace the load F. Thus, we account for an additional torque component to overcome the friction force in the thrust collar. Figure 8.5 shows a schematic of a power screw assembly, showing the power screw thread, the nut, the thrust collar, and the frame. We assume that the entire load F is concentrated at the mean collar diameter d_c, with μ_c the friction coefficient between the collar and the frame. Thus, we calculate the torque required to overcome the friction force between the collar and the frame as

$$M_{t,c} = \frac{F\mu_c d_c}{2}. \tag{8.18}$$

In Eq. (8.18), $F\mu_c$ represents the product of the normal load and the friction coefficient between the collar and the frame, which is the friction force. Hence, the product of the friction force and the leverage arm ($d_c/2$) is the torque required to overcome this friction force.

Note that the total torque to raise or lower the load F with a power screw is the sum of the torque to overcome the power screw thread friction force (see Eqs. (8.7) and (8.8)) and the torque to overcome the collar friction (Eq. (8.18)).

Example problem 8.1

Figure 8.6 shows a schematic of a car jack lifting one wheel of a car that weighs 1500 kg. Hence, the load F is a quarter of the weight of the car. A handle operates a rod with single square power screw thread with $d = 16$ mm and $p = 4$ mm. The handle rotates against a static thrust collar with $d_c = 20$ mm, and the friction coefficient between the handle and the static collar $\mu_c = 0.06$. The power screw engages with a nut that is $s = 6$ mm wide. The friction coefficient between the nut and the thread $\mu = 0.05$. Linkages with $L = 200$ mm connect the collar, nut, base, and load platform via pin joints. The distance between the top of the load platform and the hinges and between the hinges and the ground $c = 20$ mm. The distance between the joints

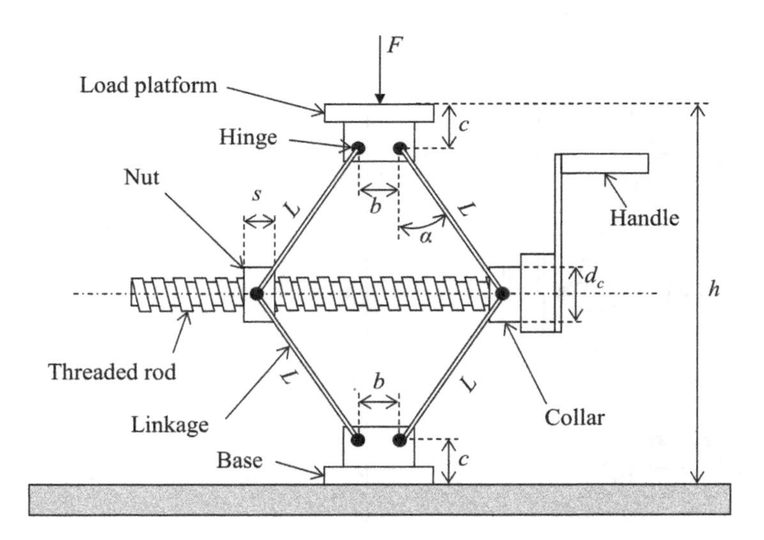

Figure 8.6 Car jack to manually raise and lower a car.

$b = 28$ mm. The height between the ground and the top of the load platform $h_{min} \leq h \leq h_{max}$, with $h_{min} = 115$ mm and $h_{max} = 410$ mm. The angle α is a function of h.

(a) Calculate the screw thread geometry: d_r, d_m.
(b) Calculate and graph the total torque required to raise and lower the load F as a function of h (for $h_{min} \leq h \leq h_{max}$).
(c) Determine whether the jack is self-locking.

Solution

(a) Referring to Figure 8.1a we write that

$d_r = d - 2p/2 = 12$ mm, and
$d_m = d - 2p/4 = 14$ mm.

(b) We determine the load in the axial direction of the power screw. Figure 8.7 schematically shows the forces along the linkages of the jack.
Thus, $F_A = F/(2\cos\alpha)$, and

$F_B = 2F_A \cos(90° - \alpha) = 2F_A \sin\alpha$, or
$F_B = F \sin\alpha / \cos\alpha = F \tan\alpha$, with $\alpha = f(h)$.

We determine $\alpha = f(h)$.

$a = (h - 2c)/2$,
$\alpha = \arccos(a/L) = \arccos\left(\frac{h/2-c}{L}\right)$.

Thus, for instance

$$\alpha_{min} = \arccos\left(\frac{h_{max}/2-c}{L}\right) = 22.33°,$$

$$\alpha_{max} = \arccos\left(\frac{h_{min}/2-c}{L}\right) = 79.19°.$$

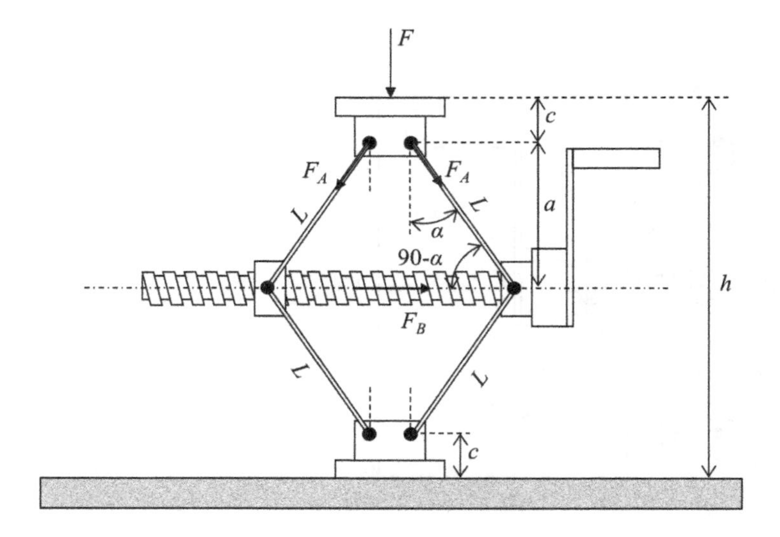

Figure 8.7 Car jack load and geometry.

We use Eqs. (8.7) and (8.8) to calculate the torque required to raise ($M_{t,r}$) or lower ($M_{t,l}$) the load F. Furthermore, we use Eq. (8.18) to calculate the torque required to overcome the friction force in the thrust collar ($M_{t,c}$).

Finally, $M_{t,r,total} = M_{t,r} + M_{t,c}$, and $M_{t,l,total} = M_{t,l} + M_{t,c}$.

Figure 8.8 shows the total torque required to raise and lower the load F as a function of the height of the car jack.

(c) Because the total torque required to lower the load F is positive, independent of the height h, the power screw is self-locking. Since this is a car jack application, self-locking is a desirable safety feature.

The collar torque is $M_{t,c} = \frac{F\mu_c d_c}{2} = \frac{3750 \times 0.06 \times 0.02}{2} = 2.25$ Nm.

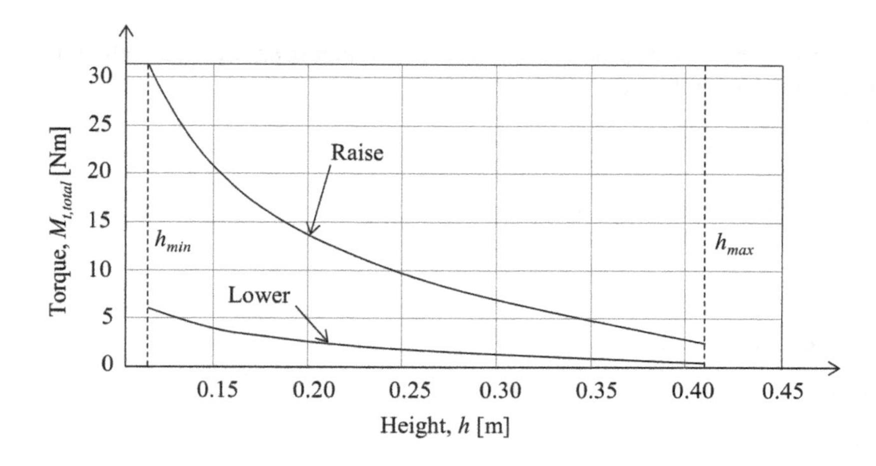

Figure 8.8 Total torque required to raise and lower the load F as a function of the height of the car jack h.

Thus, subtracting 2.25 Nm from the torque curve to lower the load F (see Figure 8.8), we observe that for $h > 0.25$ the power screw thread itself might not be self-locking, but the friction force in the thrust collar renders it self-locking.

8.3 Fasteners

Fasteners are a category of screw thread specifically designed to fasten two or more mechanical elements together. There are many types of fasteners and many types of screw threads. Here, we focus on basic metric fasteners, and how to calculate the strength of bolted joints.

Calculations of bolted joints are important because sometimes the failure of a mechanical assembly depends on the strength of a bolted joint. Furthermore, economics are important too; bolts are expensive mechanical elements. Thus, if by performing a detailed calculation we can reduce the size of a bolt or reduce the number of bolts, we can reduce cost. This is often very important in mass-produced products or assemblies, where small savings per product are multiplied by a large number of products and, thus, become substantial.

8.3.1 Screw Thread Nomenclature and Geometry

Similar to power screws, we must first understand the geometry of fastener screw thread. Figure 8.9 schematically shows the geometry of a metric bolt. We indicate the pitch p, the crest angle α, the major diameter d, the root (minor) diameter d_r, and the pitch diameter d_p. The pitch diameter is located in between d_r and d and is the diameter of a cylindrical surface, concentric with the thread, which intersects the solid thread flanks such that the distance between these points is exactly half the pitch distance.

We indicate metric screw thread with the letter M, followed by its major diameter in mm. For instance, an $M12$ bolt is a metric bolt with a major diameter $d = 12$ mm. With no other specification, this refers to the standard $M12$ bolt with standard (coarse) pitch $p = 1.75$ mm. However, metric bolts also exist with fine pitch. In that case, we specify the fine pitch after the diameter. For instance $M12 \times 1.25$ refers to a metric bolt with a major diameter $d = 12$ mm, and with a pitch $p = 1.25$ mm.

Table 8.1 compares the coarse- and fine-pitch series of metric bolts. The left column shows the major diameter of the bolt. Then, the two groups of two columns show the pitch and tensile stress area for the coarse-pitch series and the fine-pitch series, respectively. When no pitch is indicated, we refer to the coarse-pitch series. When the (fine) pitch is indicated, we refer to the fine-pitch series.

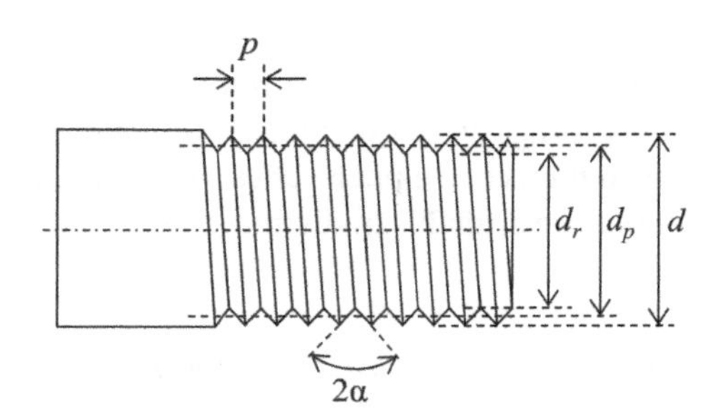

Figure 8.9 Geometry of metric fastener screw thread.

Table 8.1 Comparison of metric bolt dimensions, showing the coarse-pitch and fine-pitch series.

	Coarse-pitch series		Fine-pitch series	
Nominal major diameter d **(mm)**	**Pitch** p **(mm)**	**Tensile stress area** A_t **(mm²)**	**Pitch** p **(mm)**	**Tensile stress area** A_t **(mm²)**
2	0.40	2.07		
4	0.70	8.78		
6	1.00	20.10		
8	1.25	36.60	1.00	39.20
10	1.50	58.00	1.25	61.20
12	1.75	84.30	1.25	92.10
14	2.00	115.00	1.50	125.00
16	2.00	157.00	1.50	167.00
20	2.50	245.00	1.50	272.00
24	3.00	353.00	2.00	384.00
30	3.50	561.00	2.00	621.00
36	4.00	817.00	2.00	915.00

Source: Modified from ISO 724:1993. *General-purpose Metric Screw Threads – Basic Dimensions. International Organization for Standardization.*

Table 8.2 Standard bolt lengths.

Bolt length (mm)
3.0, 3.5, 4.0, 4.5, 5.0, 5.5, 6.0, 6.5, 7.0, 8.0, 9.0, 10, 11, 12, 14, 16, 18, 20, 22, 25, 28, 30, 32, 35, 40, 45, 50, 60, 80, 100, 120, 140, 160, 180, 200, 250, 300

Source: Modified from ISO 262:1998. *General Purpose Metric Screw Threads, Selected Sizes for Screws, Bolts and Nuts.* International Organization for Standardization.

We note the *tensile stress area* in Table 8.1. This is the cross-sectional area of the bolt that we use for strength calculations. The rationale is as follows. When calculating the cross-sectional area of the bolt, we could use, e.g. the major diameter, but this would overestimate the amount of material, or the root diameter, but this would underestimate the amount of material. However, in strength calculations, we always use the tensile stress area, which is the cross-sectional area calculated based on the average of the pitch and root diameter of the bolt, as defined in the ISO 898-1:2013 standard.[1]

Occasionally, the letter M that indicates metric may be followed by the letter J. This refers to metric screw thread with a rounded fillet at the root diameter, e.g. resulting from rolling rather than cutting the screw thread. For instance, $MJ16$ refers to a metric bolt with a major diameter $d = 16$ mm, coarse-pitch series, i.e. with pitch $p = 2$ mm, and rounded root diameter.

Bolts are available in discrete standard sizes. Table 8.2 lists common standard lengths of metric bolts.

---

1 ISO 898-1:2013. *Mechanical Properties of Fasteners Made of Carbon Steel and Alloy Steel.* International Organization for Standardization.

Figure 8.10 Strength category of a metric bolt.

8.3.2 Fastener Strength Category

Metric bolts indicate their strength category on the bolt head. Figure 8.10 illustrates the top view of a metric bolt head, with the strength category indicated by two numbers, separated by a period. Here, we schematically represent the two numbers as parameters X and Y. From these parameters, we estimate the mechanical properties of the bolt as follows. The first parameter (before the period) refers to the ultimate tensile stress σ_{ut} of the bolt. Specifically $\sigma_{ut} = 100 \times X$. The second parameter (after the period) refers to the yield stress of the bolt, i.e. $\sigma_y = 0.Y \times \sigma_{ut}$.

For instance, a metric bolt with strength category 10.9 has the following mechanical properties: $\sigma_{ut} = 1000$ MPa and $\sigma_y = 900$ MPa.

8.3.3 Bolt Preload

Figure 8.11 shows a cross-section of a bolt and nut that tightens two members together. Preload is a crucial concept when designing bolted joints. The preload F_i in the bolt exists as a result of tightening the nut onto the bolt. Thus, the preload is the tensile force that exists in the bolt before applying any external load P to the bolted joint.

Bolt preload is important for three reasons:

1. It creates a tensile load in the bolt and an equal compressive load in the members. This allows the members to better resist external tensile loading because the external load needs to first relieve the compressive load prior to creating a tensile load in the members.
2. It creates a friction force between the members to resist an external shear load. Indeed, the preload acts as a normal load, compressing the members. Multiplying the normal load by the friction coefficient between the members yields the maximum friction force that can exist between the members to resist an external shear load P_s.

Figure 8.11 Bolted joint in tension, showing the preload F_i, the external tensile load P, and an external shear load P_s.

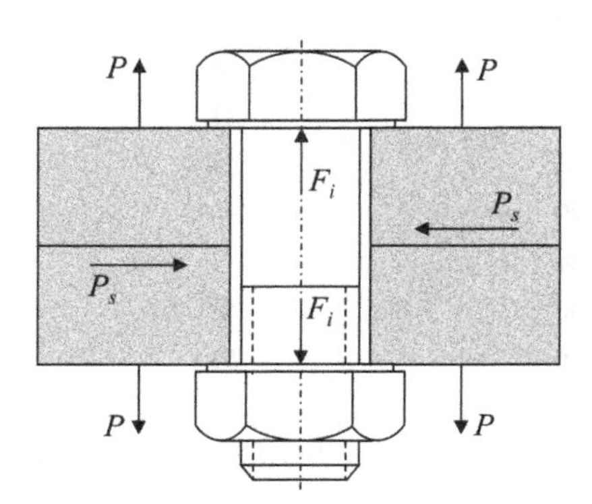

3. It improves the fatigue life of the bolt because the amplitude of a fluctuating stress in the bolt decreases with increasing preload.

8.3.4 Hexagonal Nuts

The screw thread of metric bolts matches that of metric nuts. Table 8.3 shows the width and height of standard metric hexagonal nuts.

8.3.5 Washers

We use washers to reduce the contact pressure between the members and the bolt head and the nut, respectively. The contact pressure is the total resulting load in the bolt, divided by the contact area between the members and the bolt head and the nut. If the contact pressure locally exceeds the yield stress of the members, then plastic deformation occurs, which is undesirable because it permanently changes the members, and potentially compromises the integrity of the bolted joint. Hence, washers increase the contact area between the members and the bolt head and the nut and, thus, reduce the contact pressure. Table 8.4 shows dimensions of standard plain metric washers, which allow calculating the contact area and, thus, the contact pressure.

8.3.6 Torque Requirement

The torque requirement expresses how much torque we must apply to tighten a bolt and nut to obtain the required preload F_i in the bolt. Hence, it yields practical information that we use for selecting a torque wrench setting. We calculate the torque requirement M_t as

$$M_t = KdF_i, \tag{8.19}$$

with F_i the desired bolt preload, d the major diameter of the bolt, and K a proportionality constant that is based on the bolt surface finish. Table 8.5 shows examples of different bolt surface finishes

Table 8.3 Dimensions of regular metric hexagonal nuts.

Nominal size (mm)	Width (mm)	Height (mm)
6	10	5.2
8	13	6.8
10	16	8.4
12	18	10.8
14	21	12.8
16	24	14.8
20	30	18.0
24	36	21.5
30	46	25.6
36	55	31.0

Source: Modified from ISO 262:1998. *General Purpose Metric Screw Threads, Selected Sizes for Screws, Bolts and Nuts.* International Organization for Standardization.

Table 8.4 Dimensions of regular metric plain washers.

Washer size (mm)	Min. ID (mm)	Max. OD (mm)	Max. thickness (mm)
2	2.50	6.00	0.90
4	4.70	12.00	1.40
6	6.65	18.80	1.75
8	8.90	25.40	2.30
10	10.85	28.00	2.80
12	13.30	34.00	3.50
14	15.25	39.00	3.50
16	17.25	44.00	4.00
20	21.80	50.00	4.60
24	25.60	56.00	5.10
30	32.40	72.00	5.60
36	38.30	90.00	6.40

Source: Modified from ISO 10673:1998. *Plain Washers for Screw and Washer Assemblies, Small, Normal and Large Series, Product Grade A*. International Organization for Standardization.

Table 8.5 Surface quality K for torque requirement calculation.

Bolt surface finish	K
Nonplated, black finish	0.30
Zinc-plated	0.20
Lubricated	0.18
Cadmium-plated	0.16

Source: Based on Budynas, R.G. and Nisbett, J.K. (2015). Shigley's Mechanical Engineering Design, 10e. McGraw-Hill.

and the corresponding parameter K. In the absence of specifications about the bolt surface quality, we assume $K = 0.20$ as an appropriate approximation.

Note that Eq. (8.19) is of the same form as Eq. (8.7). Indeed, the concept of tightening a bolt is similar to a power screw, and K accounts for the screw thread geometry and material properties, for which we also account when we calculate the torque to raise a load with a power screw.

8.3.7 Bolted Joints in Tension (Static)

Tightening members together with a bolt and nut creates preload in the bolt, which creates a tensile force in the bolt and a compressive load in the members. Even though the tensile and compressive load is identical, they result in different deflection of the bolt and the members because their stiffness is different. Figure 8.12 schematically illustrates this concept, and shows the load F on the vertical axis, and deflection δ on the horizontal axis. We indicate the stiffness of the bolt and

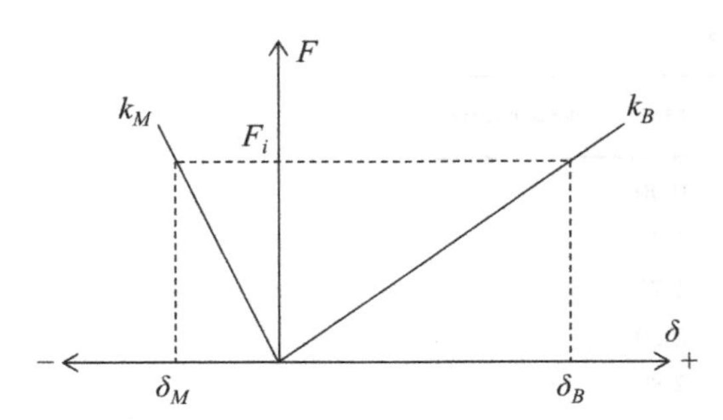

Figure 8.12 Load versus deflection diagram, graphically indicating the stiffness of the bolt and the members.

members as k_B and k_M, respectively. The stiffness relates the load to the deflection and, thus, is the tangent of the angle between the load versus deflection curves of the bolt and members. After applying the preload F_i to the bolt and members, we observe deflection δ_B of the bolt, and δ_M of the members, respectively.

We consider that the bolted joint with preload F_i is also subject to an external load P. Adding to the schematic of Figure 8.12, we recognize that an external tensile load P will increase the deflection of the bolt by $\Delta\delta_B$ and it will decrease the deflection of the members by $\Delta\delta_M$. Figure 8.13 illustrates this concept. Starting from the preload F_i and the corresponding deflection in the bolt and the members, we apply the external load P, and the arrows indicate how this affects the bolted joint. The resulting deflection in the members decreases to $\delta_M - \Delta\delta_M$, and the resulting deflection in the bolt increases to $\delta_B + \Delta\delta_B$.

Because the members do not separate as a result of the external load P (no air gap forms between them), the extra deflection in bolt and members must be identical. Thus,

$$\Delta\delta_B = \Delta\delta_M. \tag{8.20}$$

The extra deflection of the bolt and members is the result of the portion of the external load P that is borne by the bolt (P_B) and the members (P_M), respectively. Hence, we determine that

$$\Delta\delta_B = \frac{P_B}{k_B} \tag{8.21}$$

and

$$\Delta\delta_M = \frac{P_M}{k_M}. \tag{8.22}$$

Combining Eqs. (8.20), (8.21), and (8.22), we write that

$$\frac{P_B}{k_B} = \frac{P_M}{k_M}. \tag{8.23}$$

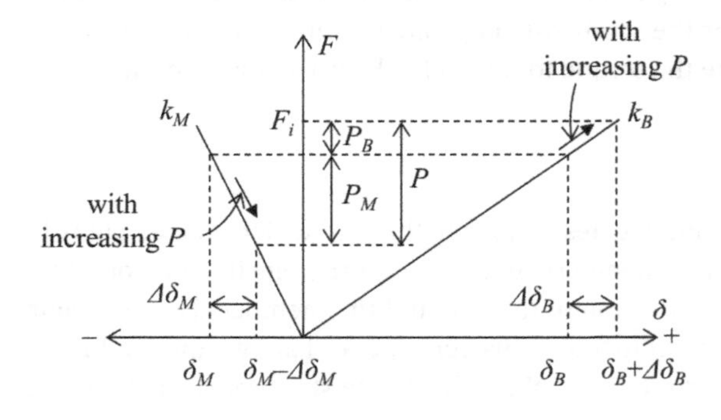

Figure 8.13 Stiffness of the bolt and the members, and graphical representation of the portion of the external load P borne by the members P_M and the bolt P_B.

The total external load P is split over the bolt and members, i.e. $P = P_B + P_M$. Hence,

$$P = k_B \frac{P_M}{k_M} + P_M = P_M \frac{k_B + k_M}{k_M}. \tag{8.24}$$

Finally, we conclude that the portion of the external load borne by the bolt P_B and by the members P_M is

$$P_B = \frac{k_B}{k_M + k_B} P \tag{8.25}$$

and

$$P_M = \frac{k_M}{k_M + k_B} P. \tag{8.26}$$

Figure 8.13 graphically illustrates the total external load P, and how it distributes over the bolt P_B and the members P_M, depending on the ratio of the bolt and member stiffness. We also observe that if the stiffness of either bolt or members changes (slope of the respective line in Figure 8.13 changes), then the portions of the external load borne by the bolt and the members change accordingly.

The total resulting load in the bolt F_B and the members F_M is the sum of the preload and the portion of the external load borne by the bolt and members, respectively. Thus,

$$F_B = F_i + P_B = F_i + \frac{k_B}{k_M + k_B} P \tag{8.27}$$

and

$$F_M = P_M - F_i = \frac{k_M}{k_M + k_B} P - F_i. \tag{8.28}$$

8.3.7.1 Determining the Preload F_i

Safe operation of a bolted joint requires meeting two criteria:

1. **Prevent member separation**, i.e. prevent that an air gap exists between the members.
2. **Prevent bolt yielding**. Note that in specific applications bolts may be preloaded beyond their yield stress. However, if that is the case, then the bolt may never be reused after disassembly. A new bolt must be used upon reassembly of the bolted joint. Here, we will only consider bolted joints that are preloaded below the yield stress.

Thus, *member separation force* $\leq F_i \leq$ *bolt yielding force*.

We first determine the force that causes member separation. Member separation occurs when the compressive force in the members is zero, i.e. $F_M = 0$. Setting Eq. (8.28) equal to zero yields

$$F_M = \frac{k_M}{k_M + k_B} P - F_i = 0 \tag{8.29}$$

or

$$F_i = \frac{k_M}{k_M + k_B} P. \tag{8.30}$$

Furthermore, substituting Eq. (8.30) into Eq. (8.27) gives

$$F_B = \frac{k_M}{k_M + k_B} P + \frac{k_B}{k_M + k_B} P = P. \tag{8.31}$$

Thus, when the members separate, the bolt bears the entire external load P. Evidently, this is an undesirable situation because it may cause the bolt to yield. Instead, we must choose the preload such that member separation does not occur and to ensure that the bolt only bears a portion of the external load P, whereas the members bear the remaining portion. To prevent member separation, we choose the preload as

$$F_i \geq \left(\frac{k_M}{k_M + k_B}\right) P. \tag{8.32}$$

We also determine the force that causes the bolt to yield, i.e. the stress in the bolt must not exceed the yield stress. Thus,

$$\frac{F_B}{A_t} \leq \sigma_y. \tag{8.33}$$

Substituting Eq. (8.27) into Eq. (8.33) gives

$$\frac{F_i}{A_t} + \frac{k_B}{k_M + k_B} \frac{P}{A_t} \leq \sigma_y, \tag{8.34}$$

or,

$$F_i \leq \sigma_y A_t - \frac{k_B}{k_M + k_B} P. \tag{8.35}$$

Thus, we select the preload to satisfy Eq. (8.35).

Accounting for both conditions (Eqs. (8.32) and (8.35)), we define the envelope within which we must choose the preload F_i as

$$\left(\frac{k_M}{k_M + k_B}\right) \leq F_i \leq \sigma_y A_t - \frac{k_B}{k_M + k_B} P. \tag{8.36}$$

Practically, we typically choose the preload as high as possible while still avoiding yielding. When the bolt is placed into service, vibrations can cause small deformations that reduce the initial preload during assembly.

8.3.7.2 Stiffness of the Bolt

The portion of the external load borne by the bolt and the members depends on the ratio of their stiffness values. For that reason, we must accurately calculate the stiffness of the bolt and the members, within the *grip* l, which we define as the entire section clamped between the face of the bolt and the face of the nut.

The stiffness of the bolt within the grip comprises two parts:

1. Unthreaded shank portion with stiffness k_d.
2. Threaded portion with stiffness k_t.

We treat those two separate sections as two springs in series. Figure 8.14 schematically shows several springs (a) in parallel and (b) in series. We calculate the equivalent spring stiffness k_{eq} for two springs in series as $1/k_{eq} = 1/k_1 + 1/k_2$.

Applying this to the unthreaded and threaded sections of the bolt within the grip, we write

$$\frac{1}{k_B} = \frac{1}{k_d} + \frac{1}{k_t} \Rightarrow k_B = \frac{k_t k_d}{k_t + k_d}, \tag{8.37}$$

with

$$k_t = \frac{EA_t}{l_t}, \tag{8.38}$$

Figure 8.14 Equivalent stiffness of multiple springs in (a) parallel and (b) series.

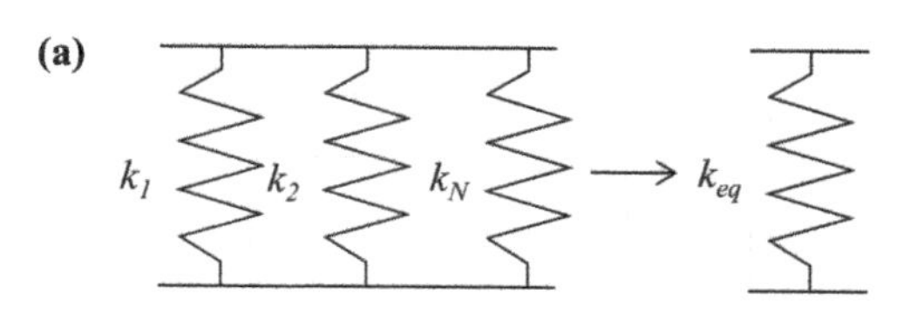

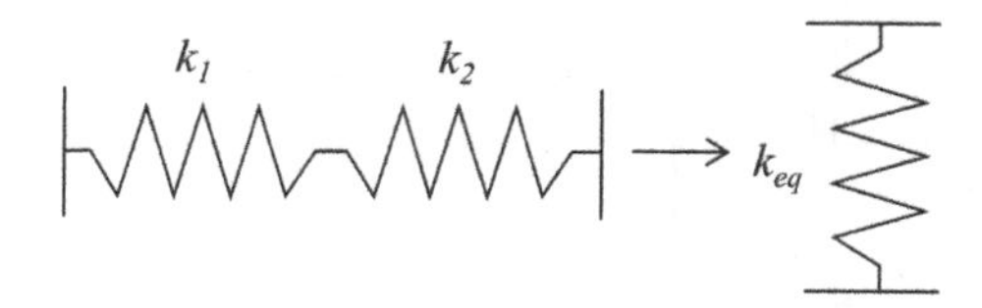

and,

$$k_d = \frac{EA_d}{l_d}. \tag{8.39}$$

Here, E is the Young's modulus of the bolt material, A_d is the cross-sectional area of the unthreaded shank section of the bolt (based on the major diameter), A_t is the tensile stress area, l_d is the length of the unthreaded shank, and l_t is the length of the threaded section of the bolt within the grip. Substituting Eqs. (8.38) and (8.39) into Eq. (8.37) yields

$$k_B = \frac{A_d A_t E}{A_d l_t + A_t l_d}. \tag{8.40}$$

Thus, to calculate k_B, we must first determine A_d, A_t, l_d, and l_t. We calculate A_d based on the major diameter of the bolt, we find A_t in Table 8.1, and we determine l_d and l_t from Figure 8.15 if the bolt goes through the members, and from Figure 8.16 if the bolt is fastened into a blind hole.

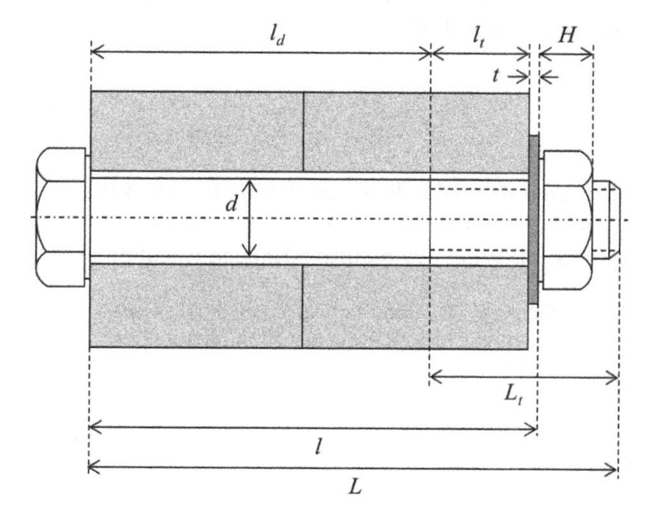

Figure 8.15 Determining bolt dimensions for a through hole with a bolt and nut. Source: Modified from Budynas, R.G. and Nisbett, J.K. (2015). *Shigley's Mechanical Engineering Design*, 10e. McGraw-Hill.

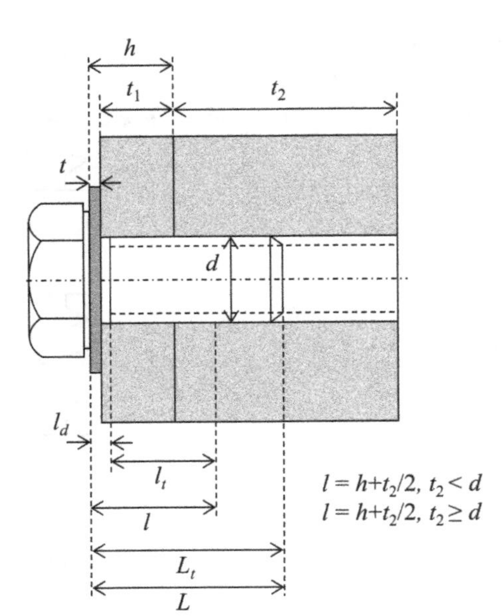

Figure 8.16 Determining bolt dimensions for a blind hole with a bolt. Source: Modified from Budynas, R.G. and Nisbett, J.K. (2015). *Shigley's Mechanical Engineering Design*, 10e. McGraw-Hill.

Note that for metric bolts of length L, the total threaded length of the bolt L_T is

$$L_T = \begin{cases} 2d + 6 \text{ mm} & L \leq 125 \text{ mm},\ d \leq 48 \text{ mm}, \\ 2d + 12 \text{ mm} & 125 \leq L \leq 200 \text{ mm}, \\ 2d + 6 \text{ mm} & L \geq 200 \text{ mm}. \end{cases} \tag{8.41}$$

8.3.7.3 Stiffness of the Members

We calculate the stiffness of the members assuming that the stress distribution in the members follows a *frustrum* shape under 45° (Figure 8.17) and that the average diameter of the washer (average between ID and OD) is approximately $1.5d$, with d the major diameter of the bolt. We also assume that both members are made of the same material. With those assumptions, the stiffness of the members is given as

$$k_M = \frac{\pi E d}{2 \ln \left[5 \left(\frac{l+0.5d}{l+2.5d} \right) \right]}, \tag{8.42}$$

where E is the Young's modulus of the members, d is the major diameter of the bolt, and l is the grip.

However, note that calculating the stiffness of the members is more complex than the substantially simplified version we describe in this section. When we require advanced calculations, we

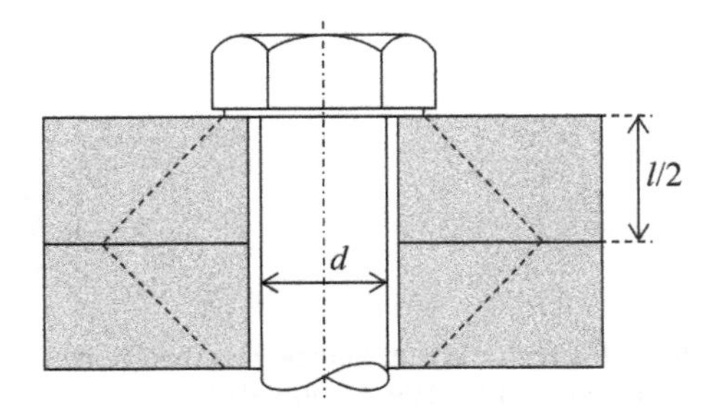

Figure 8.17 Determining the member stiffness, assuming a *frustrum-shaped* stress distribution under 45°.

refer the reader to reference works that treat the calculation of member stiffness in a more sophisticated fashion than in this introductory textbook.

8.3.7.4 Stiffness of Members with a Gasket

Sometimes we place a gasket between the members, when sealing liquids or gases is important. For instance, the valve cover gasket in a combustion engine fits between the engine block and the valve cover to seal oil and combustion gases in the combustion engine.

Figure 8.18 shows three examples of gasket geometries, including (a) a gasket compressed between both members, and (b) and (c) different sizes of gaskets that run into a groove between the members, sealing a fluid with pressure p.

When considering gaskets in bolted joint calculations, we treat the gasket as a separate spring with stiffness k_G in series with the stiffness of the members k_M. Thus, the equivalent stiffness k_{MG} of the stiffness of the gasket and members in series is

$$\frac{1}{k_{MG}} = \frac{1}{k_M} + \frac{1}{k_G}, \tag{8.43}$$

and, thus,

$$k_{MG} = \frac{k_M k_G}{k_M + k_G}. \tag{8.44}$$

Table 8.6 shows a few examples of commonly used gasket materials, and their respective Young's modulus. We observe that the Young's modulus of most gasket materials is approximately 100 times smaller than the Young's modulus of steel. Thus, $k_G \ll k_M$, and the fraction $k_M/(k_M + k_G) \cong 1$, which reduces Eq. (8.44) to

$$k_{MG} \cong k_G. \tag{8.45}$$

Therefore, using a gasket between the members has important consequences for the bolt because the portion of the external load borne by the bolt and the members depends on the ratio of their

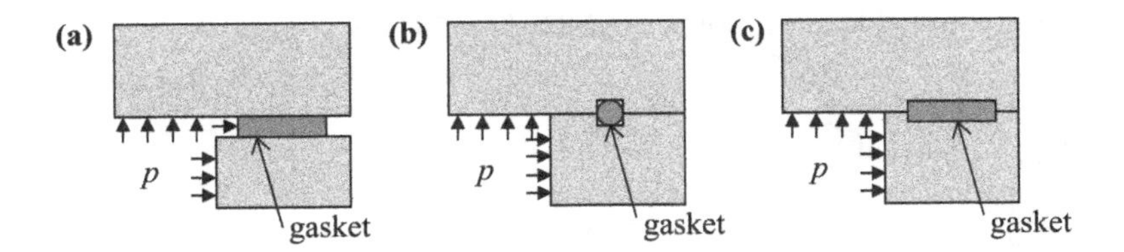

Figure 8.18 Members with gaskets.

Table 8.6 Young's modulus of common gasket materials.

Material	Young's modulus E (GPa)
Steel	210.00
Silicon	185.00
PTFE (Teflon)	0.50
Rubber	0.01–0.10
Nylon	2.00–4.00
Cork	0.80

stiffness values. When the stiffness of the members with gasket reduces to approximately the stiffness of the gasket, it significantly reduces the portion of the external load borne by the members and, consequently, increases the portion of the external load borne by the bolt. Equation (8.27) indicates the total resulting load in the bolt. Replacing k_M with k_{MG} yields

$$F_B = F_i + \frac{k_B}{k_{MG} + k_B} P. \tag{8.46}$$

Because $k_{MG} \ll k_M$, the denominator decreases, thus increasing the portion of the load borne by the bolt.

Example problem 8.2

Figure 8.19 shows a portion of a cylindrical steel pressure vessel ($E = 200$ GPa), filled with a gas maintained at $p = 2$ MPa. Figure 8.19a shows a cross-sectional view, whereas Figure 8.19b shows a top view. The inner diameter of the pressure vessel $D = 200$ mm. A steel cap is bolted on the pressure vessel with eight class 8.8 $M10$ steel bolts. The preload in each bolt $F_i = 10$ kN. A gasket is placed between the members to avoid leakage of the gas from the pressure vessel. The geometry of the vessel is as follows: $t = 50$ mm, $L_1 = 400$ mm, $L_2 = 500$ mm. The stiffness of the gasket is 100 times smaller than the stiffness of the members.

(a) Calculate the safety factor against yielding of the bolts.
(b) Calculate the safety factor against yielding of the bolts, without the gasket.

Solution

(a) We use class 8.8 bolts; $\sigma_{ut} = 800$ MPa, $\sigma_y = 640$ MPa.
We calculate the external load acting on the bolt by converting the pressure acting on the cap of the pressure vessel to a load.

$$P_{total} = p(\pi D^2/4) = 2 \cdot 10^6(\pi 0.2^2/4) = 62\ 832 \text{ N}.$$

Thus, the we estimate the external load per bolt $P = 62\ 832/8 = 7854$ N.

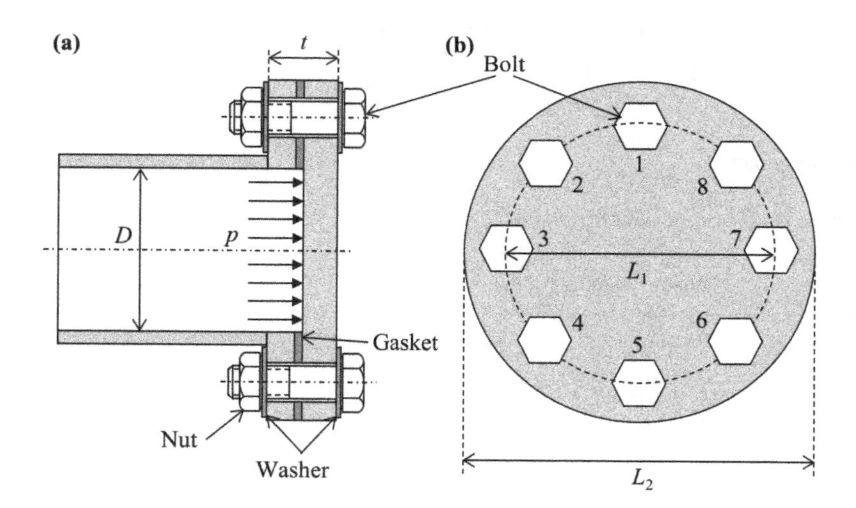

Figure 8.19 Pressure vessel with a gasket between the members.

We calculate the length of the bolt L, which is the sum of the grip l, i.e. the thickness of the members t and the thickness of two regular $M10$ washers (see Table 8.4), and the thickness of a regular hexagonal $M10$ nut (see Table 8.3). Hence,

$$L = 50 + 2 \times 2.8 + 8.4 \text{ mm} = 64 \text{ mm}.$$

Note that the grip $l = 50 + 2 \times 2.8 = 55.6$ mm.
However, 64 mm is not a standard size for metric bolts, so we round up to the next standard size, i.e. $L = 80$ mm (see Table 8.2).
We use Figure 8.15 to determine L_T, l_t, and l_d.
Equation (8.41) defines the threaded portion of the bolt $L_T = 2 \times 10 + 6 = 26$ mm.
Thus, the length of the unthreaded shank within the grip (it is always within the grip) is $l_d = L - L_T = 80 - 26 = 54$ mm.
Then, the length of the threaded portion of the bolt within the grip is $l_t = l - l_d = 55.6 - 54 = 1.6$ mm.

$A_t = 58 \cdot 10^{-6}\ \text{m}^2$ (see Table 8.1).
$A_d = \pi d^2/4 = \pi 0.01^2/4 = 78 \cdot 10^{-6}\ \text{m}^2$.

We calculate the stiffness of the bolt using Eq. (8.40), thus

$$k_B = \frac{EA_tA_d}{A_dl_t + A_tl_d} = \frac{200 \cdot 10^9 \times 58 \cdot 10^{-6} \times 78 \cdot 10^{-6}}{78 \cdot 10^{-6} \times 1.6 \cdot 10^{-3} + 58 \cdot 10^{-6} \times 54 \cdot 10^{-3}} = 2.78 \cdot 10^8\ \text{N/m}.$$

We calculate the stiffness of the members (without gasket), using Eq. (8.42), i.e.

$$k_M = \frac{\pi Ed}{2\ln\left[5\left(\frac{l+0.5d}{l+2.5d}\right)\right]} = \frac{\pi \times 200 \cdot 10^9 \times 0.010}{2\ln\left[5\left(\frac{0.0556+0.5\times0.010}{0.0556+2.5\times0.010}\right)\right]} = 2.37 \cdot 10^9\ \text{N/m}.$$

The stiffness of the gasket is 100 times smaller than the stiffness of the members. Thus,

$$K_G = 0.01K_M = 0.01 \times 2.37 \cdot 10^9 = 2.37 \cdot 10^7\ \text{N/m} \cong K_{MG}.$$

The safety factor is the total resulting load in the bolt at the inception of yielding, divided by the total resulting load in the bolt. Thus,

$n = \sigma_y A_t / F_B$.

Using Eq. (8.46), we calculate

$$F_B = F_i + \frac{k_B}{k_{MG} + k_B}P = 10\,000 + \left(\frac{2.77 \cdot 10^8}{2.77 \cdot 10^8 + 2.37 \cdot 10^7}\right)7854 = 17\,241\ \text{N}.$$

Hence, $n = 640 \cdot 10^6 \times 58 \cdot 10^{-6}/17\,241 = 2.15$.

(b) Without the gasket, we use Eq. (8.27) to calculate F_B:

$$F_B = F_i + \frac{k_B}{k_M + k_B}P = 10\,000 + \left(\frac{2.77 \cdot 10^8}{2.77 \cdot 10^8 + 2.37 \cdot 10^9}\right)7854 = 10\,822\ \text{N}.$$

Correspondingly, $n = 640 \cdot 10^6 \times 58 \cdot 10^{-6}/10\,822 = 3.43$.

Example problem 8.3

Figure 8.20 shows a traffic sign made of steel ($E = 200$ GPa), comprising a bottom portion rigidly anchored in the soil, and a top portion connected to the bottom portion with four class 8.8 $M20$ steel bolts, with a regular hexagonal nut, and a regular washer between the bolt and nut head and the members. The preload in each bolt $F_i = 100$ kN. The dimensions of the traffic sign are as follows: the distance between the centroid of the traffic sign and the connection to the bottom portion anchored in the soil $L_1 = 2$ m, bolt circle diameter $L_2 = 0.40$ m, flange diameter $L_3 = 0.60$ m, the thickness of the members $t = 0.05$ m. Assume that the wind blows from the south, and the wind load is considered a force $F = 40$ kN acting in the centroid of the traffic sign.

(a) Calculate the resulting load in each bolt.
(b) Calculate the safety factor against yielding of the heaviest loaded bolt.

Solution

(a) We use class 8.8 bolts; $\sigma_{ut} = 800$ MPa, $\sigma_y = 640$ MPa.
We first determine the external load acting on each bolt, P_1, P_2, P_3, and P_4.
The wind creates a moment on the traffic sign, which attempts to rotate the top portion about the edge of the bottom portion. The moment $M = 40\ 000 \times 2 = 80\ 000$ Nm.
Moment equilibrium about the rotation point requires that
$80\ 000 = 0.1 \times P_1 + 0.3 \times P_2 + 0.3 \times P_4 + 0.5 \times P_3$.
Each bolt is subject to an external load that is proportional to the distance from the rotation point (see Figure 8.20). Thus,

$$P_1/0.1 = P_2/0.3 = P_4/0.3 = P_3/0.5.$$

$P_2 = P_4$ because their distance to the rotation point is identical.

$$P_1 = P_2/3, P_1 = P_3/5, P_2 = 3P_3/5.$$

Combining these equations, we write that

$$80\ 000 = 0.1 \times P_1 + 2 \times 0.3 \times 3 \times P_1 + 0.5 \times 5 \times P_1 = 4.4 \times P_1.$$

Thus, each bolt is subject to the following external load:

$$P_1 = 18\ 182\ \text{N}, P_2 = P_4 = 54\ 545\ \text{N}, P_3 = 90\ 909\ \text{N}$$

Next, we calculate the stiffness of the bolt and the members, which we use to determine the total resulting load in the bolt, by combining the preload and the portion of the external load borne by the bolt.
We calculate the length of the bolt L, which is the sum of the grip l, i.e. the thickness of the members t and the thickness of two regular $M20$ washers (see Table 8.4), in addition to the thickness of a regular hexagonal $M20$ nut (see Table 8.3). Thus,

$$L = 50 + 2 \times 4.6 + 18\ \text{mm} = 77.2\ \text{mm}.$$

Note that the grip $l = 50 + 2 \times 4.6 = 59.2$ mm.
However, 77.2 mm is not a standard size for metric bolts, so we round up to the next standard size, i.e. $L = 80$ mm (see Table 8.2).
We use Figure 8.15 to determine L_T, l_t, and l_d.
Equation (8.41) defines the threaded portion of the bolt $L_T = 2 \times 20 + 6 = 46$ mm.
Thus, the length of the unthreaded shank within the grip (it is always within the grip) is $l_d = L - L_T = 80 - 46 = 34$ mm.
Then, the length of the threaded portion of the bolt within the grip is $l_t = l - l_d = 59.2 - 34 = 25.2$ mm.

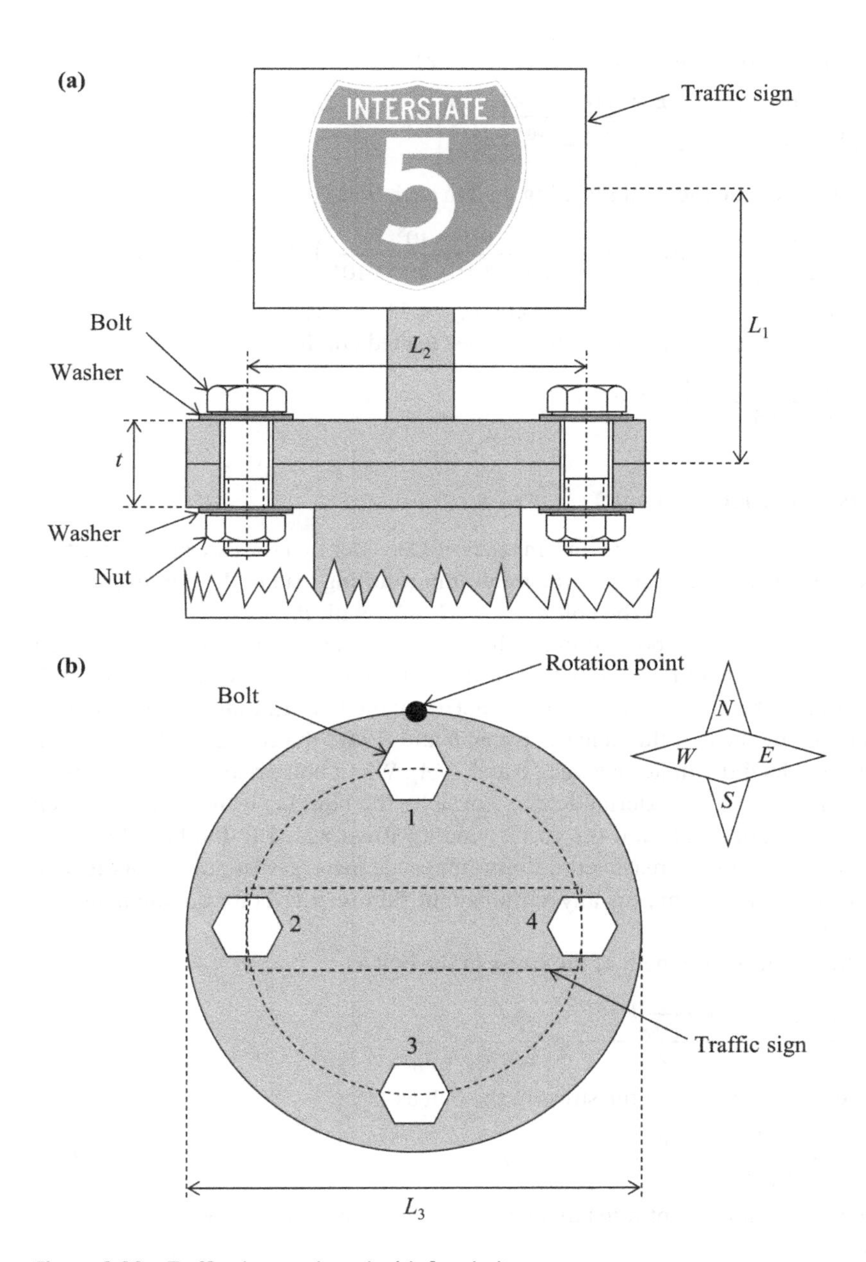

Figure 8.20 Traffic sign, anchored with four bolts.

$A_t = 245 \cdot 10^{-6}$ m^2 (see Table 8.1).

$A_d = \pi d^2/4 = \pi 0.02^2/4 = 314 \cdot 10^{-6}$ m^2.

We calculate the stiffness of the bolt using Eq. (8.40). Thus,

$$k_B = \frac{EA_tA_d}{A_dl_t + A_tl_d} = \frac{200 \cdot 10^9 \times 245 \cdot 10^{-6} \times 314 \cdot 10^{-6}}{314 \cdot 10^{-6} \times 25.2 \cdot 10^{-3} + 245 \cdot 10^{-6} \times 34 \cdot 10^{-3}} = 9.47 \cdot 10^8 \text{ N/m}.$$

We calculate the stiffness of the members using Eq. (8.42), i.e.

$$k_M = \frac{\pi E d}{2\ln\left[5\left(\frac{l+0.5d}{l+2.5d}\right)\right]} = \frac{\pi \times 200\cdot 10^9 \times 0.020}{2\ln\left[5\left(\frac{0.0592+0.5\times 0.020}{0.0592+2.5\times 0.020}\right)\right]} = 5.45\cdot 10^9 \text{ N/m}.$$

We use Eq. (8.27) to calculate the total resulting load in each bolt F_B:

$$F_{B,1} = F_i + \frac{k_B}{k_M + k_B} P_1 = 100\ 000 + \left(\frac{9.47\cdot 10^8}{9.47\cdot 10^8 + 5.45\cdot 10^9}\right) 18\ 182 = 102\ 691 \text{ N}.$$

Similarly, we find that $F_{B,2} = F_{B,4} = 108\ 073$ N, and $F_{B,3} = 113\ 455$ N.
Hence, the safety factor against yielding of the heaviest loaded bolt is

$$n = \frac{640\cdot 10^6 \times 245\cdot 10^{-6}}{113\ 455} = 1.38.$$

8.3.8 Bolted Joints in Tension (Dynamic)

We also consider dynamic loading of bolted joints in tension. Consider that the external load acting on the bolt fluctuates between 0 and P. Hence, the minimum resulting load in the bolt $F_{B,min} = F_i$, whereas the maximum resulting load in the bolt $F_{B,max} = F_i + P_B$, with $P_B = Pk_B/(k_M + k_B)$.

We re-emphasize that preload is important to maximize the fatigue life of the bolt. Figure 8.21 shows the fluctuating external load P as a function of time. If no preload exists in the bolt, or if member separation occurs, the entire external load is borne by the bolt. In that case, the portion of the external load borne by the bolt fluctuates between 0 and P, and the corresponding stress in the bolt from the external load fluctuates between 0 and P/A_t. When bolt preload prevents member separation, only a portion of the external load is borne by the bolt, i.e. it fluctuates between 0 and P_B (dashed line in Figure 8.21, and the corresponding stress added to the bolt fluctuates between 0 and P_B/A_t). As a result, decreasing the stress amplitude increases fatigue life of the bolt, which one understands by considering, e.g. any of the fatigue failure criteria for ductile materials (Goodman/Soderberg/Gerber).

We calculate the alternating component of the stress in the bolt as

$$\sigma_a = \frac{F_{B,max} - F_{B,min}}{2A_t} = \frac{F_i + \frac{k_B}{k_M+k_B}P - F_i}{2A_t} = \frac{P}{2A_t}\frac{k_B}{k_M + k_B}. \tag{8.47}$$

We calculate the steady component of the stress in the bolt as

$$\sigma_m = \frac{F_{B,min}}{A_t} + \frac{P}{2A_t}\frac{k_B}{k_M + k_B} = \frac{F_i}{A_t} + \frac{P}{2A_t}\frac{k_B}{k_M + k_B}. \tag{8.48}$$

We use the Goodman criterion for infinite fatigue life with fluctuating stress (see Eq. (6.12)):

$$\frac{1}{n} = \frac{\sigma_m}{\sigma_{ut}} + \frac{\sigma_a}{\sigma_e}, \tag{8.49}$$

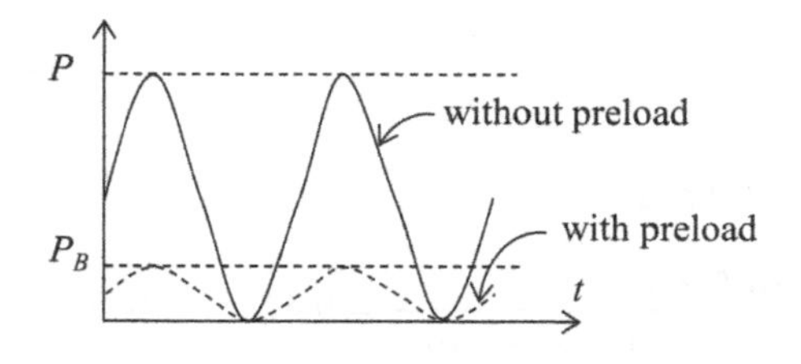

Figure 8.21 Bolted joint subject to a dynamic axial load, schematically illustrating that preload reduces the fluctuating load applied to the bolt, thus increasing its fatigue life.

Table 8.7 Endurance limit σ_e of metric bolts with rolled screw thread.

Bolt grade	Size range	Endurance limit σ_e (MPa)
ISO 8.8	$M16$–$M36$	129
ISO 9.8	$M1.6$–$M16$	149
ISO 10.8	$M5$–$M36$	162
ISO 12.9	$M1.6$–$M36$	190

with n the design (or safety) factor. Using the Goodman criterion, we can show that the safety factor against fatigue failure is given as[2]

$$n = \frac{\sigma_e \left(\sigma_{ut} - \sigma_i\right)}{\sigma_a \left(\sigma_{ut} + \sigma_e\right)}, \tag{8.50}$$

with $\sigma_i = F_i/A_t$.

When performing bolt calculations with dynamic loading, we must determine the endurance limit of the bolt. Fastener screw thread is most commonly "rolled," and we determine the endurance limit from a table that already corrects the endurance limit for stress concentrations. Table 8.7 shows the endurance limit for common bolt strength categories and sizes. Alternatively, if the fastener screw thread is "cut" using, e.g. a lathe, then we must use the endurance limit modifying factors to determine the endurance limit.

Example problem 8.4

Figure 8.22 shows a toy in the monkey cage at the zoo. A steel plate ($E = 200\,\text{GPa}$) with a half ring-shaped anchor is clamped against a fixed steel structure by means of two class 8.8 $M10$ steel bolts (rolled thread) with preload $F_i = 30$ kN. We place a regular washer on each side of the members, and we use a regular hexagonal nut to tighten each bolt. The thickness of the members

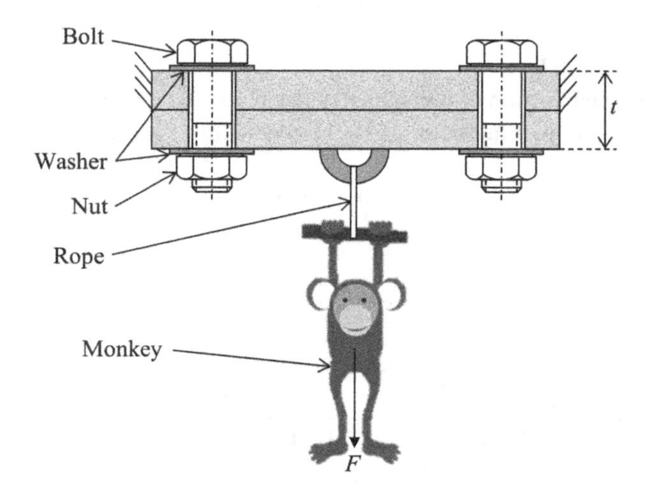

Figure 8.22 Monkey cage toy, anchored with two bolts.

2 See, e.g. Budynas, R.G. and Nisbett, J.K. (2015). *Shigley's Mechanical Engineering Design*, 10e. McGraw-Hill.

$t = 40$ mm. An in-extensible rope with a cylinder is attached to the ring from which the monkeys can hang. A single monkey weighs up to $F = 300$ kg. However, monkeys are known to bully and jump on each other while hanging from the rope. As a result the load can fluctuate between 0 and 8 monkeys at any given time. Because of the symmetric design, the load is shared equally between the bolts.

(a) Calculate the torque required to obtain a preload $F_i = 30$ kN per bolt.
(b) Calculate the stiffness of the bolt taking into account the unthreaded and threaded portion.
(c) Calculate the stiffness of the members assuming that the load spreads under a 45° angle.
(d) Calculate the minimum preload to prevent member separation.
(e) Calculate the maximum load borne by each bolt.
(f) Calculate the safety factor against yielding of each bolt.
(g) Calculate the contact pressure between the washer and the members. If the members are made from AISI 1018 HR steel, determine whether plastic deformation occurs.
(h) Determine the safety factor against fatigue failure if the external load harmonically varies between 0 and 8 monkeys.

Solution

We use class 8.8 bolts; $\sigma_{ut} = 800$ MPa, $\sigma_y = 640$ MPa.

(a) We use Eq. (8.19) to determine the torque required to obtain the prescribed preload. Furthermore, we use Table 8.5 to determine parameter K. Since the problem statement does not specifically identify the surface finish of the bolt, we use $K = 0.2$. Thus, $M_t = KdF_i = 0.2 \times 0.010 \times 30\,000 = 60$ Nm.

(b) We calculate the length of the bolt L, which is the sum of the grip l, i.e. the thickness of the members t and the thickness of two regular $M10$ washers (see Table 8.4), in addition to the thickness of a regular hexagonal $M10$ nut (see Table 8.3). Hence,

$$L = 40 + 2 \times 2.8 + 8.4 \text{ mm} = 54 \text{ mm}.$$

Note that the grip $l = 40 + 2 \times 2.8 = 45.6$ mm.
$L = 54$ mm is not a standard size for metric bolts, so we round up to the next standard size, $L = 60$ mm (see Table 8.2).
We use Figure 8.15 to determine L_T, l_t, and l_d.
Equation (8.41) defines that the threaded portion of the bolt is $L_T = 2 \times 10 + 6 = 26$ mm.
Thus, the length of the unthreaded shank within the grip (it is always within the grip) is $l_d = L - L_T = 60 - 26 = 34$ mm.
Then, the length of the threaded portion of the bolt within the grip is $l_t = l - l_d = 45.6 - 34 = 11.6$ mm.

$A_t = 58 \cdot 10^{-6}\ \text{m}^2$ (see Table 8.1).
$A_d = \pi d^2/4 = \pi 0.01^2/4 = 78 \cdot 10^{-6}\ \text{m}^2$.

We calculate the stiffness of the bolt using Eq. (8.40). Thus,

$$k_B = \frac{EA_tA_d}{A_dl_t + A_tl_d} = \frac{200 \cdot 10^9 \times 58 \cdot 10^{-6} \times 78 \cdot 10^{-6}}{78 \cdot 10^{-6} \times 11.6 \cdot 10^{-3} + 58 \cdot 10^{-6} \times 34 \cdot 10^{-3}} = 3.16 \cdot 10^8\ \text{N/m}.$$

(c) We calculate the stiffness of the members using Eq. (8.42), i.e.

$$k_M = \frac{\pi Ed}{2\ln\left[5\left(\frac{l+0.5d}{l+2.5d}\right)\right]} = \frac{\pi \times 200 \cdot 10^9 \times 0.010}{2\ln\left[5\left(\frac{0.0456+0.5\times0.010}{0.0456+2.5\times0.010}\right)\right]} = 2.46 \cdot 10^9\ \text{N/m}.$$

(d) A single monkey weighs 300 kg or approximately $F = 3000$ N. If a maximum of eight monkeys can hang from the structure at any time, then the total external load $P_t = 8 \times 3000 = 24\ 000$ N. Because the load is shared equally by the bolts, the external load per bolt is $P = 12\ 000$ N. To avoid member separation, the total resulting load in the members must exceed zero, i.e. $F_M \geq 0$. Thus,

$$F_M = P_M - F_i = \frac{k_M}{k_M + k_B} P - F_i.$$

Hence, $F_i \geq \left(\frac{2.46 \cdot 10^9}{3.16 \cdot 10^8 + 2.46 \cdot 10^9} \right) 12\ 000$, or

$F_i \geq 10\ 634$ N.

(e) We calculate the maximum load per bolt as follows:

$$F_B = F_i + P_B = F_i + \frac{k_B}{k_M + k_B} P. \text{ Thus,}$$

$$F_B = 30\ 000 + \left(\frac{3.16 \cdot 10^8}{3.16 \cdot 10^8 + 2.46 \cdot 10^9} \right) 12\ 000 = 31\ 366 \text{ N.}$$

(f) The safety factor against yielding is the total resulting load in the bolt at the inception of yielding, divided by the maximum total resulting load in the bolt, i.e.

$n = \sigma_y A_t / F_B$, or
$n = 640 \cdot 10^6 \times 58 \cdot 10^{-6} / 31\ 366 = 1.18.$

(g) We calculate the contact pressure between the washer and the members as $p_{contact} = F_B / A_{washer}$, where $A_{washer} = (OD^2 - ID^2)\pi/4$, with OD and ID the outer and inner diameter of the metric washer, respectively, as tabulated in Table 8.4.

$A_{washer} = (28^2 - 10.85^2)\pi/4 = 523.3$ mm^2,

$p_{contact} = 31\ 366 / 523.3 \cdot 10^{-6} = 59.9$ MPa.

Table 2.2 shows that the yield stress of AISI 1018 HR steel is $\sigma_y = 220$ MPa. Hence, no plastic deformation of the members occurs because $p_{contact} \ll \sigma_y$.

(h) The Goodman criterion is expressed in Eq. (6.12). We first determine the alternating normal stress σ_a and the steady normal stress σ_m in the bolt, in addition to the endurance limit of the bolt σ_e.

$\sigma_a = P_B / 2A_t = \frac{k_B}{k_M + k_B} \frac{P}{2A_t} = \left(\frac{3.16 \cdot 10^8}{3.16 \cdot 10^8 + 2.46 \cdot 10^9} \right) \frac{12\ 000}{2 \times 58 \cdot 10^{-6}} = 11.8$ MPa,

$\sigma_m = F_i / A_t + P_B / 2A_t = 30\ 000 / 58 \cdot 10^{-6} + 11.8 \cdot 10^6 = 529$ MPa,

$\sigma_i = F_i / A_t = 30\ 000 / 58 \cdot 10^{-6} = 517$ MPa.

Since the bolts have rolled threads, we use $\sigma_e = 140$ MPa from Table 8.7.
We use Eq. (8.50) to calculate the safety factor against fatigue failure. Thus,

$n = \left[\sigma_e \left(\sigma_{ut} - \sigma_i\right)\right] / \left[\sigma_a \left(\sigma_{ut} + \sigma_e\right)\right]$, or
$n = [140\,(800 - 517)] / [11.8\,(800 + 140)] = 3.58.$

8.3.9 Bolted Joints in Shear

We must avoid any shear stress in the bolt because the cross-section of a bolt is small and, thus, it can fail easily. Hence, we design the bolted joint to avoid shear stress in the bolt itself. To accomplish this, we use the preload in the bolt to compress the members. Thus, the preload is a normal force,

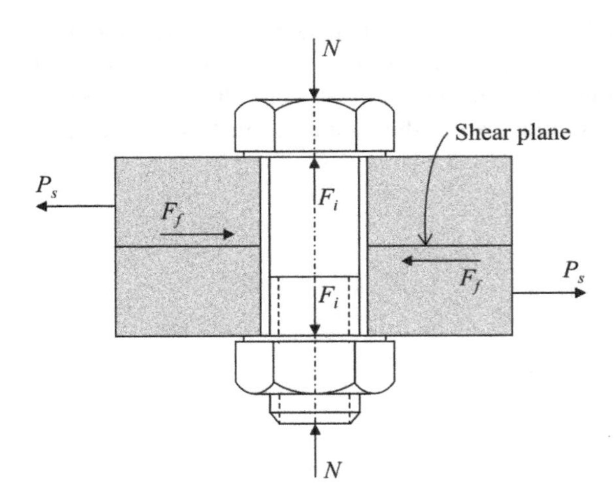

Figure 8.23 Bolted joint subject to an external shear load.

which creates a friction force between the members that counteracts a shear force acting on the bolted joint. Figure 8.23 shows a schematic of a bolted joint, subject to an external shear load P_S. The friction force between the members is equal in magnitude and direction and opposite in sense to the external shear load. The friction force increases when the external shear load increases such that no slip occurs between the members and no shear stress occurs in the bolt.

However, the maximum friction force that can exist between the members is $F_f = \mu N$ (assuming Coulomb friction), where μ is the friction coefficient between the members, and N is the normal load between the members, which is equal to the preload in the bolt F_i. If the external shear load exceeds the maximum friction force between the members, then the members slip (relative motion between the members occurs), and the bolt will be subject to shear stress.

Example problem 8.5

Figure 8.24 shows a schematic of a lap joint comprising three steel plates, bolted together using two class 8.8 metric steel bolts. All plates are 10 mm thick and made of steel with $\sigma_y = 240$ MPa. $P = 200$ kN, and the friction coefficient between the plates is $\mu = 0.50$.

(a) Calculate the required major diameter of the bolts to avoid shear stress in the bolts and consider a design factor $n_D = 1.25$.
(b) Calculate the torque required to tighten the bolts.

Solution

(a) We use class 8.8 bolts; $\sigma_{ut} = 800$ MPa, $\sigma_y = 640$ MPa.
We preload the two bolts such that we create a maximum friction force that is equal to P, additionally considering the required design factor.
Thus, $F_f = P \times n_D = \mu \times (\text{total preload})$

$$P \times n_D = \mu \times F_i \times n_b \times i,$$

where F_i is the preload in each bolt, n_b is the number of bolts, and i is the number of shear planes, i.e. interfaces between the members where a friction force is created as a result of the preload. Hence,

$$F_i = \frac{n_D \times P}{n_b \times i \times \mu} = \frac{1.25 \times 200 \cdot 10^3}{2 \times 2 \times 0.50} = 125 \cdot 10^3 \text{ N}.$$

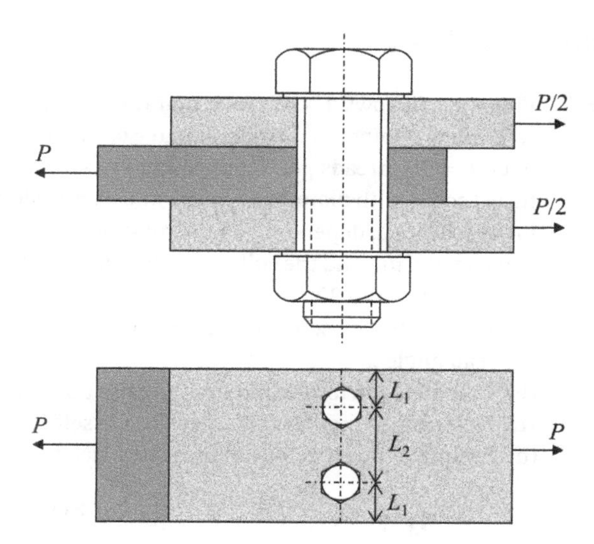

Figure 8.24 Lap joint with two bolts and three plates.

From the required preload, we can calculate the bolt diameter. We relate the preload and the yield stress to the tensile stress area, which we can reverse look-up in Table 8.1, to determine the required bolt diameter.

$$F_i = \sigma_y A_t \Rightarrow A_t = F_i/\sigma_y = 125\ 000/640 \cdot 10^6 = 195 \cdot 10^{-6}\ \text{m}^2.$$

From Table 8.1, we determine that we must select $M20$ bolts ($A_t = 245\ \text{mm}^2$).

(b) The torque to tighten the bolts is $M_t = KdF_i$ (see Eq. (8.19)). Furthermore, we use Table 8.5 to determine parameter K. Since the problem statement does not specifically identify the surface finish of the bolt, we use $K = 0.2$.

$$M_t = 0.20 \times 0.020 \times 125 \cdot 10^3 = 500\ \text{Nm}.$$

8.4 Key Takeaways

1. Two categories of screw thread exist: power screw thread, which is designed to facilitate motion between two mechanical elements, and fastener screw thread, which is designed to tighten two mechanical elements together.
2. A power screw converts a rotary to a linear motion, and we relate the torque required to raise and lower a load with the power screw to the geometry, material properties, and operating conditions of the power screw.
3. Preloading a bolted joint is crucial because (i) it reduces the portion of the external load borne by the bolt, and also places the members in compression, thus increasing their ability to withstand tensile forces. (ii) It avoids shear stress in the bolt, by means of the friction force created between the members. (iii) It increases fatigue life of the bolt by reducing the stress amplitude resulting from dynamic external loading, applied to the bolted joint.
4. We always design bolted joints such that the bolt is not subject to shear stress because a bolt has a small cross-section and, thus, does not withstand shear stress well.

8.5 Problems

8.1 Figure 8.25 shows a screw jack that raises and lowers a non-rotating mass $m = 200$ kg. The screw jack comprises a single square screw thread with a major diameter $d = 30$ mm and a pitch $p = 10$ threads per 5 cm. A rotating screw nut driven by a leverage arm moves a plain thrust collar with diameter $d_c = 50$ mm axially along the screw thread. This in turn raises and lowers the load, depending on the rotation direction of the screw nut. The friction coefficient between the nut and the collar $\mu_c = 0.25$ and the friction coefficient between the screw thread and the nut $\mu = 0.20$.

(a) Determine the screw thread pitch p, lead l, minor diameter d_r, mean diameter d_m, and lead angle λ.
(b) Calculate the torque required to raise and lower the mass m with the screw jack.
(c) Determine whether this screw jack is self-locking.
(d) Estimate the efficiency of the screw jack screw thread.

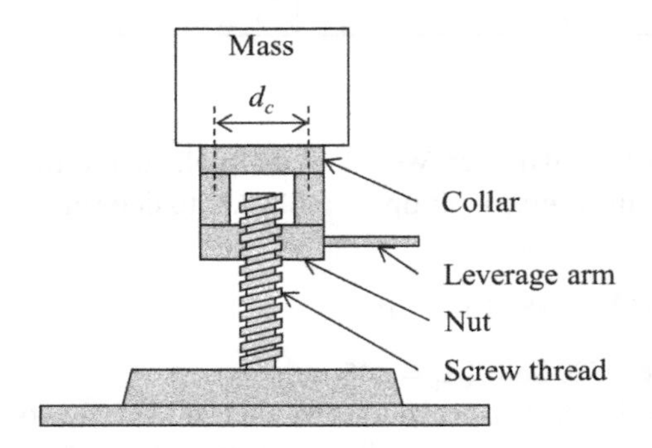

Figure 8.25 Screw jack.

8.2 Determine the yield stress σ_y and the ultimate tensile stress σ_{ut} of the following metric bolt strength classes.

Strength class	σ_y (MPa)	σ_{ut} (MPa)
6.8		
8.8		
9.8		
10.9		

8.3 Figure 8.26 shows a cross-sectional view of a flange clutch, clamped together by four class 10.9 $M16$ steel bolts (only two are shown in Figure 8.26). $L_1 = 200$ mm, $L_2 = 240$ mm,

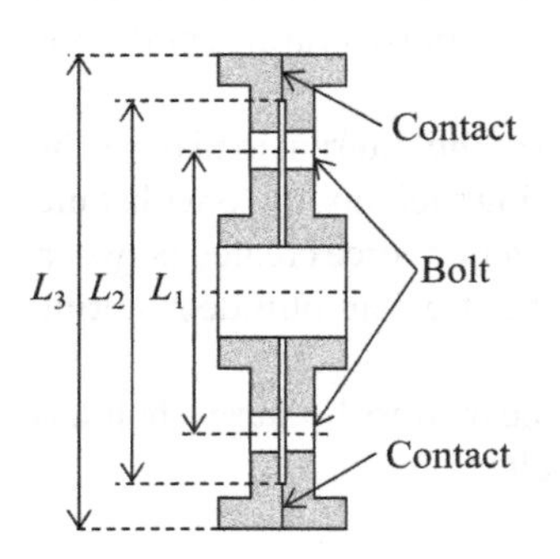

Figure 8.26 Flange clutch.

$L_3 = 260$ mm. The friction coefficient between the contacting surfaces of the clutch $\mu = 0.25$. Calculate the maximum torque moment $M_{t,max}$ that the clutch can transmit.

8.4 Figure 8.27 shows a bracket that attaches to a rigid structure with three class 8.8 metric steel bolts. The friction coefficient between the bracket and the rigid structure $\mu = 0.4$. $F =$ 24 kN acts in the shear plane between the bracket and the rigid structure, $L_1 = 100$ mm, $L_2 = 300$ mm, $L_3 = 150$ mm.

(a) Calculate the preload per bolt to ensure that no shear stress occurs in the bolts and consider a design factor $n_D = 4$.

(b) Calculate the metric bolt size needed for this application.

Figure 8.27 Bracket and rigid structure.

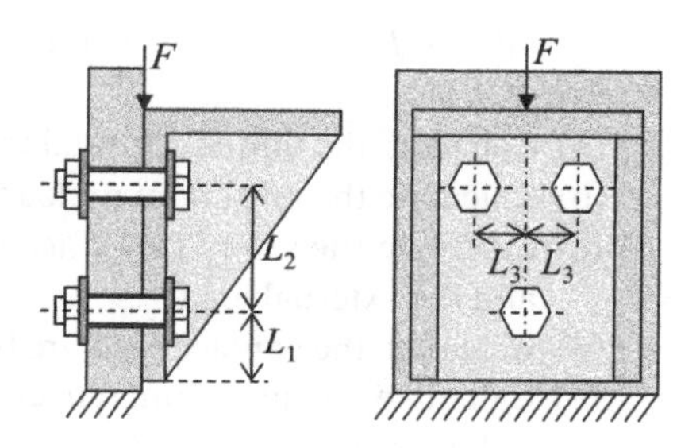

8.5 Figure 8.28 shows a machine frame that comprises two vertical plates connected by a U-shaped horizontal plate with four class 6.8 metric steel bolts. All plates are made from AISI 1060 HR steel. The friction coefficient between the plates $\mu = 0.4$. $F = 30$ kN, $L = 1$ m, $W = 300$ mm, $l = 10$ mm.

(a) Calculate the preload per bolt required to ensure that no shear stress occurs in the bolts. Use a design factor $n_D = 3$.

(b) Calculate the metric bolt size needed for this application.

Figure 8.28 Machine frame.

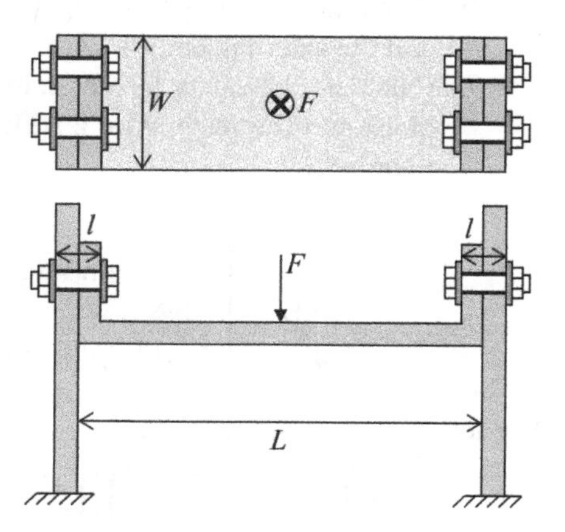

8.6 Figure 8.29 shows a bracket that attaches to a plate by means of two class 10.9 $M10$ steel bolts. Both the bracket and plate are made from AISI 1050 HR steel. The preload in each bolt

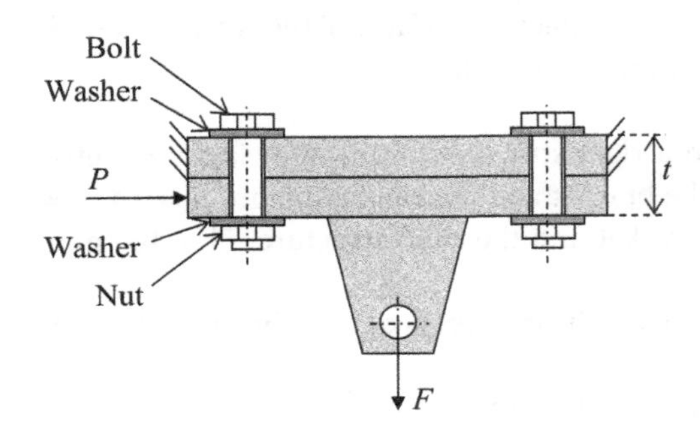

Figure 8.29 Bracket.

$F_i = 10$ kN. $F = 10$ kN. $t = 60$ mm. The friction coefficient between the bracket and the plate is $\mu = 0.3$.

(a) Calculate the torque required to obtain the bolt preload F_i.
(b) Calculate the minimum preload required to avoid member separation.
(c) Calculate the safety factor against yielding of the bolt, considering both the preload F_i and the external load F.
(d) Calculate the contact pressure between the washers and the bracket or plate.
(e) Calculate the maximum tangential force P that can act on the structure without creating a shear stress in the bolts.
(f) Calculate the safety factor for infinite fatigue life when considering that the external load F varies harmonically between 0 and $F = 10$ kN. Assume the bolts have rolled screw thread.

8.7 Figure 8.30 shows a pillow block that supports a rotating shaft. The pillow block is bolted to a plate by means of two class 10.9 $M12$ steel bolts. Both the pillow block and the plate are made from AISI 1010 HR steel. The rotating shaft creates a force $F = 20$ kN. $t = 60$ mm. The preload in each bolt $F_i = 5000$ N.

(a) Calculate the torque required to obtain the bolt preload F_i.
(b) Calculate the safety factor against yielding of the bolts, and consider both the preload F_i and the external load F.
(c) Calculate the safety factor against yielding after inserting a gasket between the pillow block and the plate, with a stiffness that is 100 times smaller than the stiffness of the members.

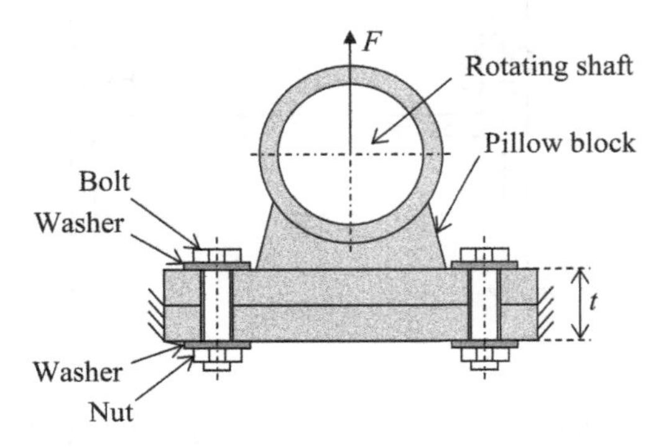

Figure 8.30 Rotating shaft supported in a pillow block.

8.8 Figure 8.31 shows a bracket that attaches to a plate by means of two class 10.9 $M12$ steel bolts. Both bracket and plate are made from AISI 1010 HR steel. The preload in each bolt $F_i = 10$ kN. $F = 20$ kN. $L_1 = 30$ mm, $L_2 = 40$ mm, $\alpha = 45°$.

(a) Calculate the torque required to obtain the bolt preload F_i.
(b) Calculate the minimum preload required to avoid member separation.
(c) Calculate the safety factor against yielding of the bolt and consider both the preload F_i and the external load F.
(d) Calculate the contact pressure between the washers and the bracket or plate.

Figure 8.31 Tilted bracket.

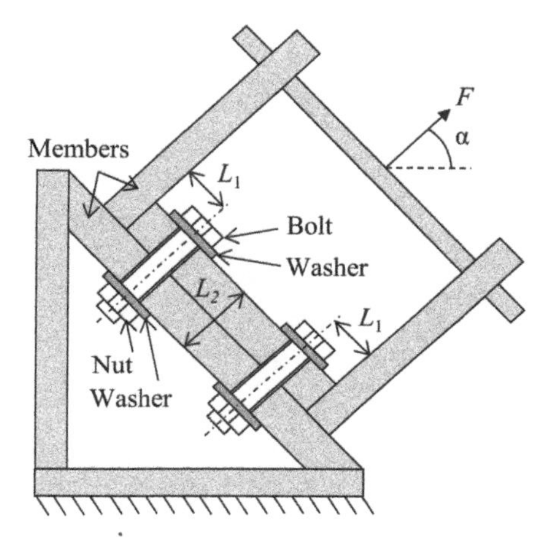

8.9 Figure 8.32 shows an arm structure attached to a frame by means of four class 8.8 $M8$ steel bolts. Both arm and frame are made from AISI 1010 HR steel. The preload in each bolt $F_i = 10$ kN. $P = 1000$ N, $F = 2000$ N. $L_1 = 30$ mm, $L_2 = L_3 = 200$ mm, $L_4 = 20$ mm, $L_5 = 500$ mm.

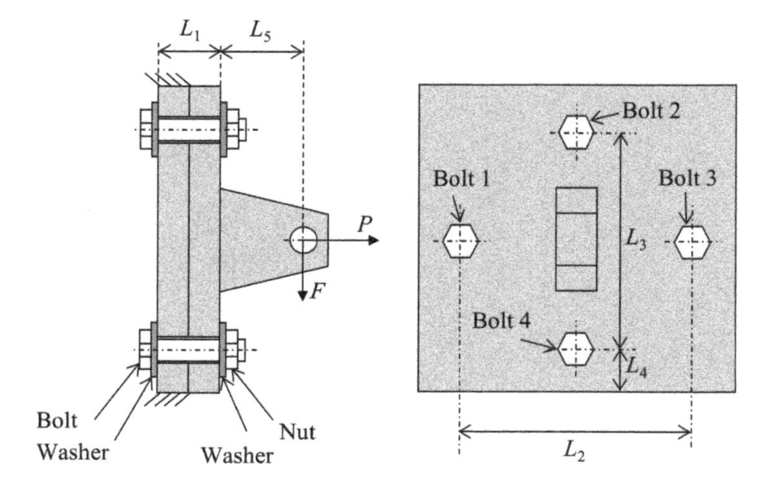

Figure 8.32 Arm structure.

(a) Determine which bolt experiences the heaviest external load.
(b) Calculate the minimum preload required to avoid member separation.
(c) Calculate the torque required to obtain the bolt preload F_i calculated in (b).
(d) Calculate the safety factor against yielding of the bolt, considering both the preload F_i and the external load F.

9

Welded Joints

9.1 Introduction

In this chapter, we will use the knowledge of the chapter on design for static strength and apply it to the design of welded joints. Welded joints are connections between mechanical elements in which we melt part of the elements we try to join and simultaneously add a filler material. Upon solidification, the mechanical elements are joined. Welded joints cannot be disassembled nondestructively, which contrasts, e.g. bolted joints that can be disassembled and re-assembled.

9.1.1 Welding Versus Brazing

We start this chapter by recognizing the difference between *welding* and *brazing*.

Brazing is a joining method that is similar to welding, but we do not melt the mechanical elements we attempt to join. Instead, we only melt the filler material that forms the connection between the mechanical elements. Therefore, brazing requires the melting temperature of the filler material to be substantially lower than that of the mechanical elements. However, because we only melt the filler material, the mechanical strength of a brazed connection is negligible (tensile strength on the order of 5 MPa).

Welding is a joining method in which we melt part of both mechanical elements we attempt to join, in addition to melting the filler material. This melting pool allows both mechanical elements to become "unified" as one new mechanical element. Therefore, the melting temperature of the filler material must be close to that of the mechanical elements.

Note that in this textbook we only consider welded (not brazed) joints because we are interested in evaluating structural connections between mechanical elements.

9.1.2 Techniques and Materials

A wide variety of welding techniques exists, some intended for broad use in different applications and with different materials. However, specialized welding techniques developed for specific applications also exist. Some examples of widely used welding techniques include shielded metal arc welding (so-called "stick welding" because the welder will use a stick-like electrode with filler material, sometimes also referred to as an "electrode rod"), gas tungsten arc welding ("TIG welding," T = Tungsten, IG = Inert Gas), gas metal arc welding ("MIG welding," M = Metal, IG = Inert Gas),

Design of Mechanical Elements: A Concise Introduction to Mechanical Design Considerations and Calculations,
First Edition. Bart Raeymaekers.

Companion website: www.wiley.com/go/raeymaekers/designofmechanicalelements

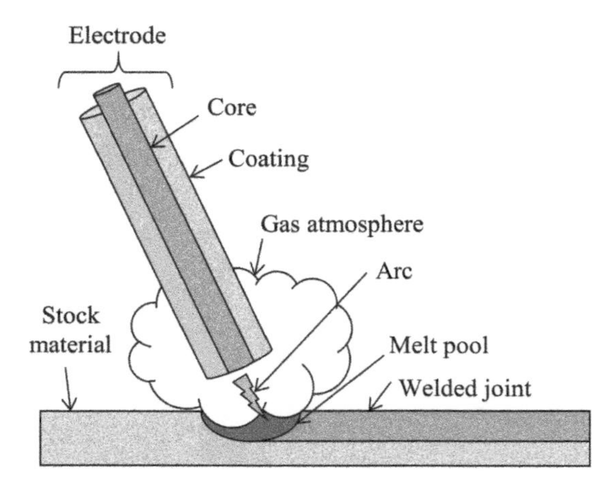

Figure 9.1 Schematic illustration of an arc welding process.

flux-cored arc welding (very similar to MIG welding), laser welding, and friction welding, among other techniques.

Welding is based on the principle that an electric current strikes an arc between the mechanical element and the consumable electrode rod, which is made of filler material (typically steel) and is covered with a coating that creates a protective carbon dioxide (CO_2) gas atmosphere during welding to protect the melt pool from oxidation and contamination. The electrode core itself acts as filler material, making a separate filler unnecessary. Figure 9.1 schematically illustrates an arc welding process.

The melting temperature of steel is approximately 1400 °C and, thus, welding is a process that involves high temperature. Inevitably, a temperature gradient forms as a result of thermal conduction from the melt pool outward into the stock material. Thus, the area around the welded joint also experiences high temperatures (just below the melting point), which might alter its crystal structure and, correspondingly, its mechanical properties. We refer to this area as the *heat affected zone*. As a result, we sometimes observe failure just adjacent to the welded joint, rather than failure of that welded joint itself. Figure 9.2 illustrates a welded joint and the heat affected zone resulting from the welding process.

Not all materials are easily weldable. For instance, low-carbon steels typically show good weldability, whereas increasing carbon content reduces the weldability of steel. Furthermore, alloying elements may reduce weldability, prompting the need for special welding precautions. We refer the reader to materials supplier catalogs to obtain a qualitative understanding of the weldability of a material. Thus, when selecting a material for a specific mechanical element, we must consider how the different elements will be assembled and verify the weldability of the material if we will use welding to join elements.

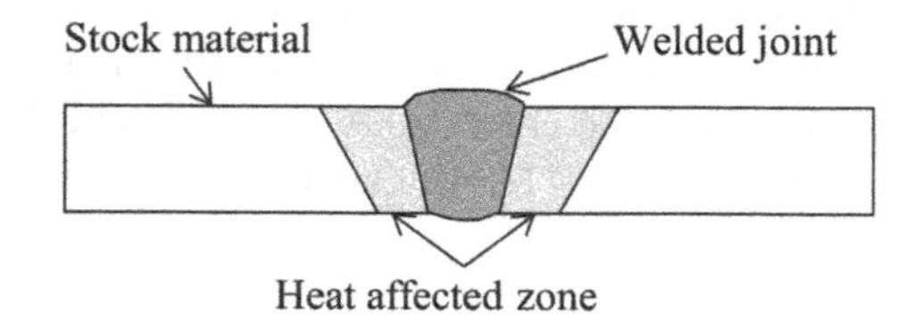

Figure 9.2 Heat affected zone adjacent to the welded joint, resulting from thermal conduction from the melt pool outward into the stock material.

Another consequence of welding is that either deformation or residual stresses exist in the mechanical elements after solidification of the welded joints. Indeed, when creating a melt pool, the materials will shrink upon solidification as a result of thermal contraction. If the mechanical elements are unconstrained, the thermal contraction may create deformation. Alternatively, if we constrain the mechanical elements during the welding process, then we physically constrain any deformation that may occur during solidification of the melt pool and thermal contraction. However, this will create a residual stress in the welded mechanical elements. Such a residual stress is not visible and is superimposed on any stress resulting from external loading. Thus, we must account for the residual stress when calculating the strength of welded joints or ensure that no residual stress exists in the welded mechanical elements by performing heat treatment after welding to relax the residual stress. Figure 9.3 schematically illustrates (a) the intended welded part, (b) the welded part after deformation resulting from the welding process, and (c) the welded part with residual stress resulting from the welding process.

Two types of welded joints exist, depending on the orientation of the two mechanical elements: *butt welds* and *fillet welds*.

A *butt weld* joins two mechanical elements oriented in the same plane. Figure 9.4a–d shows different examples of butt welds and illustrate the different ways in which we can prepare the mechanical elements to accommodate the butt weld. Even though no preparation of the mechanical elements is generally acceptable (Figure 9.4a), we obtain the best results when preparing the mechanical elements by creating one or multiple V-shaped grooves to accommodate the melt pool and filler material (Figure 9.5b–d).

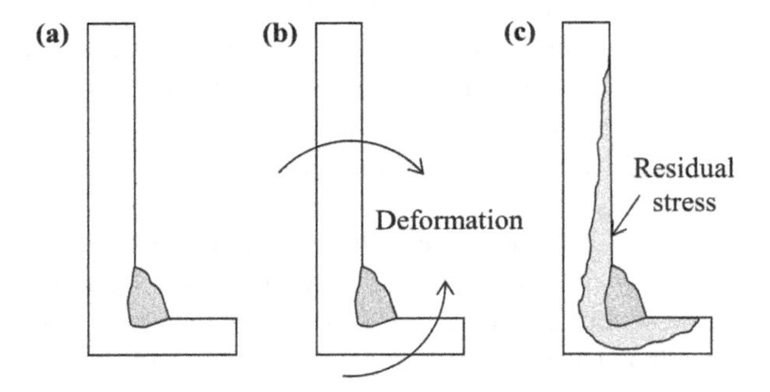

Figure 9.3 Instead of obtaining (a) the intended assembly of mechanical elements, welding causes either (b) deformation or (c) residual stress in the mechanical elements that are joined.

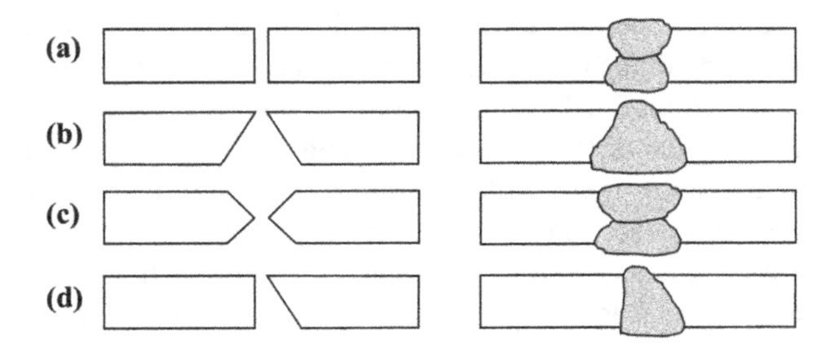

Figure 9.4 Examples of butt welds, illustrating different ways to prepare the mechanical elements on the left and the corresponding welded joints on the right.

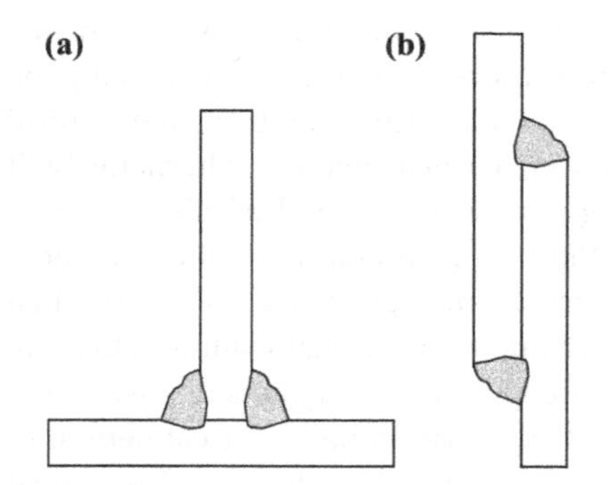

Figure 9.5 Examples of fillet welds, showing (a) a tee joint and (b) a lap joint.

A *fillet weld* joins two mechanical elements oriented in orthogonal (tee joint) or parallel planes (lap joint). Figure 9.5a,b illustrate two examples of fillet welds.

9.2 Welded Joint Geometry

Figure 9.6 illustrates the geometry of the cross-section of a welded joint. Figure 9.6a illustrates a butt weld, whereas Figures 9.6b–d depict different fillet welds. We note that the geometry of a welded joint comprises two parts; the *throat section* and the *reinforcement*. The throat section is the plane in which failure of the welded joint occurs. We use the thickness of the throat section h as the thickness of the welded joint in strength calculations and do not consider the reinforcement, despite its name. In the case of a butt weld, the thickness of the throat section h of the welded joint is equal to the thickness of the mechanical elements it joins together, whereas in the case of a fillet weld, we determine the thickness of the throat section h as illustrated in Figures 9.6b–d.

Figure 9.7a shows a fillet weld of two orthogonally oriented plates, whereas Figure 9.7b only displays the fillet weld, rotated with the right angle upward. The right angle coincides with the corner where the two plates join in Figure 9.7a. The presentation of Figure 9.7b visualizes the plane of the throat section (gray shaded) along the longitudinal axis of the welded joint and allows identifying the different stress components within the throat section of the welded joint.

There are two normal stress components that can cause failure in the throat section of the welded joint. The first one is oriented parallel to the longitudinal axis of the welded joint, i.e. $\sigma_{\parallel}$. This normal stress component causes failure of the welded joint in a plane perpendicular to the plane of the throat section, but the stress vector is oriented parallel to the longitudinal axis of the welded joint. Second, a normal stress can exist perpendicular to the longitudinal axis of the welded joint, i.e. $\sigma_{\perp}$, and normal to the plane of the throat section. Furthermore, there are two shear stress components that can cause failure in the throat section of the welded joint. The first one is oriented parallel to the longitudinal axis of the welded joint, i.e. $\tau_{\parallel}$. Additionally, a shear stress can exist perpendicular to the longitudinal axis of the throat section, i.e. $\tau_{\perp}$. Both shear stress components are located in the plane of the throat section and cause failure by shearing the plane of the throat section along the direction parallel or perpendicular to the longitudinal axis of the welded joint, respectively.

The different stress components combine into an equivalent normal stress σ_{eq} according to a failure criterion. For welded joints, the following failure criterion is commonly used, which is based

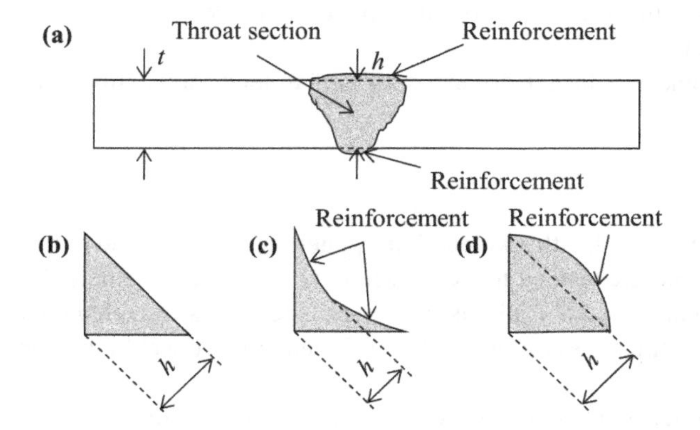

Figure 9.6 Thickness h of the throat section of a welded joint showing (a) a butt weld and (b)–(d) different fillet welds.

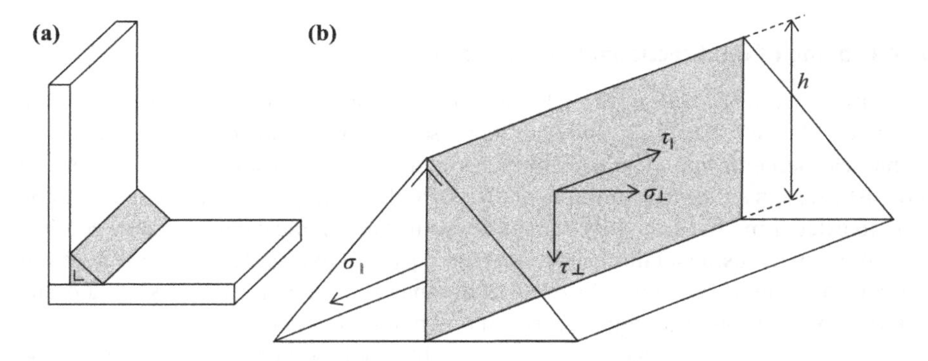

Figure 9.7 Four stress components in the throat section of a welded joint $\sigma_{\parallel}$, $\sigma_{\perp}$, $\tau_{\parallel}$, and $\tau_{\perp}$.

on the Von Mises criterion[1]

$$\sigma_{eq} = \sqrt{\sigma_{\perp}^2 + 3(\tau_{\perp}^2 + \tau_{\parallel}^2)}. \tag{9.1}$$

We note that $\sigma_{\parallel}$ is mostly not present in welded joints, or not accounted for in the strength calculation. We also emphasize that other failure criteria exist for welded joint calculations. However, in this textbook, we illustrate the basic concepts of strength calculations of welded joints by means of the failure criterion of Eq. (9.1).

9.3 Calculation of Welded Joints

When calculating the strength of welded joints, it is common to adopt the following assumptions:

1. The mechanical properties of the filler material are at least the same as those of the base material (measured after cool down).

1 See, e.g. European Standard EN 1993-1-8. *Eurocode 3: Design of Steel Structures – Part 1-8: Design of Joints.* European Committee for Standardization.

2. Residual stresses are negligible, either due to the order of finishing different welded joints or as a result of heat treatment after welding.
3. The loading is static. In this textbook, we do not discuss design of welded joints subject to dynamic loading.

9.3.1 Butt Welded Joints

When calculating a *full penetration* butt weld, i.e. the welded joint penetrates through the entire thickness of the mechanical elements that are welded together (see Figure 9.4), we calculate the strength of the welded joint assuming that the welded joint is as strong as the mechanical elements it joins. Thus, we select $h = t$, with t the thickness of the mechanical elements we weld together (see Figure 9.6a).

For instance, if the mechanical elements are loaded with axial force F and create a uniform normal stress σ in the welded joint, then we calculate the stress as the load divided by the area of the throat section of the welded joint, which is the product of the length L and the thickness of the throat section h of the welded joint. Thus, the normal stress $\sigma = F/hL$.

9.3.2 Simple Loading of Unidirectional Fillet Welded Joints

We discuss six cases of simple loading (three forces, three moments), to which a fillet weld can be subject as a result of external loading. To illustrate these six different loading cases, we use the same two mechanical elements that we join together using two fillet welds into a T-shaped element. However, in realistic design situations, multiple of these simple loading cases can exist simultaneously. Similarly to calculating the strength of shafts and other mechanical elements, we combine different stress components using a failure criterion (see Eq. (9.1)) to calculate an equivalent normal stress that we compare to, e.g. the yield stress of the material. In doing so, we solve a design problem if the thickness of the welded joint h is unknown and we attempt to achieve a specific design factor n_D or, conversely, we solve a verification problem when the thickness of the welded joint h is known *a priori*, and we verify the safety factor n of the design.

9.3.2.1 Case 1: Axial Load

Figure 9.8 shows two mechanical elements (plates) that we weld together using two fillet welds along the length of the mechanical element L. Figure 9.8a displays a front view showing both fillet welds on either side of the mechanical element, and Figure 9.8b shows a side view, indicating the axial load F. Figure 9.8c shows a top view of the welded joints.

First, we create a free-body diagram by separating the welded joints through their respective throat sections, and we determine the reaction force in each throat section. Figure 9.8c illustrates this process. We observe an $F/2$ reaction force in each fillet weld, resulting from the external axial load F.

We note that the force vector $F/2$ is oriented under 45° with respect to the throat section of the welded joint. However, we know that failure occurs in the plane of the throat section, and therefore, we convert this force vector in components that are oriented within or perpendicular to the plane of the throat section. Then, we convert those forces to one or multiple of the stress components that can occur in the throat section of the welded joint, as illustrated in Figure 9.7. Thus, we project the force vector $F/2$ into the plane and perpendicular to the plane of the throat section, and find force components:

$$N = T = \frac{F}{2}\frac{\sqrt{2}}{2}, \tag{9.2}$$

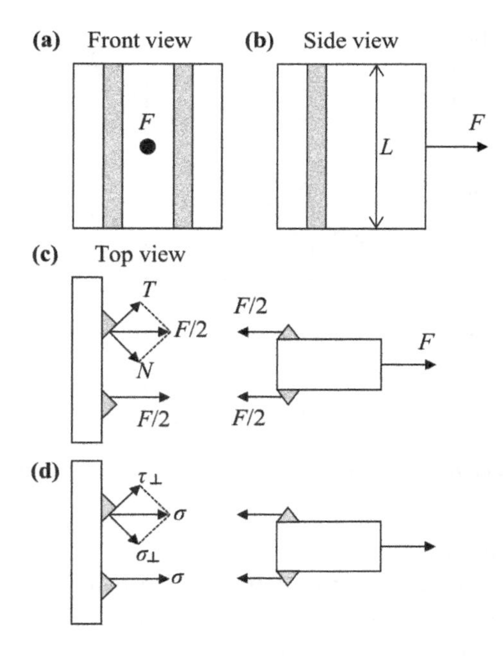

Figure 9.8 Unidirectional fillet welds subject to an axial load F.

as illustrated in Figure 9.8c. Note that the vector sum of N and T yields $F/2$, and that $\cos(45°) = \sin(45°) = \sqrt{2}/2$.

We then determine the stress components that correspond to the respective force vectors by dividing the force by the area over which it is distributed. This is the area of the throat section of the fillet weld, i.e. the product of the thickness of the throat section h and the length of the welded joint L. Thus,

$$\sigma_{\perp} = \tau_{\perp} = \frac{F}{2}\frac{\sqrt{2}}{2}\frac{1}{hL} = \frac{\sqrt{2}F}{4hL}. \tag{9.3}$$

Figure 9.8e shows the two stress components in the throat section of the welded joint, as calculated in Eq. (9.3). Thus, an axial load results in $\sigma_{\perp}$ and $\tau_{\perp}$ stress components in the throat section of the welded joint.

Example problem 9.1

Figure 9.9 shows a schematic representation of a lap joint comprising two steel plates welded together using two fillet welded joints. Both plates are t = 10 mm thick, w = 200 mm wide (orthogonal to the page), and made of steel with σ_y = 240 MPa. Calculate the thickness of the welded joints, and consider a design factor n_D = 1.4.

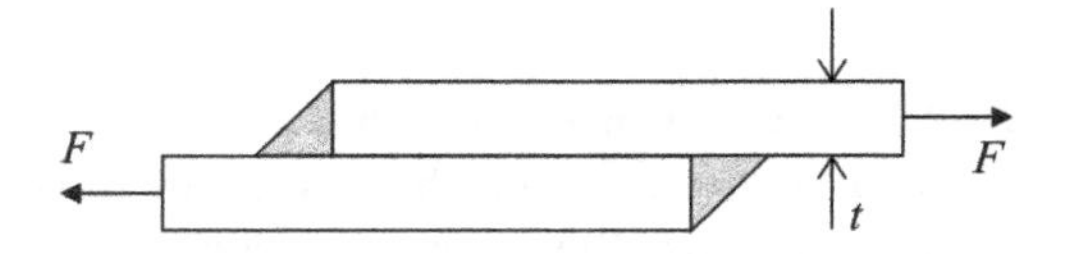

Figure 9.9 Lap joint with two fillet welded joints, subject to an axial load.

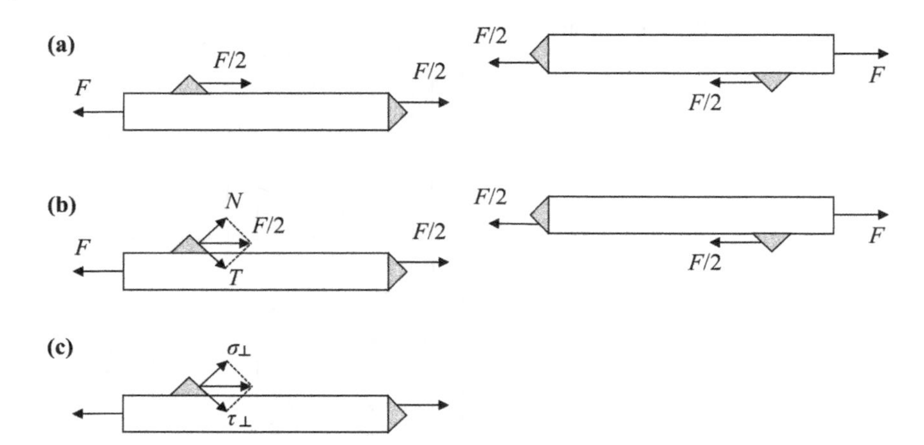

Figure 9.10 Lap joint with two fillet welded joints, showing the force and stress components in the throat section.

Solution

We note that the magnitude of the load F is not specified in the problem statement. In that case, it is appropriate to design the welded joints such that they are as strong as the mechanical elements they join. Considering the axial loading of the plates, the load F_{max} at the inception of yielding is:

$$F_{max} = \sigma_y\ t\ w = 240 \cdot 10^6 \times 10 \cdot 10^{-3} \times 200 \cdot 10^{-3} = 480 \cdot 10^3 \text{ N}.$$

Figure 9.10a shows the free-body diagram of the welded joints, after separating them in the throat section. It also identifies the reaction forces in the throat section of the welded joints. Figure 9.10b shows the free-body diagram after we project the reaction force in the plane and perpendicular to the plane of the throat section of the welded joint. Finally, Figure 9.10c shows the stress components in the throat section of the welded joints.

We calculate $N = T = F/2 \times \sqrt{2}/2$.

We convert the forces in the plane and perpendicular to the plane of the throat section to stresses, by dividing the forces by the area over which they are distributed, which is the area of the throat section of the welded joint. Thus,

$$\sigma_\perp = \tau_\perp = N/hw = T/hw, \text{ or}$$

$$\sigma_\perp = \tau_\perp = \sqrt{2}F/(4hw) = \sqrt{2} \times 480 \cdot 10^3/(4 \times h \times 200 \cdot 10^{-3}) = 848\ 528/h \text{ Pa}$$

The throat section of the welded joint is subject to combined loading because it experiences both a normal and a shear stress. Thus, we determine an equivalent normal stress σ_{eq} using Eq. (9.1). We calculate

$$\sigma_{eq} = \sqrt{\sigma_\perp^2 + 3\tau_\perp^2} = \sqrt{(848\ 528/h)^2 + 3 \times (848\ 528/h)^2} = 1\ 697\ 056/h \text{ Pa}.$$

We define the design factor $n_D = \sigma_y/\sigma_{eq}$

Finally, $1.4 = 240 \cdot 10^6 \times h/1\ 697\ 056$, or $h = 10$ mm.

9.3.2.2 Case 2: Longitudinal Load

Figure 9.11 shows two mechanical elements (plates) that we weld together using two fillet welds along the length of the mechanical element L. Figure 9.11a displays a front view showing both fillet

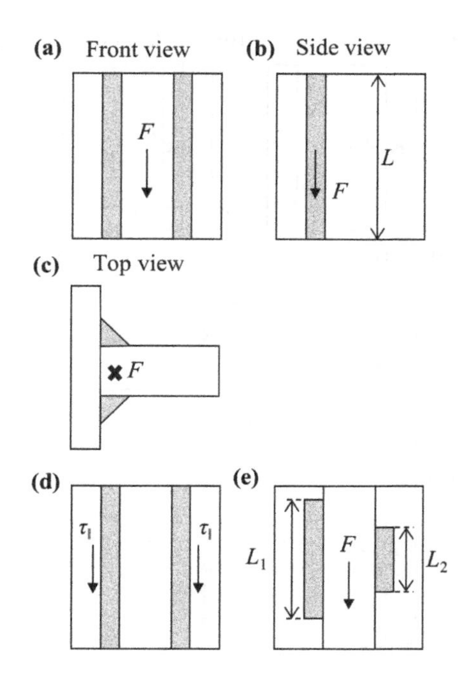

Figure 9.11 Unidirectional fillet welds subject to a longitudinal load *F*.

welds on either side of the mechanical element, and Figure 9.11b shows a side view, indicating the longitudinal load F. Figure 9.11c shows a top view of the welded joints.

First, we create a free-body diagram by separating the welded joints through their respective throat sections, and we determine the reaction force in each throat section. The symmetry of the fillet welds with respect to the external load F creates an $F/2$ reaction force in each fillet weld.

We note that the force vector $F/2$ is oriented in the plane of the throat section and parallel to the longitudinal axis of the welded joint. We calculate the stress components corresponding to the respective force vector by dividing the force by the area over which it is distributed, which is the area of the throat section of the fillet weld, i.e. the product of the thickness of the throat section h and the length of the welded joint L. Thus,

$$\tau_\parallel = \frac{F}{2}\frac{1}{hL} = \frac{F}{2hL}. \tag{9.4}$$

Figure 9.11d shows the stress components in the throat section of the welded joint, as calculated in Eq. (9.4). Thus, a longitudinal load results in a $\tau_\parallel$ stress component in the throat section of the welded joint.

We also note that if the fillet welds have a different length as shown in Figure 9.11e, we calculate that $L = L_1 + L_2$ in Eq. (9.4) and thus,

$$\tau_\parallel = \frac{F}{2h(L_1 + L_2)}. \tag{9.5}$$

9.3.2.3 Case 3: Transverse Load

Figure 9.12 shows two mechanical elements (plates) that we weld together using two fillet welds along the length of the mechanical element L. Figure 9.12a displays a front view showing both fillet welds on either side of the mechanical element and indicating the transverse load F. Figure 9.12b shows a side view and Figure 9.8c shows a top view of the welded joints.

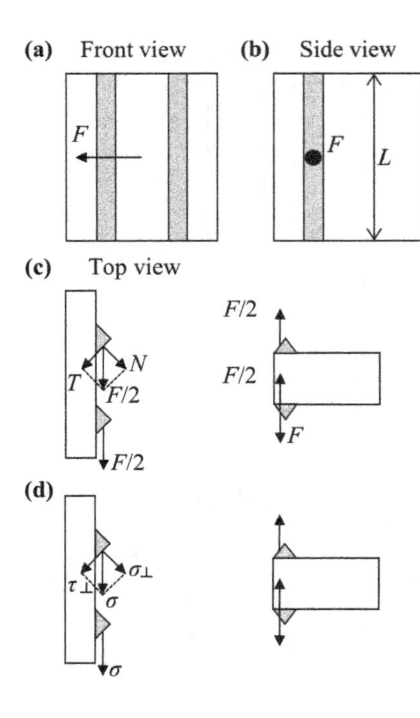

Figure 9.12 Unidirectional fillet welds subject to a transverse load F.

First, we create a free-body diagram by separating the welded joints through their respective throat sections, and we determine the reaction force in each throat section. Figure 9.12c illustrates this process. We observe an $F/2$ reaction force in each fillet weld, resulting from the transverse external load F.

We note that the force vector $F/2$ is oriented under 45° with respect to the throat section of the welded joint. However, we know that failure occurs in the plane of the throat section and, therefore, we convert this force vector in components that are oriented within or perpendicular to the plane of the throat section. Then, we convert those forces to one or multiple of the stress components that can occur in the throat section of the welded joint, as illustrated in Figure 9.7. Thus, we project the force vector $F/2$ into the plane and perpendicular to the plane of the throat section, and determine force components:

$$N = T = \frac{F}{2}\frac{\sqrt{2}}{2}, \tag{9.6}$$

as illustrated in Figure 9.12c. Note that the vector sum of N and T yields $F/2$, and that $\cos(45°) = \sin(45°) = \sqrt{2}/2$.

We determine the stress components corresponding to the respective force vectors by dividing the force by the area over which it is distributed, which is the area of the throat section of the fillet weld, i.e. the product of the thickness of the throat section h and the length of the welded joint L. Thus,

$$\sigma_{\perp} = \tau_{\perp} = \frac{F}{2}\frac{\sqrt{2}}{2}\frac{1}{hL} = \frac{\sqrt{2}F}{4hL}. \tag{9.7}$$

Figure 9.12d shows the two stress components in the throat section of the welded joint, as calculated in Eq. (9.7). Thus, an axial load results in a $\sigma_{\perp}$ and $\tau_{\perp}$ stress component in the throat section of the welded joint.

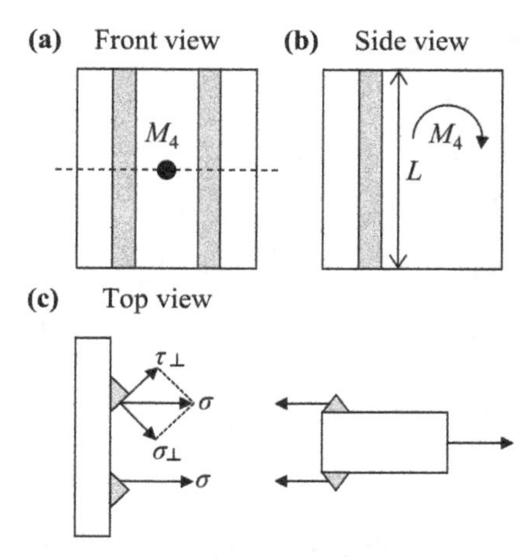

Figure 9.13 Unidirectional fillet welds subject to an in-plane bending moment M_4.

9.3.2.4 Case 4: In-plane Bending Moment

Figure 9.13 shows two mechanical elements (plates) that we weld together using two fillet welds along the length of the mechanical element L. Figure 9.13a displays a front view showing both fillet welds on either side of the mechanical element, and Figure 9.13b shows a side view, indicating the in-plane bending moment M_4. The horizontal dashed line indicates the neutral axis, which runs through the centroid of the throat section of the welded joints. Figure 9.13c shows a top view of the welded joints.

First, we create a free-body diagram by separating the welded joints through their respective throat sections. We recognize that in the case of a bending moment, the stress in the welded joint is a function of the distance from the neutral axis. Figure 9.13c illustrates the stress components in the throat section of the welded joint at a distance y from the neutral axis. We indicate the normal stress resulting from bending as σ.

$$\sigma = \frac{M_4 y}{I}, \tag{9.8}$$

with $I = hL^3/12$, the area moment of inertia of the throat section of the welded joint, about the neutral axis. Since there are two welded joints, the total area moment of inertia of the throat section of all welded joints is $I = 2hL^3/12$. The maximum normal stress from bending occurs at the top and bottom of the welded joint, i.e. the maximum distance from the neutral axis $y = L/2$. Thus,

$$\sigma = \frac{12M_4(L/2)}{2hL^3} = \frac{3M_4}{hL^2}. \tag{9.9}$$

We note that σ is oriented under 45° with respect to the throat section of the welded joint. However, we know that failure occurs in the plane of the throat section and, therefore, we convert this stress vector in components that are oriented within or perpendicular to the plane of the throat section, corresponding to one or multiple of the stress components that can occur in the throat section of the welded joint, as illustrated in Figure 9.7. Thus, we project the vector σ into the plane and perpendicular to the plane of the throat section, and determine

$$\sigma_\perp = \tau_\perp = \frac{3\sqrt{2}M_4}{2hL^2}, \tag{9.10}$$

as illustrated in Figure 9.13c. Note that the vector sum of $\sigma_\perp$ and $\tau_\perp$ yields σ and that $\cos(45°) = \sin(45°) = \sqrt{2}/2$. Thus, an in-plane bending moment results in a $\sigma_\perp$ and $\tau_\perp$ stress component in the throat section of the welded joint.

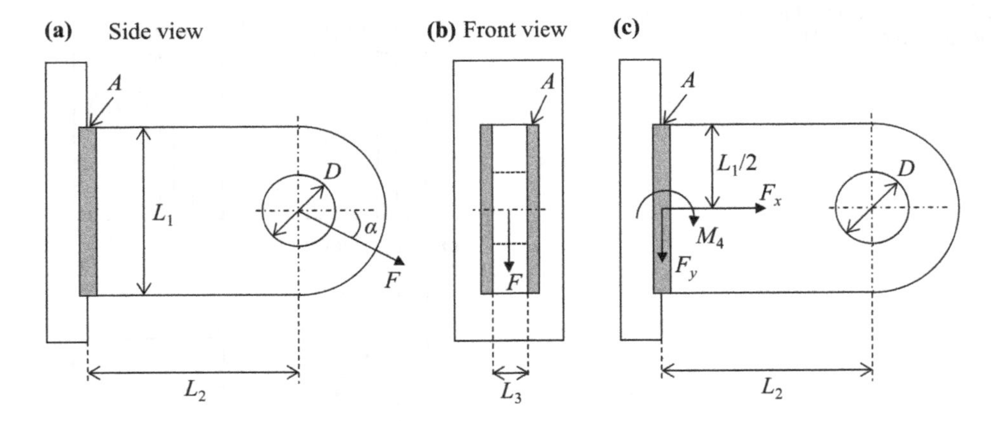

Figure 9.14 Bracket attached to a rigid structure with two fillet welded joints, showing the stress components in the throat section.

Example problem 9.2

Figure 9.14 shows a schematic representation of a bracket welded to a rigid frame using two fillet welded joints. Figure 9.14a shows a side view, whereas Figure 9.14b shows a front view. All parts are made from AISI 1080 HR steel, with $\sigma_{ut} = 770$ MPa and $\sigma_y = 420$ MPa. $F = 50$ kN with $\alpha = 30°$. $L_1 = 200$ mm, $L_2 = 300$ mm, $L_3 = 20$ mm. The thickness of the welded joint $h = 6$ mm. Calculate the safety factor against yielding in point A.

Solution

We first replace the external load with the equivalent combination of simple loading cases (Figure 9.14c). Here, F breaks down in a horizontal component F_x and a vertical component F_y. We move the vertical component such that both F_x and F_y run through the centroid of the throat section of the welded joints. However, when moving F_y, we must compensate with an in-plane bending moment M_4. Thus,

$$F_x = F\cos\alpha = 50\ 000\cos 30° = 43\ 301\ \text{N (Case 1)},$$

$$F_y = F\sin\alpha = 50\ 000\sin 30° = 25\ 000\ \text{N (Case 2)},$$

$$M_4 = F_y \times (L_2 - \sqrt{2}h/4) = 25\ 000 \times (0.3 - \sqrt{2} \times 0.006/4) = 7447\ \text{Nm (Case 4)}.$$

We emphasize that the rotation axis of the bending moment runs through the middle of the throat section h. Thus, the leverage arm of the force F_y is $L_2 - \sqrt{2}h/4$, which we can calculate in a verification calculation when h is known. However, when performing a design calculation, h is not known *a priori* and, thus, we approximate (and slightly overestimate) the leverage arm as L_2.

Figure 9.15 shows the free-body diagram of the welded joints, after we separate them in their respective throat sections, and we show the stress components in the throat section of the welded joints, representative of both Case 1 and Case 4 in point A. Case 2 results in a shear stress component $\tau_{\|}$, parallel to the longitudinal axis of the welded joint.

Case 1: We calculate $\sigma_\perp = \tau_\perp = \sqrt{2}F_x/(4hL_1) = \sqrt{2} \times 43\ 301/(4 \times 0.006 \times 0.2) = 12.76$ MPa.

Case 2: We calculate $\tau_{\|} = F_y/(2hL_1) = 25\ 000/(2 \times 0.006 \times 0.2) = 10.42$ MPa.

Case 4: We calculate $\sigma = \frac{M_4 L_1/2}{2hL_1^3/12} = \frac{3M_4}{hL_1^2}$.

Hence, $\sigma_\perp = \tau_\perp = \frac{3\sqrt{2}M_4}{2hL_1^2} = \frac{3\sqrt{2}}{2}\frac{7447}{0.006 \times 0.2^2} = 65.82$ MPa.

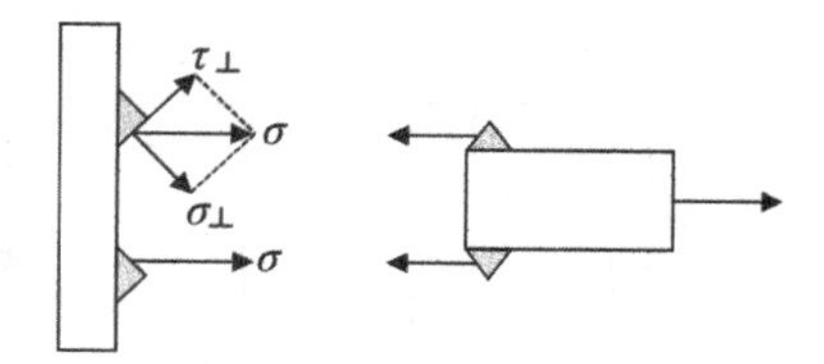

Figure 9.15 Bracket attached to a rigid structure with two fillet welds.

The throat section of the welded joint is subject to combined loading because it experiences both a normal and a shear stress. Thus, we determine the equivalent normal stress σ_{eq} using Eq. (9.1), i.e.

$$\sigma_{eq} = \sqrt{\sigma_\perp^2 + 3(\tau_\perp^2 + \tau_\parallel^2)} = \sqrt{(65.82 + 12.76)^2 + 3[(65.82 + 12.76)^2 + 10.42^2)]} = 158.2 \text{ MPa}.$$

We define the safety factor $n = \sigma_y/\sigma_{eq}$ and thus $n = 420/158.2 = 2.65$.

9.3.2.5 Case 5: Out-of-plane Bending Moment

Figure 9.16 shows two mechanical elements (plates) that we weld together using two fillet welds along the length of the mechanical element L. Figure 9.16a displays a front view showing both fillet welds on either side of the mechanical element, and Figure 9.16b shows a side view, indicating the force F that creates the out-of-plane bending moment M_5. Figure 9.16c shows a top view of the welded joints, again indicating the out-of-plane bending moment M_5.

First, we create a free-body diagram by separating the welded joints through their respective throat sections, and we determine the reaction force in each throat section. Figure 9.16c illustrates this process. We observe that the out-of-plane bending moment creates a *force couple* in the welded joints, with a reaction force F in each fillet weld, oriented in the same direction but opposite sense, derived from the sense of the out-of-plane bending moment M_5. We quantify the force couple as the product of the force vectors and the distance between them. Here, we determine the distance between both force vectors located at the mid-point of the throat section of the welded joints. Thus, the distance between the force vectors is the thickness of the mechanical element d, with half of the thickness of the throat section projected onto the vertical, i.e. $\sqrt{2}h/4$, on either side of the mechanical element. The distance between both force vectors F is $d + \sqrt{2}h/2$. Therefore,

$$F = \frac{M_5}{d + \sqrt{2}h/2}. \tag{9.11}$$

We note that the force vector F is oriented under 45° with respect to the throat section of the welded joint. However, we know that failure occurs in the plane of the throat section and, therefore, we convert this force vector in components that are oriented within or perpendicular to the plane of the throat section. Then, we convert those forces to one or multiple of the stress components that may occur in the throat section of the welded joint, as illustrated in Figure 9.7. Thus, we project the force vector F into the plane and perpendicular to the plane of the throat section, and determine force components:

$$N = T = \frac{\sqrt{2}F}{2}, \tag{9.12}$$

as illustrated in Figure 9.16c. Note that the vector sum of N and T yields F, and that $\cos(45°) = \sin(45°) = \sqrt{2}/2$.

We calculate the stress components corresponding to the respective force vectors by dividing the force by the area over which it is distributed, which is the area of the throat section of the fillet

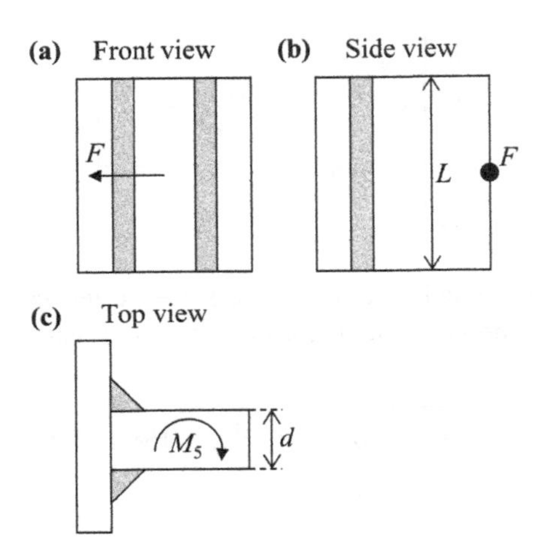

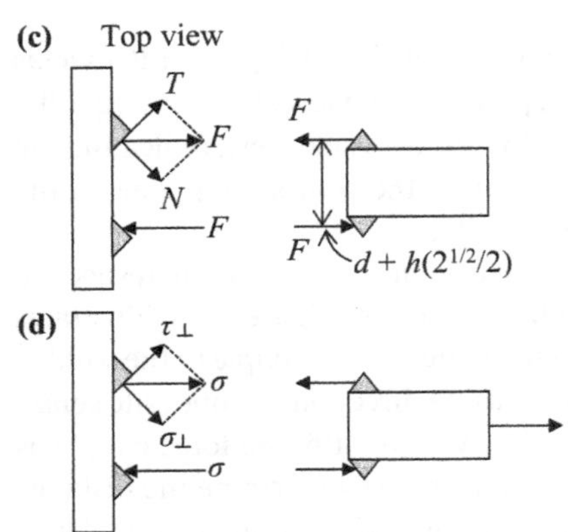

Figure 9.16 Unidirectional fillet welds subject to out-of-plane bending moment M_5.

weld, i.e. the product of the thickness of the throat section h and the length of the welded joint L. Thus,

$$\sigma_\perp = \tau_\perp = \frac{\sqrt{2}F}{2}\frac{1}{hL} = \frac{\sqrt{2}M_5}{2hL(d + \sqrt{2}h/2)}. \tag{9.13}$$

Figure 9.16d shows the two stress components in the throat section of the welded joint, as calculated in Eq. (9.13). Thus, an out-of-plane bending moment results in a $\sigma_\perp$ and $\tau_\perp$ stress component in the throat section of the welded joint.

9.3.2.6 Case 6: Torque Moment

Figure 9.17 shows two mechanical elements (plates) that we weld together using two fillet welds along the length of the mechanical element L. Figure 9.17a displays a front view showing both fillet welds on either side of the mechanical element and indicates the torque moment M_6. Figure 9.17b shows a side view.

First, we create a free-body diagram by separating the welded joints through their respective throat sections, and we determine the reaction force in each throat section.

We observe that the torque moment M_6 creates a *force couple* in the welded joints, with a reaction force F in each fillet weld, oriented in the same direction but opposite sense, derived from the

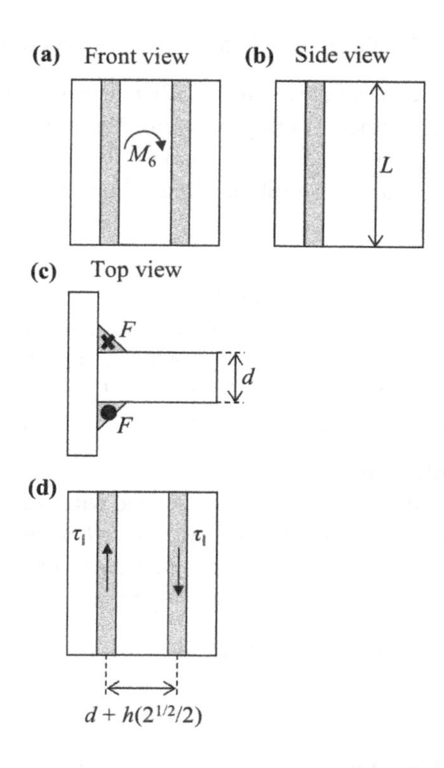

Figure 9.17 Unidirectional fillet welds subject to a torque moment M_6.

sense of the torque moment M_6, and illustrated in Figure 9.17c. We characterize the force couple as the product of the force vectors and the distance between them. Here, we determine the distance between both force vectors located at the mid-point of the throat section of the welded joints. Thus, the distance between the force vectors is the thickness of the mechanical element d, with half of the thickness of the throat section projected on the vertical, i.e. $\sqrt{2}h/4$, on either side of the mechanical element. The distance between both force vectors F is $d + \sqrt{2}h/2$. Therefore,

$$F = \frac{M_6}{d + \sqrt{2}h/2}. \tag{9.14}$$

We note that the force vector F is oriented in the plane of the throat section and parallel to the longitudinal axis of the welded joint. We then determine the stress components corresponding to the respective force vectors by dividing the force by the area over which it is distributed. This is the area of the throat section of the fillet weld, i.e. the product of the thickness of the throat section h and the length of the welded joint L. Thus,

$$\tau_\| = \frac{F}{hL} = \frac{M_6}{hL(d + \sqrt{2}h/2)}. \tag{9.15}$$

Figure 9.17d shows the stress components in the throat section of the welded joint, as calculated in Eq. (9.15). Thus, a torque moment results in a $\tau_\|$ stress component in the throat section of the welded joint.

9.3.3 Combined Loading of Unidirectional Fillet Welded Joints

We can combine any number of cases of simple loading with unidirectional fillet welds into a combined loading problem. We break down the combined loading into individual cases of simple loading, and calculate the stress in the throat section of the welded joints as a result of each case of simple loading. Then, we combine all stress components into the failure criterion of Eq. (9.1).

Example problem 9.3

Figure 9.18 shows a schematic of two parts welded together with two fillet welded joints. Figure 9.18a shows a front view, whereas Figure 9.18b shows a side view. All parts are made from steel with $\sigma_y = 140$ MPa. $F = 3000$ N. $L_1 = 220$ mm, $L_2 = 90$ mm, $L_3 = 20$ mm. Calculate the thickness of the welded joint h, and consider a design factor $n_D = 3$.

Solution

We first replace the external load with the equivalent combination of simple loading cases. Here, F breaks down as follows. We move F such that it runs through the centroid of the throat section of the welded joints, where it acts as a transverse load. However, when moving F we must compensate with an out-of-plane bending moment M_5 and a torque moment M_6. Thus,

$$F = 3000 \text{ N (Case 3).}$$

$$M_5 = F \times (L_2 - \sqrt{2}h/4) = 3000 \times (0.09 - \sqrt{2}h/4) \cong 270 \text{ Nm (Case 5).}$$

We approximate the out-of-plane bending moment because the thickness of the welded joint h is unknown. However, we note that the approximation overestimates the moment and, thus, it is a conservative approximation.

$$M_6 = F \times (L_2/2) = 3000 \times 0.11 = 330 \text{ Nm (Case 6).}$$

Figure 9.19 shows the free-body diagram of the welded joints, after separating them in the throat sections. Figure 9.19a illustrates Case 3, Figure 9.19b illustrates Case 5, and Figure 9.19 illustrates Case 6.

Case 3: We calculate $\sigma_\perp = \tau_\perp = \sqrt{2}F/(4hL_1) = \sqrt{2} \times 3000/(4 \times 0.22 \times h) = 4821/h$.

Case 5: We calculate $M_5 = P\,(L_3 + 2 \times \sqrt{2}h/4) \Rightarrow P = M_5/(L_3 + \sqrt{2}h/2)$

$$\sigma_\perp = \tau_\perp = \frac{\sqrt{2}P}{2hL_1} = \frac{M_5}{(L_3 + \sqrt{2}h/2)} \frac{\sqrt{2}}{2hL_1}.$$

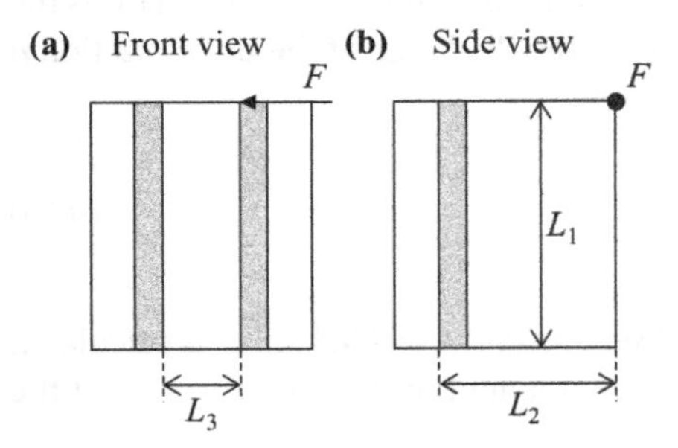

Figure 9.18 Two steel parts welded together with two fillet welds.

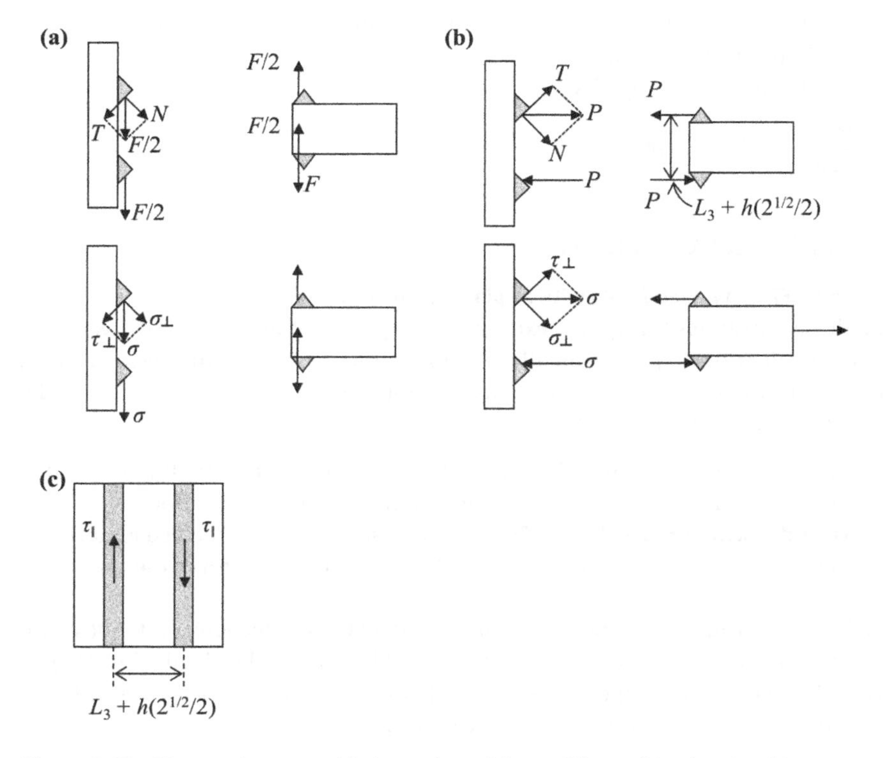

Figure 9.19 Two steel parts welded together with two fillet welds, showing (a) transverse load (Case 3), (b) out-of-plane bending moment (Case 5), and (c) torque moment (Case 6).

Because the higher order terms of h are small compared to the first order term, we neglect them. Thus,

$$\sigma_\perp = \tau_\perp \cong \frac{M_5}{(L_3 \times h \times L_1)}\frac{\sqrt{2}}{2} = \frac{270}{(0.02 \times h \times 0.22)}\frac{\sqrt{2}}{2} = 43\ 391/h.$$

Case 6: We calculate the longitudinal force in the welded joint $V = \frac{M_6}{L_3+\sqrt{2}h/2}$. Note that we use V to avoid confusing with F, which is the external load.

$$\tau_\parallel = V/hL_1 = \frac{M_6}{L_3 + \sqrt{2}h/2}\frac{1}{hL_1}.$$

Because the higher order terms of h are small compared to the first order term, we neglect them. Thus,

$$\tau_\parallel = \frac{M_6}{L_3 \times h \times L_1} = \frac{330}{0.02 \times h \times 0.22} = 75\ 000/h.$$

The throat section of the welded joint is subject to combined loading, as it experiences both a normal and a shear stress. We calculate an equivalent normal stress σ_{eq} using the failure criterion of Eq. (9.1). Thus,

$$\sigma_{eq} = \sqrt{\sigma_\perp^2 + 3(\tau_\perp^2 + \tau_\parallel^2)},$$

$$\sigma_{eq} = \sqrt{(4821/h + 43\ 391/h)^2 + 3((4821/h - 43\ 391/h)^2 + 75\ 000^2)} = 153\ 826/h.$$

Note that the $\tau_\perp$ vectors that result from Case 3 and Case 5 have opposite sense see Figure 9.19a, b. Hence, we subtract them in the calculation of σ_{eq}.

We define the design factor $n_D = \sigma_y/\sigma_{eq}$. Hence,

$$n_D = \frac{140 \cdot 10^6 \times h}{158\ 826} \Rightarrow h = 3.4 \text{ mm}.$$

9.3.4 Multidirectional Fillet Welded Joints

9.3.4.1 Multidirectional Fillet Welded Joints with In-plane Load, No Bending

Figure 9.20 schematically illustrates this type of external loading. It is a combination of Case 2 and Case 3, but the load is not distributed proportionally to the strength of the individual welded joints, due to limited elastic deformation of the material. Instead, the load distribution depends on the length of the individual welded joints.

- **No weld at the free edge and $L_2 > 1.5L_1$:** The entire load F is borne by the longitudinal welded joints (L_2). The transverse welded joint is not considered in the strength calculation.
- **No weld at the free edge and $0.5L_1 < L_2 < 1.5L_1$:** The total strength of the welded joints is the strength of the longitudinal welded joints (L_2) increased by 1/3 of the strength of the transverse welded joint (L_1).
- **No weld at the free edge and $L_2 < 0.5L_1$:** The total strength of the welded joints is the strength of the transverse weld (L_1) increased by 1/3 of the strength of longitudinal welded joints (L_2).
- **Weld at the free edge**: The total strength of the welded joints is the strength of the welded joint at the free edge increased by 1/3 of the strength of the other welded joints.

9.3.4.2 Multidirectional Fillet Welded Joints with In-plane Load and Bending

Figure 9.21 schematically illustrates an example of this type of external loading, showing an I-beam welded to a rigid frame, with welded joints in two orthogonal directions. Figure 9.21a displays a front view with an in-plane load F, whereas Figure 9.21b shows a side view with the in-plane load F and an in-plane bending moment M_4. This is a combination of Cases 2, 3, and 4, but we make additional assumptions to perform the strength calculation.

First, we calculate the stress resulting from bending in the welded joints following the method of Case 4, taking into account all welded joints, both longitudinal (in the direction of the in-plane load) and transverse (perpendicular to the direction of the in-plane load). Second, for the in-plane load, we only consider the strength of the longitudinal welded joints (Case 2), and we neglect the strength of the transverse welded joints (Case 3). In the example of Figure 9.21, this means that the

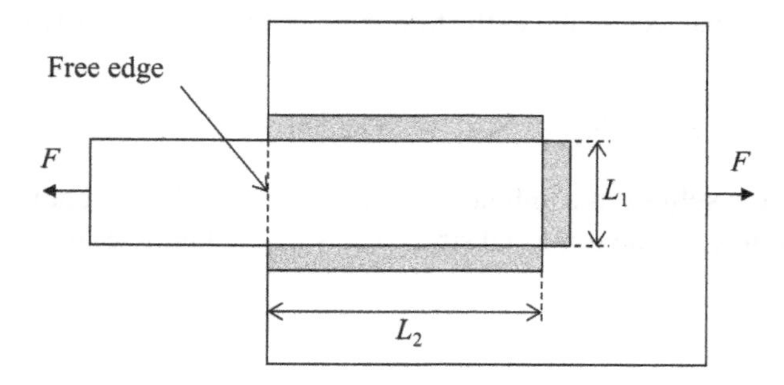

Figure 9.20 Multidirectional fillet welded joints with in-plane loading, no bending.

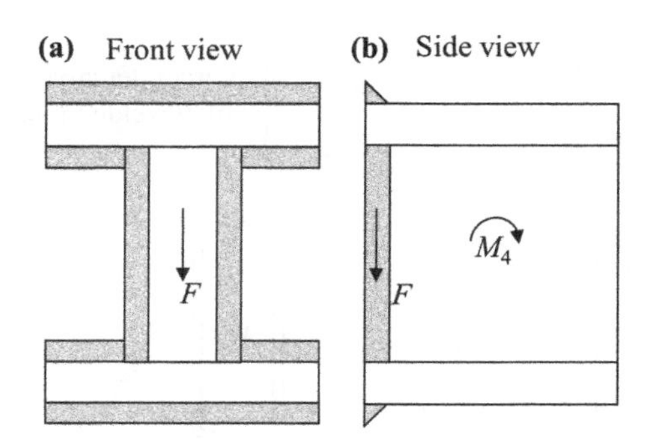

Figure 9.21 Multidirectional fillet welded joints subject to in-plane loading and bending.

horizontal welded joints are not considered in the calculation of stress resulting from the in-plane load (they are considered in the stress resulting from bending!). The in-plane load only creates a longitudinal shear stress in the vertical welded joints (Case 2).

9.3.4.3 Multidirectional Fillet Welded Joints with Torque Moment

Figure 9.22a schematically illustrates an example of this type of external loading, showing a circular tube welded to a rigid structure with a circumferential fillet welded joint, and subject to a torque moment M_6. To calculate the strength of the welded joint, we first recognize that this problem is similar to Case 2, i.e. the torque moment M_6 creates a force in the welded joint that runs along the circumference of the welded joint. This force converts into a $\tau_{\|}$ stress component that follows the direction and sense of the force in the welded joint, i.e. it is tangential to the longitudinal axis of the welded joint (Figure 9.22b). An alternative way to think about the shear stress in multidirectional welded joints subject to a torque moment is to consider the *hydrodynamic analogy*. Indeed, consider the welded joints to be a channel filled with a liquid. Then, consider the torque moment to be a force that drives the liquid through the channel; the direction in which the liquid flows is identical to the direction and sense of the local shear tress $\tau_{\|}$. We calculate the shear stress as

$$\tau_{\|} = \frac{M_6}{2hA}, \tag{9.16}$$

where A is the area within the contour through the middle point of the throat section of the welded joint shown in Figure 9.22b by the dash-dot line. Hence, $A = \pi(d_o + h)^2/4$ for the circular tube.

Figure 9.23a schematically illustrates another example of this type of external loading, showing a rectangular tube welded to a rigid frame with a circumferential fillet welded joint, and subject to

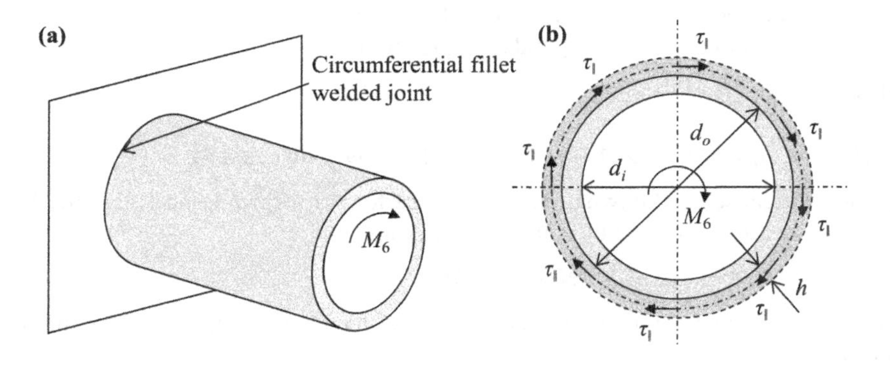

Figure 9.22 Multi-directional fillet welded joints around a circular tube subject to a torque moment M_6.

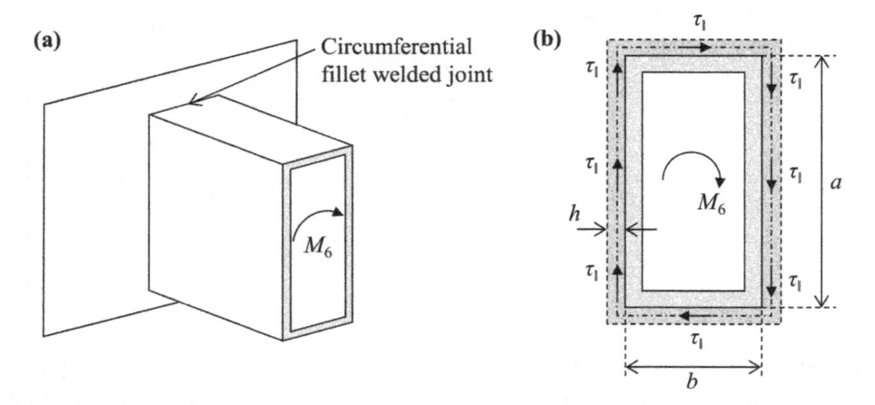

Figure 9.23 Multidirectional fillet welded joints around a rectangular tube subject to a torque moment M_6.

a torque moment M_6. We calculate $\tau_{\|}$ using Eq. (9.16), where A again is the area within the contour through the middle point of the throat section of the welded joint shown in Figure 9.23b by the dash-dot line. Thus, $A = (a + h) \times (b + h)$ for the rectangular tube.

Physically, one can think of Eq. (9.16) as the torque moment M_6 divided by an "average leverage arm" equal to A divided by the perimeter of the mean contour through the throat section of the welded joint, in units of (m). This yields the tangential force in the welded joint resulting from the torque moment. Then, this tangential force divided by the area of the throat section of the welded joint, i.e. the product of the perimeter of the mean contour through the throat section of the welded joint and the thickness of the throat section h, yields the shear stress $\tau_{\|}$.

Example problem 9.4

Figure 9.24 shows a schematic representation of two steel parts welded together using fillet welded joints Figure 9.24a shows a front view, whereas Figure 9.24b shows a side view. All parts are made from steel with $\sigma_y = 600$ MPa. $F = 5000$ N. $L_1 = 100$ mm, $L_2 = 10$ mm, $L_3 = 1$ m, and $L_2 = 45$ mm. The thickness of the welded joints $h = 5$ mm. Calculate the safety factor against yielding in points A and B.

Solution

We first replace the external loading with the equivalent combination of simple loading cases. Here, F breaks down as follows. We move F such that it runs through the centroid of the throat section of the welded joints, where it acts as a longitudinal load. However, when moving F, we must compensate with an in-plane bending moment M_4 and a torque moment M_6. Thus,

$$F = 5000 \text{ N (Case 2).}$$

$$M_4 = F \times (L_3 - \sqrt{2}h/4) = 5000 \times (1 - \sqrt{2} \times 0.005/4) = 4991 \text{ Nm (Case 4).}$$

Here, we account for h because it is known *a priori*, as we perform a verification calculation. We do not approximate the bending moment by ignoring an unknown h.

$$M_6 = F \times (L_1/2) = 5000 \times 0.05 = 250 \text{ Nm (Case 6).}$$

We first determine the centroid of the throat section of the welded joints. The area of the throat section of all welded joints is:

$$A = (5 \times 100) + 2 \times (5 \times 100) + 2 \times (5 \times 40) + (5 \times 10) = 1950 \text{ mm}^2,$$

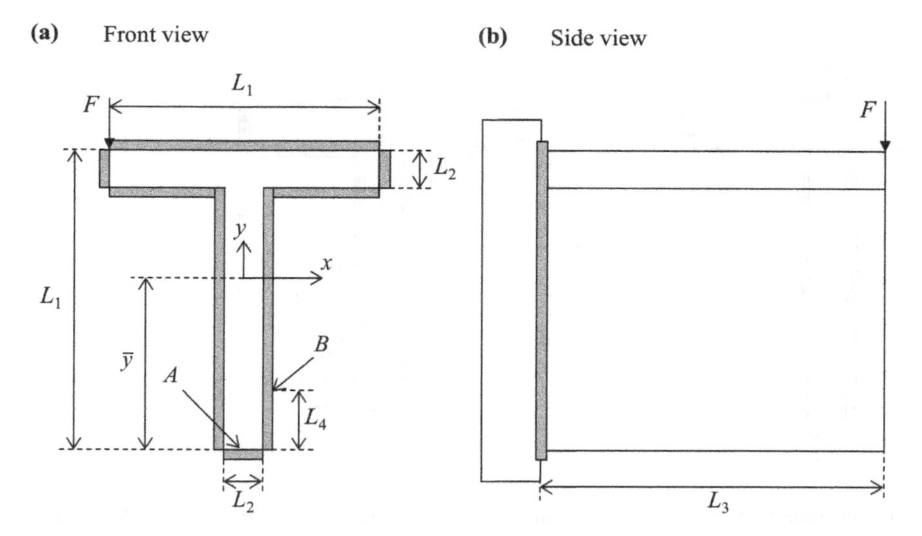

Figure 9.24 Two steel parts welded together with combined longitudinal and transverse fillet welded joints.

$$\bar{y}A = \Sigma y_i A_i,$$

$$\bar{y}A = (5 \times 100) \times 102.5 + 2 \times (5 \times 100) \times 87.5 + 2 \times (5 \times 40) \times 50 - (5 \times 10) \times 2.5,$$

$\bar{y} = 69.8$ mm. We measure $\bar{y}$ from the bottom of the T-section.

We calculate the area moment of inertia around the horizontal axis x through the centroid because this is the neutral axis (bending). The area moment of inertia represents the resistance of the throat section of all welded joints against bending. Hence,

$$I_{xx} = 2 \times \left(\frac{5 \times 100^3}{12} + 100 \times 5 \times 19.8^2 \right) + \left(\frac{100 \times 5^3}{12} + 100 \times 5 \times 32.7^2 \right)$$
$$+2 \times \left(\frac{40 \times 5^3}{12} + 40 \times 5 \times 17.7^2 \right) + \left(\frac{10 \times 5^3}{12} + 10 \times 5 \times 72.3^2 \right) = 2\ 148\ 678 \text{ mm}^4.$$

We then calculate the stress in the throat section at points A and B, for each case of simple loading.
Case 2: We calculate $\tau_{\parallel} = \frac{F}{2hL_1} = \frac{5000}{2 \times 0.005 \times 0.1} = 5$ MPa.
Note that in the case of multidirectional welded joints with in-plane loading and bending, the longitudinal load is only borne by the longitudinal (vertical) welded joints. Figure 9.25a shows the free-body diagram of the welded joints, indicating the location, direction, and sense of $\tau_{\parallel}$ resulting from the longitudinal load.
Case 4: We calculate $\sigma = \frac{M_4 y}{I_{xx}}$.

$$\sigma_{\perp,A} = \tau_{\perp,A} = \frac{\sqrt{2}\sigma}{2} = \frac{\sqrt{2}}{2}\frac{M_4 y}{I_{xx}} = \frac{\sqrt{2}}{2}\frac{4991 \times 0.0698}{2\ 148\ 678 \cdot 10^{-12}} = 114.64 \text{ MPa (compression)},$$

$$\sigma_{\perp,B} = \tau_{\perp,B} = \frac{\sqrt{2}\sigma}{2} = \frac{\sqrt{2}}{2}\frac{M_4 y}{I_{xx}} = \frac{\sqrt{2}}{2}\frac{4991 \times 0.0248}{2\ 148\ 678 \cdot 10^{-12}} = 40.73 \text{ MPa (compression)}.$$

Case 6: We calculate the area within the mean perimeter of the throat section of all welded joints. Figure 9.25b shows the free-body diagram of the welded joints, indicating the location, direction,

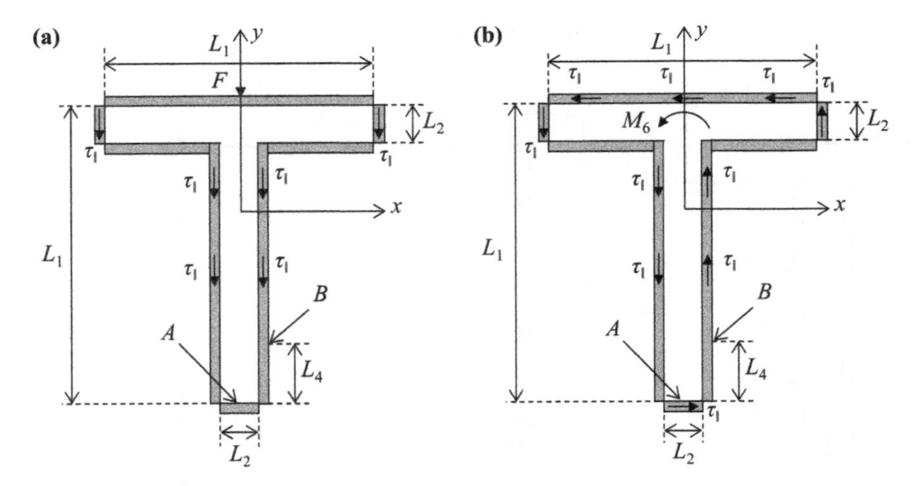

Figure 9.25 Location, direction, and sense of $\tau_{\|}$ resulting from (a) longitudinal load (Case 2) and (b) torque moment (Case 6).

and sense of $\tau_{\|}$ resulting from the longitudinal load. Thus,

$$A = (15 \times 90) + (105 \times 15) = 2925 \ \text{mm}^2.$$

$$\tau_{\|} = \frac{M_6}{2hA} = \frac{250}{2 \times 0.005 \times 2925 \cdot 10^{-6}} = 8.55 \text{ MPa}.$$

The throat section of the welded joint is subject to combined loading because it experiences both a normal and a shear stress. Thus, we must calculate an equivalent normal stress σ_{eq} using Eq. (9.1), i.e.

$$\sigma_{eq} = \sqrt{\sigma_{\perp}^2 + 3(\tau_{\perp}^2 + \tau_{\|}^2)}.$$

$$\sigma_{eq,A} = \sqrt{114.64^2 + 3(114.64^2 + 8.55^2)} = 229.76 \text{ MPa}.$$

$$\sigma_{eq,B} = \sqrt{40.73^2 + 3[40.73^2 + (-8.55 + 5)^2]} = 81.69 \text{ MPa}.$$

Note that in point B, we account for the additional $\tau_{\|}$ resulting from the longitudinal load, which is only borne by the longitudinal welded joints. Furthermore, the $\tau_{\|}$ resulting from the longitudinal load is oriented in the same direction but opposite sense than the $\tau_{\|}$ resulting from the torque moment (in B). Thus, they superimpose (same direction), but subtract from each other (opposite sense). In point A, the only $\tau_{\|}$ component results from the torque moment. Figure 9.25a,b graphically illustrate the superposition of $\tau_{\|}$.

We define the safety factor $n = \sigma_y/\sigma_{eq}$.

$$n_A = \sigma_y/\sigma_{eq,A} = 600/229.76 = 2.61 \text{ and}$$

$$n_B = \sigma_y/\sigma_{eq,B} = 600/81.69 = 7.34.$$

9.4 Key Takeaways

1. The thickness of a welded joint is measured in its throat section.
2. Welding results in either deformation or residual stresses. Residual stresses can be removed by heat treatments of the welded parts.
3. We describe six cases of simple loading, and we demonstrate how to convert an external load into a stress in the throat section of the welded joint, for unidirectional and multidirectional welded joints.
4. A realistic strength calculation of welded joints requires breaking down the external loading in multiple cases of simple loading, and calculating an equivalent normal stress based on the individual stress components resulting from the different cases of simple loading.

9.5 Problems

9.1 Figure 9.26 shows two plates welded together with a fillet weld. The yield stress of the plates $\sigma_y = 160$ MPa. $L_1 = 200$ mm, $\theta = 30°$. $F = 30$ kN. Calculate the thickness of the fillet weld, and consider a design factor $n_D = 4$.

Figure 9.26 Two plates welded together.

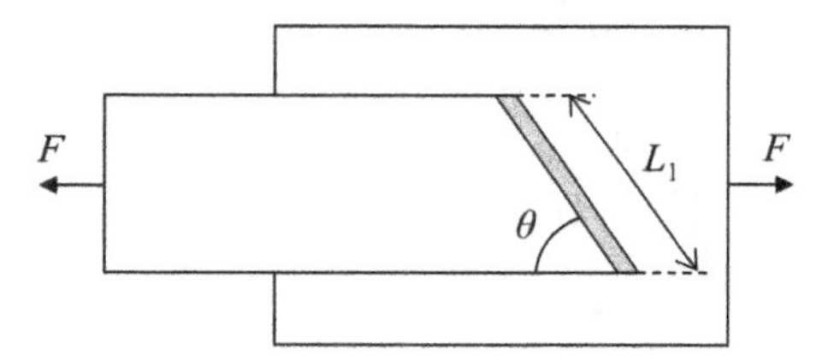

9.2 Figure 9.27 shows a fork lift that holds a load $F = 10$ kN. The fillet weld that joins the vertical and horizontal parts of the fork assembly runs along the entire width of the fork, i.e. $L_1 = L_2 = 0.5$ m. $t = 0.05$ m. The thickness of the welded joint $h = 5$ mm. The yield stress of the material $\sigma_y = 160$ MPa. Calculate the safety factor against yielding of the welded joint.

Figure 9.27 Fork lift.

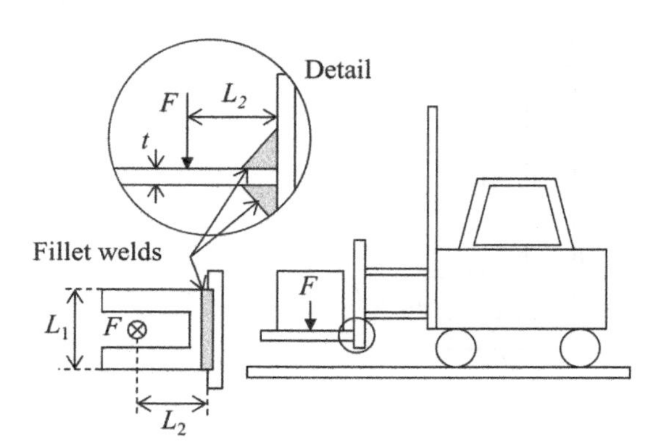

9.3 Figure 9.28 shows a square steel tube welded to a rigid frame with a circumferential fillet weld. The yield stress of the material $\sigma_y = 140$ MPa. $L_1 = 5$ mm, $L_2 = 50$ mm. The tube is subject to a clockwise torque moment $M_t = 2000$ Nm. Calculate the thickness of the welded joint, and consider a design factor $n_D = 2$.

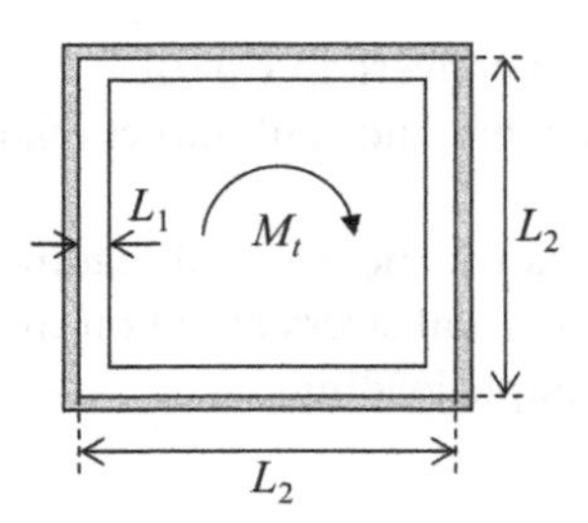

Figure 9.28 Square tube welded to a rigid frame.

9.4 Figure 9.29 shows a triangular tube welded to a rigid frame with a circumferential fillet weld. All parts are made from AISI 1010 HR steel. $L_1 = L_2 = 100$ mm, $L_3 = 10$ mm. The tube is subject to a clockwise torque moment $M_t = 1000$ Nm. Calculate the thickness of the welded joint, and consider a design factor $n_D = 2$.

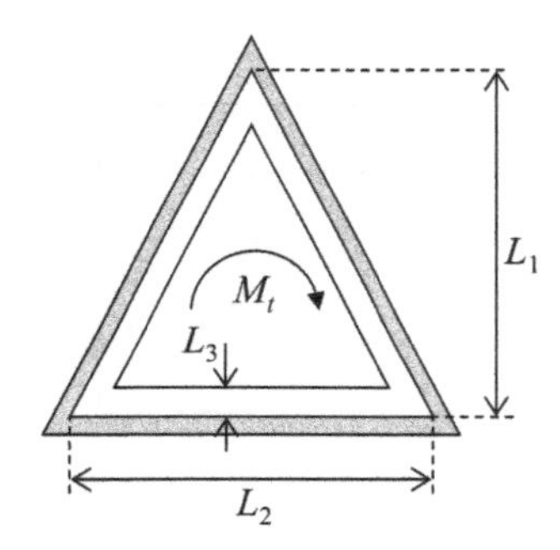

Figure 9.29 Triangular tube welded to a rigid frame.

9.5 Figure 9.30 shows an L-shaped element welded to a rigid frame with a circumferential fillet weld. All parts are made from AISI 1040 HR steel. $L_1 = 200$ mm, $L_2 = 300$ mm, and $L_3 = 20$ mm. The element is subject to a clockwise torque moment $M_t = 1000$ Nm. Calculate the thickness of the welded joint, and consider a design factor $n_D = 5$.

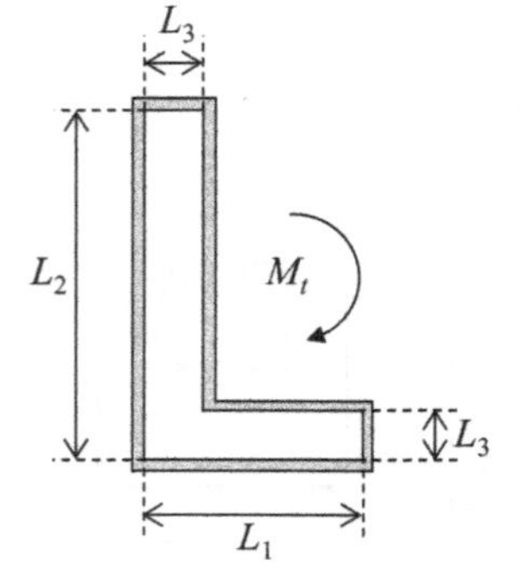

Figure 9.30 L-shaped element welded to a rigid frame.

9.6 Figure 9.31 shows a bracket welded to a rigid frame with two fillet welds. All parts are made from AISI 1010 HR steel. $L_1 = 10$ mm, $L_2 = 300$ mm, $L_3 = 100$ mm, $L_4 = 150$ mm, $L_5 = 200$ mm, and $L_6 = 300$ mm. The thickness of the welded joints $h = 5$ mm. $F = P =$ 7.5 kN. Calculate the safety factor against yielding in point A.

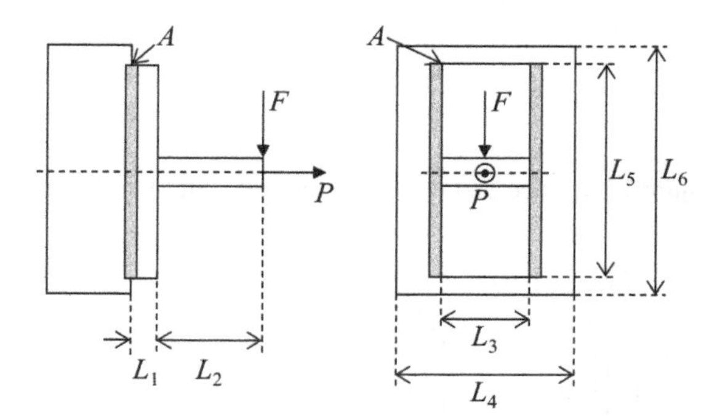

Figure 9.31 Bracket welded to a rigid frame.

9.7 Figure 9.32 shows a bracket welded to a rigid frame with two fillet welds. All parts are made from AISI 1080 HR steel. $L_1 = 100$ mm, $L_2 = 300$ mm, $L_3 = 150$ mm, $L_4 = 20$ mm, and $\alpha = 30°$. The thickness of the welded joints $h = 6$ mm. $F = 50$ kN. Calculate the safety factor against yielding in points A and B.

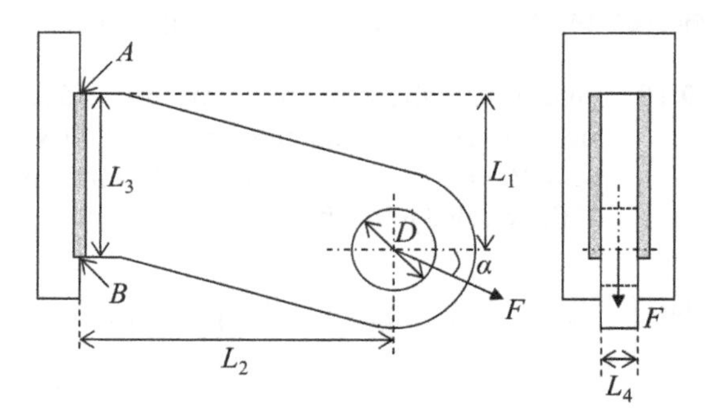

Figure 9.32 Bracket welded to a rigid frame.

9.8 Figure 9.33 shows two steel plates welded together with two fillet welds. All parts are made from AISI 1010 HR steel. $L_1 = 300$ mm, $L_2 = 100$ mm, $L_3 = 400$ mm, $L_4 = 30$ mm, $D =$ 50 mm, and $\alpha = 45°$. $F = 10$ kN. Calculate the thickness of the welded joint and consider a design factor $n_D = 5$.

9.9 Figure 9.34 shows a U-shaped beam welded to a rigid frame. The yield stress of the material $\sigma_y = 600$ MPa. $L_1 = 1$ m, $L_2 = 80$ mm, $L_3 = 100$ mm, and $L_4 = 10$ mm. $F = 5$ kN. The thickness of the welded joints $h = 5$ mm. Calculate the safety factor against yielding in points A and B.

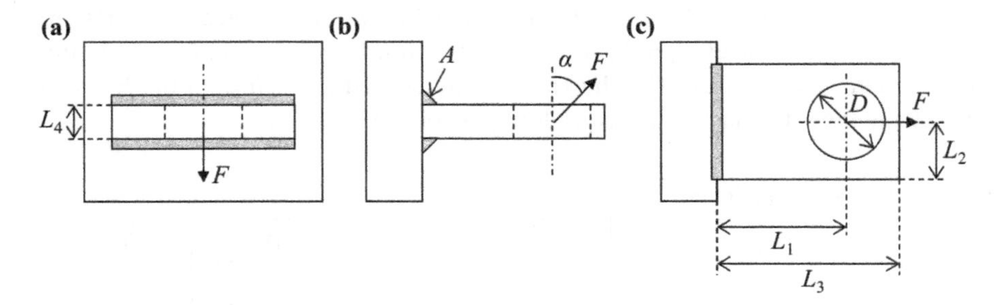

Figure 9.33 Two plates welded together with two fillet welds.

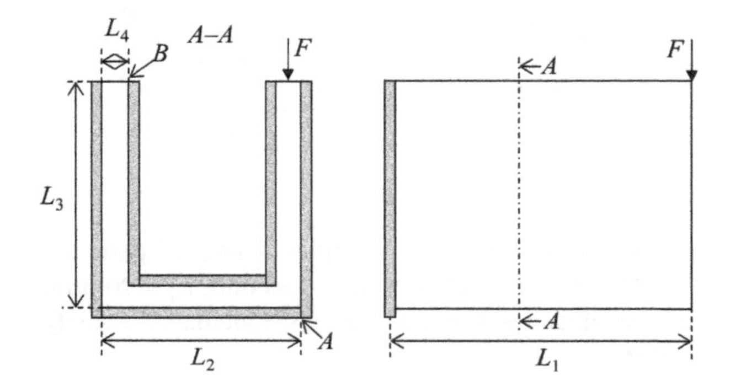

Figure 9.34 U-shaped beam welded to a rigid frame.

9.10 Figure 9.35 shows a steel bracket welded to a rigid frame with two fillet welds. The yield stress of all elements is $\sigma_y = 700$ MPa. $L_1 = 20$ mm, $L_2 = 300$ mm, $L_3 = 50$ mm, $L_4 = 100$ mm, $L_5 = 40$ mm, $D = 45$ mm, and $\alpha = 45°$. $F = 50$ kN. Calculate the thickness of the welded joints h, and consider a design factor $n_D = 2$.

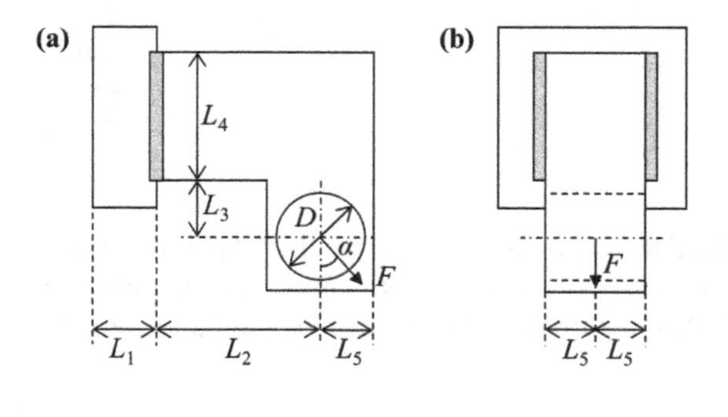

Figure 9.35 Bracket welded to a rigid frame.

10

Rolling Element Bearings

10.1 Introduction

10.1.1 Definition

A rolling element bearing facilitates rotary motion between two mechanical elements. Specifically, it uses rolling elements between two races to carry the bearing load and allow motion with low friction and wear. A rolling element bearing is different from a sleeve bearing, where two components rotate relative to each other without rolling elements, instead separated by lubricant.

10.1.2 Terminology and Geometry

Figure 10.1 shows an isometric view of a rolling element bearing, identifying its components. A rolling element bearing typically comprises an outer ring and an inner ring, with rolling elements in between. The contact surfaces between the rolling elements and the inner and outer ring are the inner and outer races. The rolling elements maintain a uniform spatial distribution along the perimeter of the bearing by means of a separator, i.e. a cage-like structure that maintains the spacing between the different rolling elements. Often, rolling element bearings are sealed, i.e. a seal is integrated between the inner and outer ring, which prevents dirt from entering the bearing and acting as a third-body abrasive between the rolling elements and the races, thus accelerating wear and, ultimately, failure. Furthermore, the seal prevents bearing grease to leave the bearing, again contributing to reducing failure.

We characterize the geometry of a rolling element bearing by its inner diameter because this dimension must match the diameter of the shaft with which it mates, and by its outer diameter, which must match the housing or bore in which it mounts. Also, the width is important when allocating space on a shaft for a bearing. The face of the bearing is the surface that mates with, e.g. the bearing housing, or the shoulder of a shaft. Bearing manufacturers commonly prescribe specific requirements for the shaft surface where a bearing will seat, including surface quality and hardness, as these parameters are important for the bearing to perform its function and achieve its manufacturer-predicted life.

10.1.3 Design Parameters

Bearing manufacturers provide design "recipes" in their respective catalogs, using empirical equations derived from experimental data sets. However, the design parameters and principles that underscore these empirical equations are the same for different bearing manufacturers.

Design of Mechanical Elements: A Concise Introduction to Mechanical Design Considerations and Calculations,
First Edition. Bart Raeymaekers.

Companion website: www.wiley.com/go/raeymaekers/designofmechanicalelements

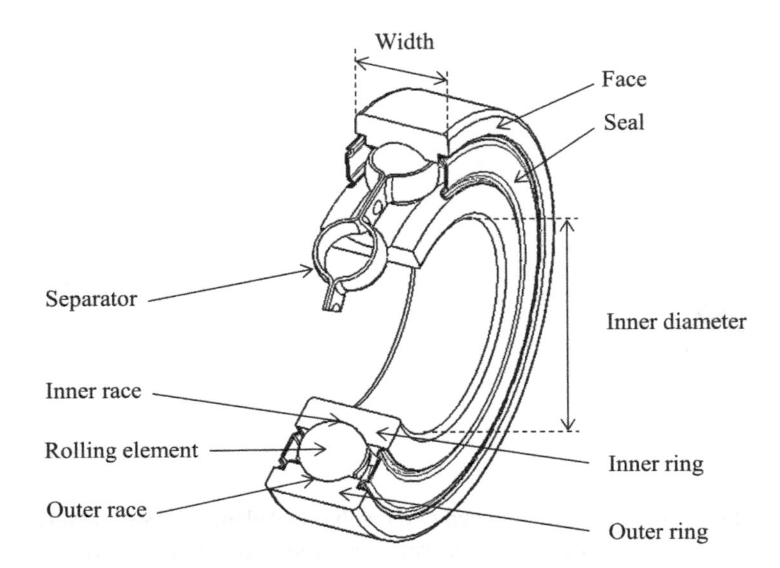

Figure 10.1 Rolling element bearing terminology.

Thus, the design parameters we must consider when designing a rolling element bearing include the geometry (width, inner and outer diameter), the magnitude and direction of the load, the rotational speed of the bearing, bearing life, and bearing reliability, among additional application-specific parameters such as corrosion resistance, tolerances, heat, or friction forces.

10.2 Types of Rolling Element Bearings

This section does not intend to give a comprehensive overview of all different types of rolling element bearings and their function. Instead, we try to provide an appreciation for the myriad of rolling element bearing options and solutions that exist, and for how these options can help address different design specifications and trade-offs.

Different types of bearings exist to serve specific purposes. One of the main parameters by which to categorize bearings is the direction of the external load for which they are intended. Specific bearing designs exist to withstand either a radial or an axial load. Figure 10.2 shows a schematic of (a) a radial bearing and (b) an axial bearing. We indicate the radial load F_r and the axial load F_a, and recognize intuitively that a radial bearing is suited to withstand a radial load but not an axial load, and vice versa.

Bearing manufacturers offer a multitude of bearing types not just for a radial or an axial load, but also for combined loading, and for loads of different magnitude. A deep-groove ball bearing provides *point contact* between the rolling elements and the inner and outer races, whereas a cylindrical or needle rolling element bearing provides *line contact* between the rolling elements and the inner and outer races. Line contact allows a higher load-carrying capacity than point contact because the load is distributed over a larger contact area, thus avoiding yielding of the rolling elements and race materials. Figure 10.3 illustrates the difference between point and line contact.

Thus, to withstand a large radial load, we may opt to use multiple deep-groove ball bearings, or we can select a double-row deep-groove ball bearing (two parallel races and sets of rolling elements on

Figure 10.2 (a) Radial and (b) axial rolling element bearing.

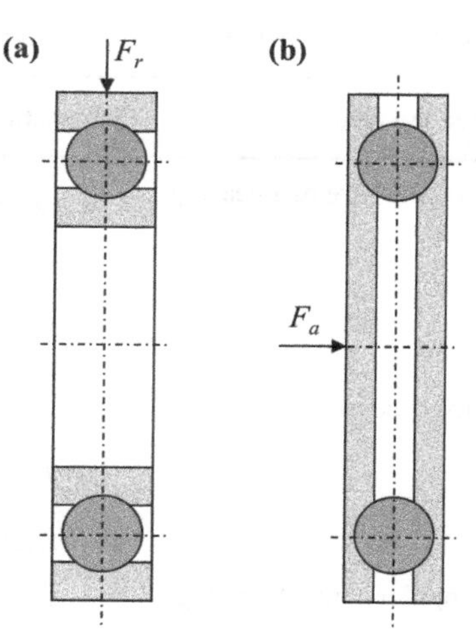

Figure 10.3 (a) Deep-groove ball bearing with point contact and (b) cylindrical rolling element bearing with line contact.

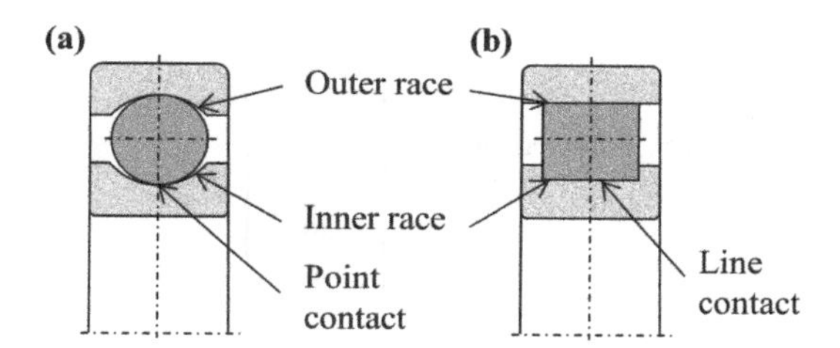

the inner and outer ring), or select a cylindrical rolling element bearing. A double-row deep-groove ball bearing and a cylindrical rolling element bearing are narrower than two deep-groove ball bearings adjacent to each other, and the practicalities of assembly/disassembly and alignment are also easier to handle.

Table 10.1 provides an overview of a few commonly used bearing types, a schematic, and a qualitative comparison between the suitability of each bearing type to perform selected key bearing functions, including carrying a radial load, an axial load, absorb misalignment, operate at high rotating speed, operate with low run-out, and operate with low friction. The qualitative scale ranges between + + + (the very best) to − − − (the least good).

10.3 Hertz Contact Stress

When we load two bodies with curved surfaces against each other, we create a *Hertz* contact stress in both bodies. As a result of this stress, and its corresponding strain, the theoretical point or line contact expands into a finite contact area. If we consider contact between two spheres, we note that point contact exists when the spheres first make contact. However, the size of the point contact increases to a finite contact area when we load the spheres against each other. We emphasize that Hertz contact stress theory only applies to elastic contact between curved bodies.

Contact between curved bodies exists between the rolling elements and the inner and outer races of a rolling element bearing. Other than in rolling element bearings, Hertz contact stresses are important to describe e.g. the stress resulting from mating gear teeth, between a train wheel and rail, cam and follower mechanism, or contact between asperities of two rough surfaces.

Table 10.1 Different rolling element bearing types and their function.

Bearing type	Schematic	Radial *F*	Axial *F*	Align	Speed	Run-out	Friction
Deep-groove ball bearing	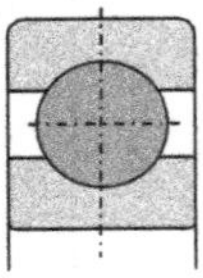	+	+	−	+++	+++	+++
Insert bearing	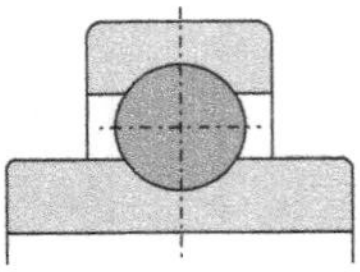	+	+	++	++	+	++
Angular contact bearing	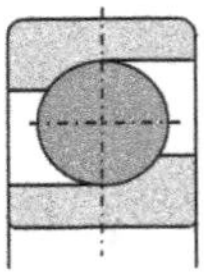	+	+	−	++	+++	+
Self-aligning ball bearing	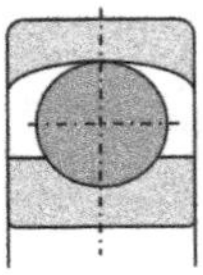	+	−	+++	++	++	+++
Cylindrical roller bearing	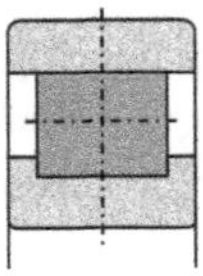	++	−−	−	+++	+++	+++
Needle roller bearing	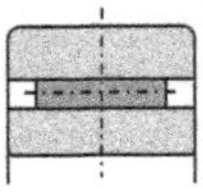	++	−−	−	++	+	+
Tapered roller bearing	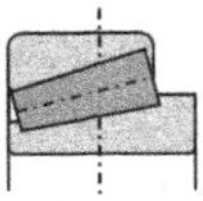	+++	++	−	++	+++	+
Spherical roller bearing	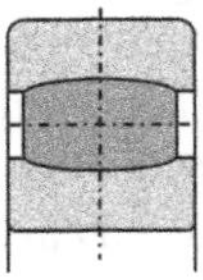	+++	−−	++	++	+++	+
Thrust bearing	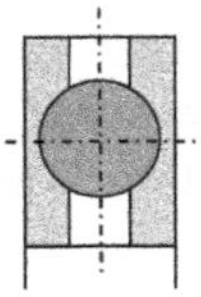	−−	+	−−	−	++	+

10.3.1 Hertz Contact Stress Between Spherical Bodies

Figure 10.4 schematically illustrates contact between two spherical bodies with Young's moduli and Poisson coefficients E_1, ν_1 and E_2, ν_2, respectively. We axially load both spheres against each other with force F, thus creating elastic strain in both spheres. We indicate a Cartesian coordinate system, the diameters d_1 and d_2 of the two spherical bodies, and the diameter of the circular contact area $2a$, where a is given as

$$a = \left\{ \frac{3F}{8} \frac{\left[(1-\nu_1^2)/E_1\right] + \left[(1-\nu_2^2)/E_2\right]}{1/d_1 + 1/d_2} \right\}^{1/3}. \tag{10.1}$$

Note that in case of contact between a spherical and a flat body, $d_2 = \infty$, and in case of contact with an internal diameter, $d_2 < 0$. Figure 10.5 illustrates that the spatial contact stress distribution within each sphere is semi-elliptical and that the maximum stress occurs at the center of the contact area, and is given as

$$\sigma_{max} = \frac{3F}{2\pi a^2}. \tag{10.2}$$

Figure 10.6 shows the magnitude of the different stress components of the triaxial Hertz contact stress state, σ_x, σ_y, σ_z, and τ_{max}, as a function of the orthogonal distance from the contact point at the interface between the two contacting bodies (z-direction), expressed in terms of the contact radius a. We observe that the maximum normal stress occurs at the contact surface, whereas the maximum shear stress occurs away from the surface at $z = 0.48a$.

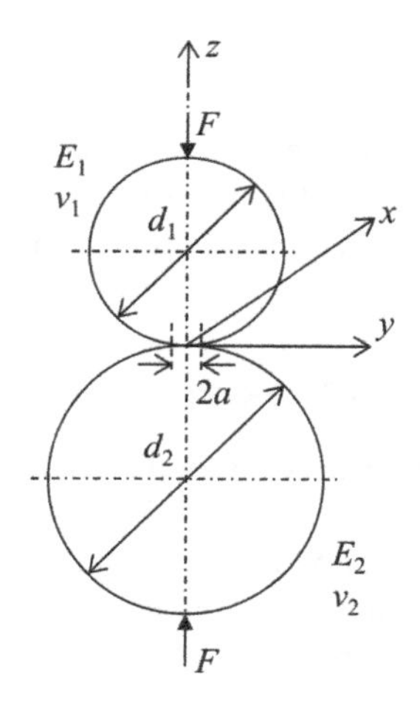

Figure 10.4 Hertz contact stress between two spherical bodies.

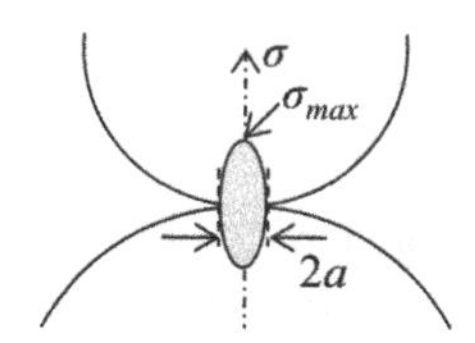

Figure 10.5 Schematic illustration of the semi-elliptical Hertz contact stress distribution between two spherical bodies.

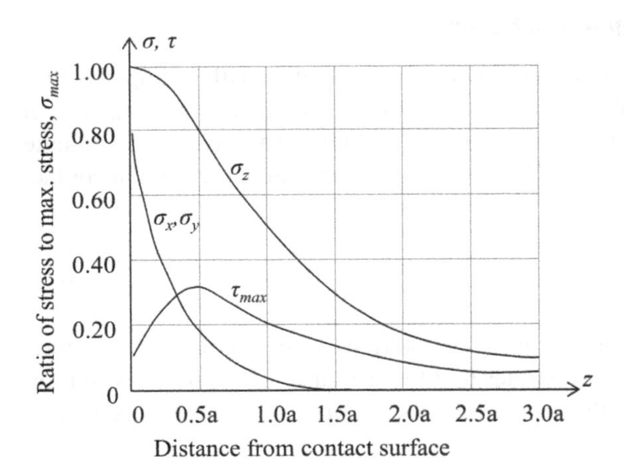

Figure 10.6 Ratio of the stress magnitude and the maximum stress versus orthogonal distance from the contact point at the interface between the contacting bodies, normalized with the contact radius a.

The expressions of the different stress components are[1]

$$\sigma_x = \sigma_y = -\sigma_{max}\left\{\left[1 - \frac{z}{a}\arctan\left(\frac{a}{z}\right)\right](1+\nu) - \frac{1}{2\left(1+\frac{z^2}{a^2}\right)}\right\}, \tag{10.3}$$

$$\sigma_z = \frac{-\sigma_{max}}{1+\frac{z^2}{a^2}}, \tag{10.4}$$

$$\tau_{xz} = \tau_{yz} = \frac{\sigma_x - \sigma_z}{2} = \frac{\sigma_y - \sigma_z}{2}. \tag{10.5}$$

Furthermore, $\sigma_x = \sigma_y \Rightarrow \tau_{xy} = \frac{\sigma_x - \sigma_y}{2} = 0$.

Note that we use Eqs. (10.3)–(10.5) to calculate the stress components in either sphere. However, when both spheres are made of different materials, we use the appropriate Poisson coefficient ν to account for the correct material of the sphere in which we calculate the stress components.

10.3.2 Hertz Contact Stress Between Cylindrical Bodies

Figure 10.7 schematically illustrates contact between two cylindrical bodies with Young's moduli and Poisson coefficients E_1, ν_1 and E_2, ν_2, respectively. We axially load both cylinders against each other with force F, thus creating elastic strain in both cylinders. We indicate a Cartesian coordinate system, the diameters d_1 and d_2 of the two cylindrical bodies, respectively, and the length l and width $2b$ of the rectangular contact area. l is the length over which the cylinders make contact and b is given as

$$b = \left\{\frac{2F}{\pi l}\frac{\left[(1-\nu_1^2)/E_1\right] + \left[(1-\nu_2^2)/E_2\right]}{1/d_1 + 1/d_2}\right\}^{1/2}. \tag{10.6}$$

Note that in case of contact between a cylindrical and a flat body, $d_2 = \infty$, and in case of contact with an internal diameter, $d_2 < 0$. The spatial contact stress distribution within each cylinder is semi-elliptical, and the maximum stress occurs at the center,

$$\sigma_{max} = \frac{2F}{\pi b l}. \tag{10.7}$$

1 See, e.g. Popov, V.L., Hess, M., Willert, E. (2019). *Handbook of Contact Mechanics: Exact Solutions of Axisymmetric Contact Problems*. Springer-Verlag.

Figure 10.7 Hertz contact stress between two cylindrical bodies.

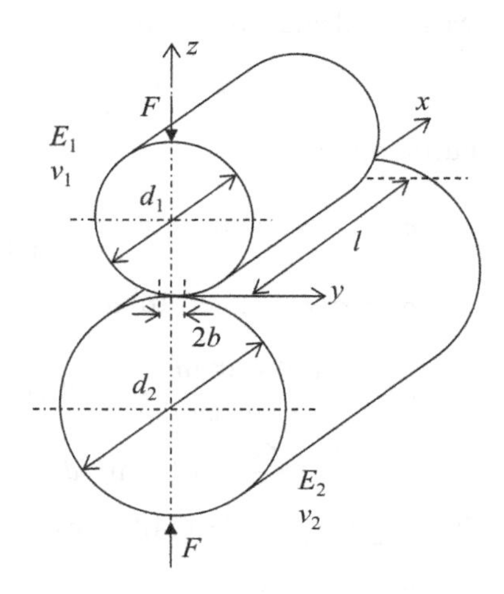

The expressions of the different stress components are[2]

$$\sigma_x = -2\nu\sigma_{max}\left(\sqrt{1+\frac{z^2}{b^2}}-\frac{z}{b}\right), \tag{10.8}$$

$$\sigma_y = -\sigma_{max}\left\{\left[2-\frac{1}{1+\frac{z^2}{a^2}}\right]\sqrt{1+\frac{z^2}{b^2}}-2\frac{z}{b}\right\}, \tag{10.9}$$

$$\sigma_z = \frac{-\sigma_{max}}{\sqrt{1+\frac{z^2}{b^2}}}. \tag{10.10}$$

Furthermore, $\tau_{xy} = \frac{\sigma_x-\sigma_y}{2}$, $\tau_{xz} = \frac{\sigma_x-\sigma_z}{2}$, $\tau_{yz} = \frac{\sigma_y-\sigma_z}{2}$.

Note that we use Eqs. (10.8)–(10.10) to calculate the stress components in either cylinder. However, when both cylinders are made of different materials, we use the appropriate Poisson coefficient ν to account for the correct material of the sphere in which we calculate the stress components.

Example problem 10.1

We load a steel sphere against a steel flat with $F = 100$ N. Calculate the minimum diameter of the sphere required to avoid that the maximum shear stress in the sphere exceeds 180 MPa.

Solution

We start from Eq. (10.1) to calculate the radius of the contact area between the sphere and the flat, i.e.

$$a = \left\{\frac{3F}{8}\frac{\left[(1-\nu_1^2)/E_1\right]+\left[(1-\nu_2^2)/E_2\right]}{1/d_1+1/d_2}\right\}^{1/3}.$$

$$a = \left\{\frac{3\times 100}{8}\frac{\left[(1-0.30^2)/210\cdot 10^9\right]+\left[(1-0.30^2)/210\cdot 10^9\right]}{1/d+1/\infty}\right\}^{1/3} = 6.875\cdot 10^{-4}\, d^{1/3}.$$

2 See, e.g. Popov, V.L., Hess, M., and Willert, E. (2019). *Handbook of Contact Mechanics: Exact Solutions of Axisymmetric Contact Problems*. Springer-Verlag.

We also calculate the maximum contact stress using Eq. (10.2), i.e.

$$\sigma_{max} = 3F/2\pi a^2 = \frac{3 \times 100}{2\pi(6.875 \cdot 10^{-4}\, d^{1/3})^2} = 1.01 \cdot 10^8/d^{2/3}.$$

Furthermore,

$$\sigma_x = \sigma_y = -(1.01 \cdot 10^8/d^{2/3}) \left\{ \left[1 - 0.48 \text{ arctan}\left(\frac{1}{0.48}\right)\right](1+0.30) - \frac{1}{2\left(1+0.48^2\right)} \right\},$$

$$\sigma_x = \sigma_y = -1.946 \cdot 10^7\, d^{-2/3},$$

$$\sigma_z = \frac{-1.01 \cdot 10^8/d^{2/3}}{1+0.48^2} = -8.210 \cdot 10^7\, d^{-2/3},$$

$$\tau_{xz} = \frac{\sigma_x - \sigma_z}{2} = 3.13 \cdot 10^7\, d^{-2/3}.$$

Since $\tau_{xz} \leq \tau_{max} = 180$ MPa, we calculate

$$d \geq 0.0725 \text{ m}.$$

10.4 Bearing Calculations

10.4.1 Bearing Life

We define bearing life as the number of hours, or the number of revolutions, at constant velocity, before failure occurs. Two common metrics exist to quantify bearing life; the *rating life* and the *average life*.

The *rating life* is the number of hours, or the number of revolutions, at constant velocity that 90% of a group of bearings complete or exceed before failure occurs. We refer to this metric as the L_{10} life. Note that each manufacturer may define L_{10} life differently. For instance, SKF and Schaeffler define their L_{10} life as 10^6 revolutions, whereas The Timken Company defines it at as 3000 hours at 500 RPM.

The *average life* is the number of hours, or the number of revolutions, at constant velocity that 50% of a group of bearings complete or exceed before failure occurs. The average life actually describes the median rather than the average. However, average life is the terminology that is often (erroneously) used.

The Weibull distribution (under constant load and constant velocity) describes the life of rolling element bearings because its three parameters offer flexibility to obtain a good fit to experimental data of bearing life. Figure 10.8 shows the percentage of bearings (of a group representing 100%) in operation as a function of bearing life. We indicate both the L_{10} life and the average life.

10.4.2 Bearing Load

We define bearing load as the radial force that causes bearing failure at constant bearing velocity. Commonly, we quantify bearing load using the *basic (dynamic) load rating*, or C_{10} rating, which is the radial load that causes 10% of a group of bearings to fail at the bearing manufacturer's rating life, i.e. its L_{10} life. Note that the ISO 281:2007 standard[3] specifies methods of calculating the basic dynamic load rating of rolling element bearings.

---

3 ISO 281:2007. *Rolling Bearings – Dynamic Load Ratings and Rating Life. International Organization for Standardization.*

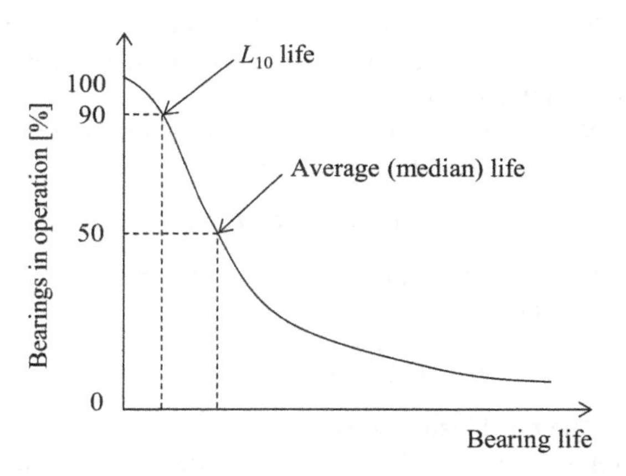

Figure 10.8 Schematic representation of bearing life.

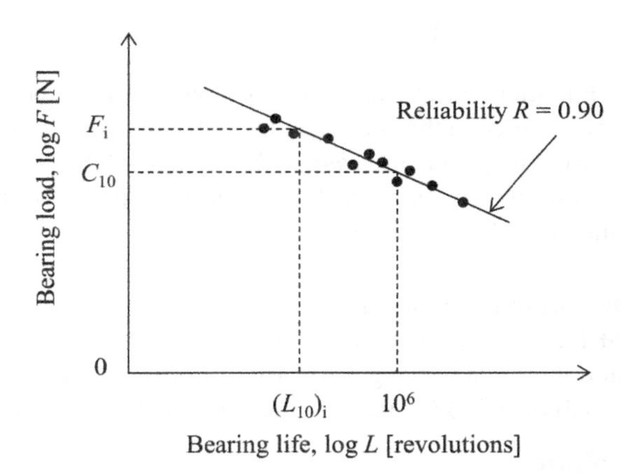

Figure 10.9 Log–log presentation of bearing load versus bearing life.

Another metric is the C_0 basic (static) load rating, which is the load that produces a total permanent deformation in the race and rolling element at any contact point of 0.0001 times the diameter of the rolling element. Both the static and dynamic load rating appear in the bearing manufacturer's catalog.

Figure 10.9 shows a double-logarithmic presentation of bearing load as a function of bearing life. We schematically show a cloud of data points and a linear curve fit through the data cloud. The linear curve fit shows different L_{10} life values, $(L_{10})_i$ on the horizontal axis, and their corresponding bearing load on the vertical axis. As an example, we indicate the $L_{10} = 10^6$ revolutions (SKF and Schaeffler), where the bearing load corresponds to the basic load rating C_{10}. Note also that because each data point in Figure 10.9 documents a group of bearings where 10% of the group has failed, each data point implicitly represents a reliability $R = 90\%$.

The linear curve fit of the logarithmic data shows that

$$FL^{1/a} = \text{constant}, \tag{10.11}$$

with $a = 3$ for a ball bearing, and $a = 10/3$ for any other rolling element bearing. Thus, from Eq. (10.11), we write that

$$F_r L_r^{1/a} = F_d L_d^{1/a}, \tag{10.12}$$

where the subscripts *r* and *d* refer to *rated* and *design*, respectively. We indicate *rated* for the values that appear in the manufacturer's catalog, whereas we indicate *design* for the values for which we design the bearing.

We convert bearing life in number of revolutions L to bearing life in number of hours $\overline{L}$, which is more practical to use. Hence,

$$L \text{ (revolutions)} = 60\, N\, \overline{L} \text{ (hours)}, \tag{10.13}$$

where N is the number of rotations per minute (RPM). Thus, combining Eqs. (10.12) and (10.13), we find that

$$F_r\left(60\, N_r\, \overline{L_r}\right)^{1/a} = F_d\left(60\, N_d\, \overline{L_d}\right)^{1/a}. \tag{10.14}$$

The rated load F_r is equal to the C_{10} basic load rating, given that all data points in Figure 10.9 represent a reliability $R = 90\%$. Hence, we conclude that

$$C_{10} = F_d\left(\frac{60\, N_d\, \overline{L_d}}{60\, N_r\, \overline{L_r}}\right)^{1/a}. \tag{10.15}$$

Equation (10.15) allows converting a design bearing load, life, and velocity to the rated value documented in the bearing manufacturer's catalog. Indeed, tabulating all possible combinations of bearing life, load, and velocity is impractical. For that reason, we use Eq. (10.15) to determine the load rating we must select from a bearing manufacturer catalog, for specific design requirements and operating parameters.

Example problem 10.2

SKF rates its bearings at 10^6 revolutions. We attempt to select a ball bearing for a specific design and require a bearing life of 10 000 hours at 1000 RPM, supporting a radial load of 1000 N, and a reliability of 90%. Calculate the basic load rating to select from the SKF catalog.

Solution

We start from Eq. (10.15), i.e.

$$C_{10} = F_d\left(\frac{60\, N_d\, \overline{L_d}}{60\, N_r\, \overline{L_r}}\right)^{1/a}.$$

Introducing the L_{10} life of 10^6 revolutions,

$$C_{10} = F_d\left(\frac{60\, N_d\, \overline{L_d}}{10^6}\right)^{1/a},$$

$$C_{10} = 1000\left(\frac{60 \times 1000 \times 10\,000}{10^6}\right)^{1/3} = 8434 \text{ N}.$$

10.4.3 Bearing Reliability

We use the Weibull distribution to express the reliability of a bearing under constant load. We define *reliability* as (1 – *probability of failure*). If we consider a random variable *bearing life* L, then we write the bearing reliability as

$$R = 1 - p(L \leq T) = 1 - F(T), \tag{10.16}$$

where $F(T)$ refers to the cumulative distribution function evaluated at $L = T$. It is sometimes more convenient to nondimensionalize variables. Random variable $x = L/L_{10}$ expresses the bearing life nondimensionalized by the rating life. Hence, we write the cumulative distribution function of a Weibull distribution as a function of random variable x as

$$F(x) = 1 - \exp\left[-\left(\frac{x - x_0}{\theta - x_0}\right)^b\right], \tag{10.17}$$

where x_0, θ, and b are the Weibull parameters, which we obtain through curve fitting an experimental dataset, i.e. bearing life under constant load. Substituting Eq. (10.17) into Eq. (10.16) and accounting for the nondimensional variables gives

$$R = \exp\left[-\left(\frac{x - x_0}{\theta - x_0}\right)^b\right]. \tag{10.18}$$

Next, we relate the bearing reliability to Eq. (10.15), such that we may determine the basic load rating of a bearing, even if we desire a reliability that is different from 90%.

Figure 10.10 shows the design specifications, which we indicate with point d. We indicate the conditions tabulated in the bearing manufacturer catalog with point a. Points a and d do not overlap because the design specifications are different from the bearing operating parameters (bearing speed, load, life) that the manufacturer uses to rate its bearings. Thus, we must convert the design specification to the operating conditions that the manufacturer uses to rate its bearings, to allow selecting the correct basic load rating. To accomplish this conversion, we write along a constant reliability line in Figure 10.10 that

$$FL^{1/a} = \text{constant}, \tag{10.19}$$

or,

$$F_b x_b^{1/a} = F_d L_d^{1/a}, \tag{10.20}$$

alternatively,

$$F_b = F_d\left(\frac{x_d}{x_b}\right)^{1/a}. \tag{10.21}$$

Along a constant load line between points a and b, the Weibull parameters x_0, θ, b are valid and, thus, we write

$$R_d = R_b = \exp\left[-\left(\frac{x_b - x_0}{\theta - x_0}\right)^b\right]. \tag{10.22}$$

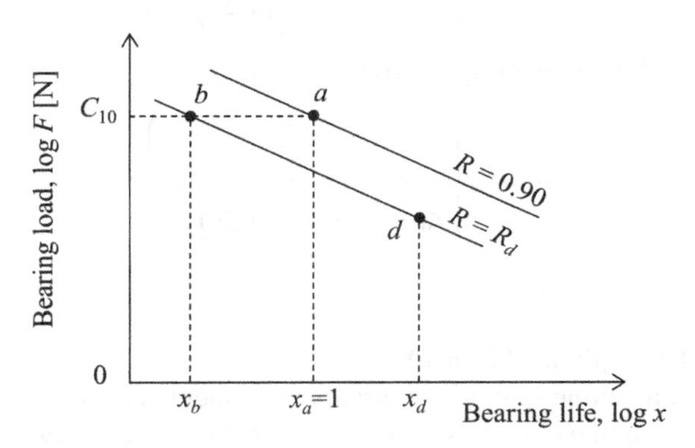

Figure 10.10 Log–log presentation of bearing load versus bearing life, for different values of the reliability R.

Table 10.2 Typical bearing application factor values a_f.

Type of loading	Application factor a_f
Precision gearing	1.0–1.1
Applications with poor bearing seals	1.2
Machinery with no impact	1.0–1.2
Machinery with light impact	1.2–1.5
Machinery with moderate impact	1.5–3.0

Source: Based on Budynas, R.G. and Nisbett, J.K. (2015). *Shigley's Mechanical Engineering Design*, 10th ed. McGraw-Hill.

Rewriting Eq. (10.22) gives

$$x_b = x_0 + (\theta - x_0)\left(\ln \frac{1}{R_d}\right)^{1/b}. \tag{10.23}$$

Next, we substitute Eq. (10.23) in Eq. (10.21), which yields

$$F_b = F_d \left(\frac{x_d}{x_0 + (\theta - x_0)\left(\ln \frac{1}{R_d}\right)^{1/b}} \right)^{1/a}. \tag{10.24}$$

We note that $F_b = C_{10}$, i.e. the manufacturer's basic load rating. Furthermore, we introduce an application factor a_f (see Table 10.2) to account for potential impact loading of the bearing, and we write that

$$C_{10} = a_f F_d \left(\frac{x_d}{x_0 + (\theta - x_0)\left(\ln \frac{1}{R_d}\right)^{1/b}} \right)^{1/a}. \tag{10.25}$$

Table 10.2 lists a few common examples of loading types and the corresponding application factor a_f. Furthermore, we modify Eq. (10.25) to make the dimensional parameters accessible for design calculations. Thus,

$$C_{10} = a_f F_d \left(\frac{L_d / L_{10}}{x_0 + (\theta - x_0)\left(\ln \frac{1}{R_d}\right)^{1/b}} \right)^{1/a}, \tag{10.26}$$

with L in revolutions. Alternatively,

$$C_{10} = a_f F_d \left(\frac{\frac{\overline{L}_d}{\overline{L}_r} \frac{N_d}{N_r}}{x_0 + (\theta - x_0)\left(\ln \frac{1}{R_d}\right)^{1/b}} \right)^{1/a}, \tag{10.27}$$

with $\overline{L}$ in hours.

Example problem 10.3

A machine shaft is supported in two identical ball bearings. The shaft is not subject to impact. The Weibull parameters we use for the bearing reliability calculations were experimentally determined as $\theta = 6.84$, $x_0 = 0$, and $b = 1.17$. The design requires a bearing life of 20000 hours at 1000 RPM with a reliability of at least 95%. Each bearing is subject to a radial load 4 kN. Calculate the basic load rating to select from the manufacturer's catalog, considering that the manufacturer rates its bearings at 3000 hours L_{10} life at 600 RPM.

Solution

We start from Eq. (10.27), i.e.

$$C_{10} = a_f F_d \left(\frac{\overline{L_d}/\overline{L_r} \times N_d/N_r}{x_0 + (\theta - x_0)\left(\ln(1/R_d)\right)^{1/b}} \right)^{1/a}.$$

$$C_{10} = 1.0 \times 4000 \left(\frac{20000/3000 \times 1000/600}{0 + (6.84 - 0)(\ln(1/0.95))^{1/1.17}} \right)^{1/3} = 13808 \text{ N}.$$

10.4.4 Combined Radial and Axial Loading

Sections 10.4.2 and 10.4.3 were entirely based on bearings subject to a radial bearing load only because the manufacturer's catalog ratings are based on radial load only. However, a bearing may experience combined axial and radial loading. Therefore, we define an equivalent load F_e that has the same effect on the bearing life than the combination of an axial and radial load.

Figure 10.11 shows two nondimensional parameters, $F_e/(VF_r)$ versus $F_a/(VF_r)$. Here, F_a is the axial or thrust load, V is the rotation factor, with $V = 1$ for rotating inner ring, and $V = 1.2$ for rotating outer ring of the bearing. We schematically illustrate experimental data as solid black circles. We approximate the data by means of two line segments, with

$$\begin{cases} F_e/(VF_r) = 1 & \text{when } F_a/(VF_r) \leq e, \\ F_e/(VF_r) = X + Y\frac{F_a}{VF_r} & \text{when } F_a/(VF_r) > e, \end{cases} \tag{10.28}$$

X is the radial factor (intercept with the vertical axis in Figure 10.11) and Y is the thrust factor (slope of the line for $F_a/(VF_r) > e$ in Figure 10.11).

Commonly, we express Eq. (10.28) as one equation

$$F_e = X_i V F_r + Y_i F_a. \tag{10.29}$$

Here, $i = 1$ when $F_a/(VF_r) \leq e$ and $i = 2$ when $F_a/(VF_r) > e$. Table 10.3 shows the different parameters required to calculate F_e using Eq. (10.29). We use F_e as the design load F_d in the calculation of the basic dynamic load rating factor C_{10}, using Eqs. (10.25)–(10.27).

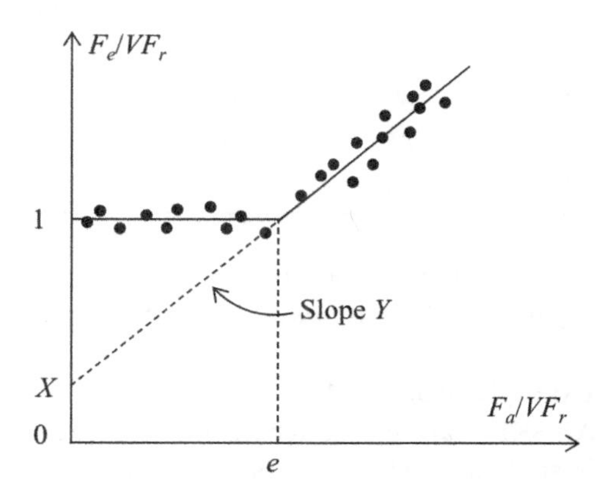

Figure 10.11 Combined loading of a bearing showing two nondimensional parameters F_e/VF_r versus F_a/VF_r.

Table 10.3 Equivalent radial load factors for ball bearings.

		$F_a/(VF_r) \le e$		$F_a/(VF_r) > e$	
F_a/C_0	e	X_1	Y_1	X_2	Y_2
0.014[a]	0.19	1.00	0	0.56	2.30
0.021	0.21	1.00	0	0.56	2.15
0.028	0.22	1.00	0	0.56	1.99
0.042	0.24	1.00	0	0.56	1.85
0.056	0.26	1.00	0	0.56	1.71
0.070	0.27	1.00	0	0.56	1.63
0.084	0.28	1.00	0	0.56	1.55
0.110	0.30	1.00	0	0.56	1.45
0.170	0.34	1.00	0	0.56	1.31
0.280	0.38	1.00	0	0.56	1.15
0.420	0.42	1.00	0	0.56	1.04
0.560	0.44	1.00	0	0.56	1.00

a) Use 0.014 if $F_a/C_0 < 0.014$.
Source: Based on Budynas, R.G. and Nisbett, J.K. (2015). *Shigley's Mechanical Engineering Design*, 10e, McGraw-Hill.

Table 10.4 Excerpt from a bearing manufacturer catalog for deep-groove ball bearings.

Bore (mm)	OD (mm)	Width (mm)	C_{10} (kN)	C_0 (kN)
10	30	9	5.07	2.24
12	32	10	6.89	3.10
15	35	11	7.80	3.55
17	40	12	9.56	4.50
20	47	14	12.70	6.20
25	52	15	14.00	6.95
30	62	16	19.50	10.00
35	72	17	25.50	13.70
40	80	18	30.70	16.60
45	85	19	33.20	18.60
50	90	20	35.10	19.60

Furthermore, Table 10.4 shows an excerpt from a bearing manufacturer catalog, illustrating three key bearing dimensions and both the C_{10} and C_0 basic load ratings for deep-groove ball bearings.

Example problem 10.4

An SKF bearing experiences a thrust load of 2000 N and a radial load of 2500 N. The outer ring is stationary. The basic (static) load rating $C_0 = 18.6$ kN and the basic (dynamic) load rating

$C_{10} = 33.2$ kN. Estimate the bearing life in hours at a rotation speed of 720 RPM, with a reliability of at least 95%. The bearing is not subject to impact. The Weibull parameters for the bearing reliability calculations are $\theta = 6.84$, $x_0 = 0$, and $b = 1.17$.

Solution

We determine $F_a/C_0 = 2000/18\,600 = 0.107$, and $V = 1$ because the inner ring rotates. Thus, we approximate $e = 0.298$ by linearly interpolating in Table 10.3.

$F_a/(VF_r) = 2000/2500 = 0.8 > 0.298$. Thus, we approximate $Y_2 = 1.46$ by linearly interpolating in Table 10.3.

We use Eq. (10.29) to calculate the equivalent radial load F_e, i.e.

$$F_e = X_2VF_r + Y_2F_a = 0.56 \times 1 \times 2500 + 1.46 \times 2000 = 4320 \text{ N}.$$

Finally, we substitute F_e for F_d in Eq. (10.27) and calculate

$$C_{10} = a_f F_d \left(\frac{\frac{\overline{L_d}}{L_r}\frac{N_d}{N_r}}{x_0 + (\theta - x_0)\left(\ln \frac{1}{R_d}\right)^{1/b}} \right)^{1/a}, \text{ or}$$

$$\overline{L_d} = \left(\frac{C_{10}}{a_f F_d}\right)^3 \left(x_0 + (\theta - x_0)\left(\ln \frac{1}{R_d}\right)^{1/b}\right) \frac{10^6}{60N_d},$$

$$\overline{L_d} = \left(\frac{33\,200}{1 \times 4320}\right)^3 \left(0 + (6.84 - 0)\left(\ln \frac{1}{0.95}\right)^{1/1.17}\right) \frac{10^6}{60 \times 720} = 5676 \text{ hours}.$$

10.5 Key Takeaways

1. It is important to select the correct rolling element bearing for each specific task. We account for the geometry, loading (direction and magnitude), environment, and temperature, among other application-specific parameters.
2. Hertz contact stress theory is a good approximation of the stresses occurring at the interface between the rolling elements and the races of the bearing, under elastic loading.
3. The basic load rating of a bearing is listed in the manufacturer's catalog for a specific L_{10}-life. However, we may calculate the required basic load rating for a rolling element bearing as a function of reliability, magnitude and direction of the load, rotational velocity, and bearing life.

10.6 Problems

10.1 An AISI 1020 steel ball with diameter $D = 30$ mm is loaded with force $F = 20$ N against an AISI 1020 flat steel plate.

(a) Calculate the maximum shear stress in the plate.

(b) Calculate the depth in the plate at which the maximum shear stress occurs.

10.2 A titanium ball with diameter $D = 20$ mm is loaded with force $F = 40$ N against a concave surface made from AISI 304 stainless steel, with radius of curvature $R = 200$ mm.

(a) Calculate the maximum shear stress in the ball.

(b) Calculate the depth in the ball at which the maximum shear stress occurs.

10.3 Reconsider the previous problem, but calculate the maximum shear stress and depth of the maximum shear stress in the concave surface.

10.4 An AISI 1018 steel sphere with diameter $D = 25$ mm is placed in between a plate made from 2024 T3 aluminum (top) and a second flat plate made from ASTM 30 gray cast iron (bottom). Calculate the maximum load that can be applied normal to this stack without exceeding a maximum shear stress of 140 MPa in any of the three elements.
Note: Assume that the maximum shear stress occurs at $z/a = 0.48$ for each of the three materials (but, use the appropriate values of the Poisson's ratio and the Young's modulus in the calculations).

10.5 A steel sphere is loaded against a steel flat with force $F = 200$ N. Calculate the minimum radius of the sphere required to avoid that the maximum shear stress in the sphere exceeds 240 MPa.

10.6 SKF rates its bearings at 10^6 revolutions. Calculate the basic load rating for a ball bearing that requires a bearing life of 20 000 hours at 2000 RPM, with a radial load of 2000 N, and a reliability of 90%.

10.7 The Timken Company rates its bearings at 3000 hours and 500 RPM. Calculate the basic load rating for a ball bearing that requires a bearing life of 20 000 hours at 2000 RPM, with a radial load of 2000 N and a reliability of 90%.

10.8 Figure 10.12 shows a shaft supported in a deep-groove ball bearing in point A and in a cylindrical bearing in point C. The bearings are rated at an L_{10}-life of 10^6 revolutions. The shaft is driven by a belt and pulley transmission, with $P = 20$ kW. $\omega = 3000$ RPM. The shaft is machined from AISI 1020 HR steel. The pulley has a diameter $D_{pulley} = 0.1$ m. The tight side tension in the belt $T_1 = 2000$ N. The belt is wrapped around half the circumference of the pulley. $L_1 = 0.2$ m, $L_3 = 0.8$ m. Calculate the basic load rating for each bearing if a bearing life of 10 000 hours with 98% reliability is required. Use an application factor of 1.5. The reliability of the bearings follows a Weibull distribution with $x_0 = 0.02$, $\theta - x_0 = 4.439$, and $b = 1.483$.

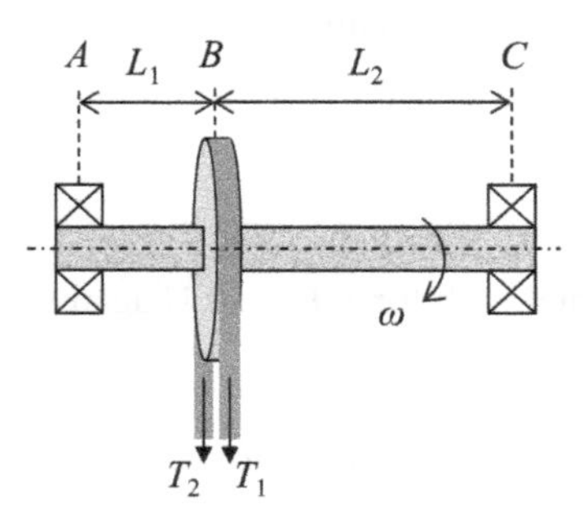

Figure 10.12 Shaft supported in two bearings.

10.9 Figure 10.13 shows a portion of a machine. A shaft is supported in two deep-groove ball bearings. The bearings are rated at an L_{10}-life of 10^6 revolutions. The shaft is driven by a belt and pulley transmission. $\omega = 1000$ RPM. The tight side tension in the belt $T_1 = 5000$ N and the friction coefficient between the belt and the pulley $\mu = 0.4$. The belt is wrapped around half the circumference of the pulley. The diameter of the pulley $D_{pulley} = 100$ mm. The weight of the flywheel $m_{flywheel} = 50$ kg. $L_1 = L_2 = 0.4$ m, $L_3 = 0.2$ m. Calculate the basic load rating for each bearing if a bearing life of 10 000 hours with 95% reliability is required. Use an application factor of 1.2. The reliability of the bearings follows a Weibull distribution with $x_0 = 0.02$, $\theta - x_0 = 4.439$, and $b = 1.483$.

Figure 10.13 Shaft supported in two bearings.

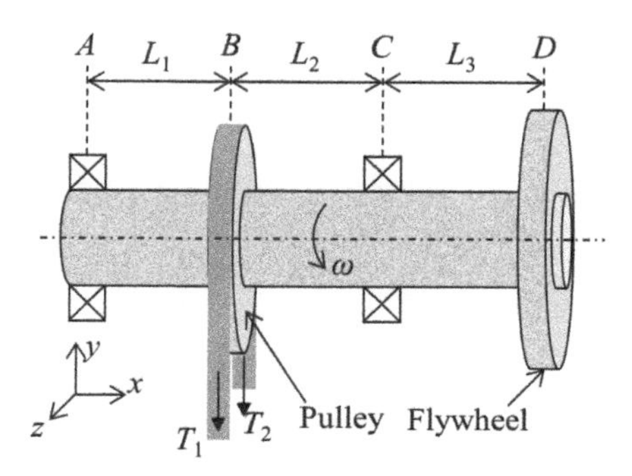

10.10 Figure 10.14 shows a portion of a machine. A shaft is supported in two deep-groove ball bearings. The shaft is driven by a belt and pulley transmission. $\omega = 2000$ RPM. The tight side tension in the belt $T_1 = 1000$ N and the friction coefficient between the belt and the pulley $\mu = 0.4$. The belt is wrapped around half the circumference of the pulley. The diameter of the pulley $D_{pulley} = 100$ mm. The weight of the flywheel $m_{flywheel} = 40$ kg. $L_1 = L_2 = 0.4$ m, $L_3 = 0.2$ m. Calculate the basic load rating for each bearing if a bearing life of 20 000 hours with 95% reliability is required. Use an application factor of 1.2. The reliability of the bearings follows a Weibull distribution with $x_0 = 0.02$, $\theta - x_0 = 4.439$, and $b = 1.483$.

Figure 10.14 Shaft supported in two bearings.

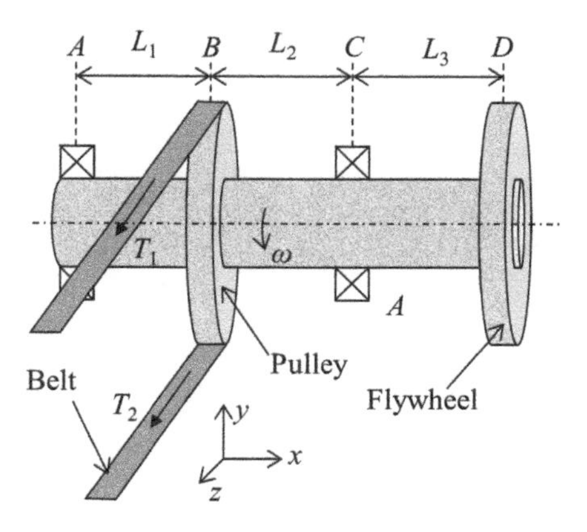

11

Gears

11.1 Introduction

A gear is a rotating mechanical element with "teeth" that engage or mesh with the "teeth" of another gear to transmit torque and rotary motion between two mechanical elements, such as two shafts. Meshing gears of different sizes enable tuning the rotational speed and torque of the shafts to which the gears are mounted. We refer to several meshing gears working in sequence as a *gear train* or a *transmission*. Figure 11.1 compares the rotation direction of (a) a belt/pulley and (b) a gear transmission. A finite amount of slip exists between the belt and pulley, which does not occur in gear transmissions. Furthermore, in a belt/pulley transmission (with noncrossed belt), the two pulleys rotate in the same direction, whereas the two meshing gears rotate in opposite direction.

11.1.1 Types of Gears

Figure 11.2 schematically illustrates the four basic categories of gears, showing (a) spur gears or "straight-cut" gears, (b) helical gears, (c) bevel gears, and (d) worm gears. We briefly describe the different gear categories in the following paragraphs.

Spur gears (Figure 11.2a) project teeth radially outward and the teeth are parallel to the axis of rotation. Thus, they transmit power and rotary motion between parallel shafts. No axial thrust loads are created between meshing spur gears because all force components act radially and tangentially to the gear. Spur gears are the simplest type of gear and, therefore, are used abundantly. They perform well at moderate velocity but may become "noisy" at high velocity because the entire gear tooth face makes contact with the meshing gear tooth at once. We primarily discuss spur gears in this textbook and use spur gears to introduce the terminology and kinematics of gears.

Helical gears (Figure 11.2b) have teeth that are inclined toward the axis of rotation and, thus, provide a gradual engagement between the meshing teeth. As a result, helical gears operate more quietly than spur gears. Thus, we use helical gears in high-speed applications, and applications that require large power transmission or where noise level is important. Helical gear combinations can transmit power and rotary motion between both parallel and nonparallel shafts. A disadvantage of helical gears compared to spur gears is that a thrust force component exists parallel to the axis of rotation, for which we must account in the shaft design and the design of the bearings that support the shaft. Alternatively, we may use double-helix or "herringbone" gears to neutralize the thrust load. Another disadvantage of helical gears is that the gradual engagement of the meshing teeth creates a friction force between them, which we typically address with lubricant.

Design of Mechanical Elements: A Concise Introduction to Mechanical Design Considerations and Calculations,
First Edition. Bart Raeymaekers.

Companion website: www.wiley.com/go/raeymaekers/designofmechanicalelements

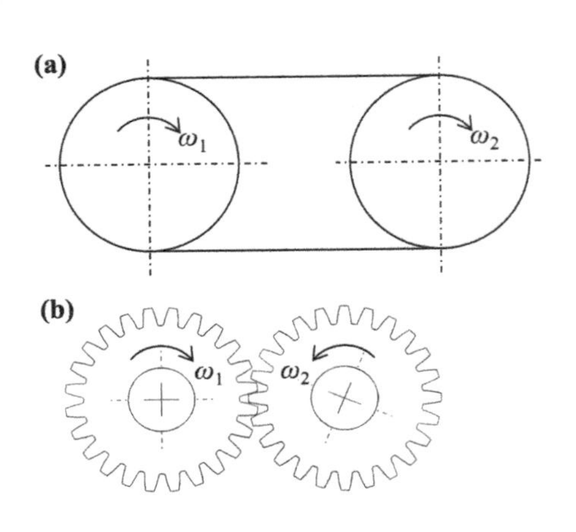

Figure 11.1 Rotation direction of (a) belt and pulley transmission, and (b) gear transmission.

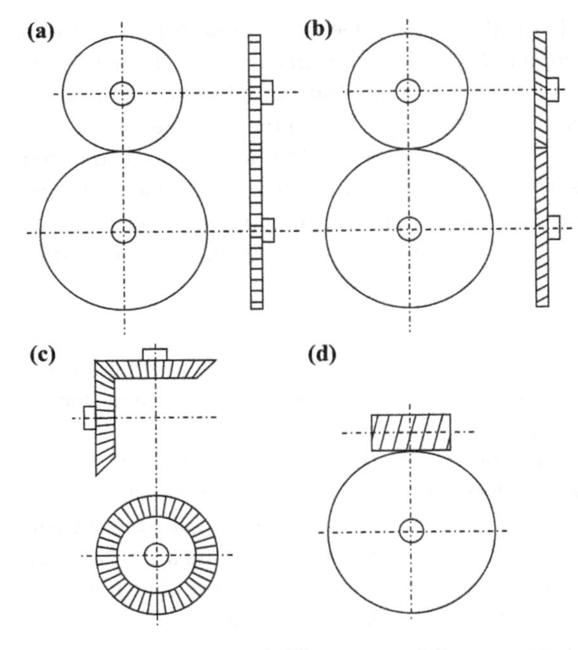

Figure 11.2 Different gear categories (a) spur gears, (b) helical gears, (c) bevel gears, and (d) worm gears.

Bevel gears (Figure 11.2c) have teeth that are manufactured on a cone-shape, in contrast to the cylinder-shape of spur and helical gears. We use bevel gears to transmit power and rotary motion between intersecting shafts. Figure 11.2c shows a bevel gear with straight teeth. Additionally, spiral bevel gears have teeth that are cut along a circular arc and engage more gradually than straight teeth bevel gears.

Finally, Figure 11.2d shows a worm gear combination. The *worm* resembles a screw and engages with the *worm gear* or *worm wheel*, which resembles a spur gear. The rotation direction of the worm gear depends on the teeth orientation (worms can be right- or left-handed, similar to screws) and the rotation direction of the worm. A worm essentially is a helical gear, but its helix angle is usually close to 90° and its body is long in the axial direction, rendering it screw-like. Note that in a worm and worm-gear combination, the worm can always drive the worm gear, but the reverse is not always true. The force component circumferential to the worm may not be sufficiently large to overcome the friction force, which particularly occurs for small lead angles.

11.1.2 Terminology

Figure 11.3 shows a section of a spur gear, identifying the different elements of a gear.[1] We refer to the top part of a gear tooth as the *top land* and the area in between gear teeth as the *bottom land*. The width of the gear tooth is the *face width*. The addendum a is the upper part of the gear tooth often called the *face*, whereas the dedendum b is the bottom part of the gear tooth, typically called the *flank*. The *pitch circle* is an imaginary circle that divides the gear tooth in face and flank. As such, the addendum is the distance between the top land and the pitch circle, and the dedendum is the distance between the bottom land and the pitch circle. All theoretical gear calculations are based on the pitch circle (and correspondingly the *pitch circle diameter d*). The pitch circles of meshing gears are tangent to each other. The *circular pitch p* is the distance, measured along the pitch circle, between any point on a gear tooth, and the corresponding point on the adjacent gear tooth. We divide the circular pitch in the *tooth thickness*, which is the solid portion of the circular pitch, and the *width of space*, which represents the empty portion of the circular pitch. Furthermore, we define the *addendum and dedendum circles* as the circles that are tangent to the top land and the bottom land of the gear tooth. The *clearance* is the spacing between the bottom land of one gear tooth and the top land of the mating gear tooth. As such, the *clearance circle* is tangent to the addendum circle of the mating gear.

We define the *module m* of a gear as the ratio of the pitch circle diameter d and the number of teeth N, typically in mm. Thus,

$$m = \frac{d}{N}. \tag{11.1}$$

Since we know that the circular pitch p is equal to the ratio of the circumference of the pitch circle and the number of teeth, i.e. $p = \pi d/N$, we write that

$$m = \frac{p}{\pi}. \tag{11.2}$$

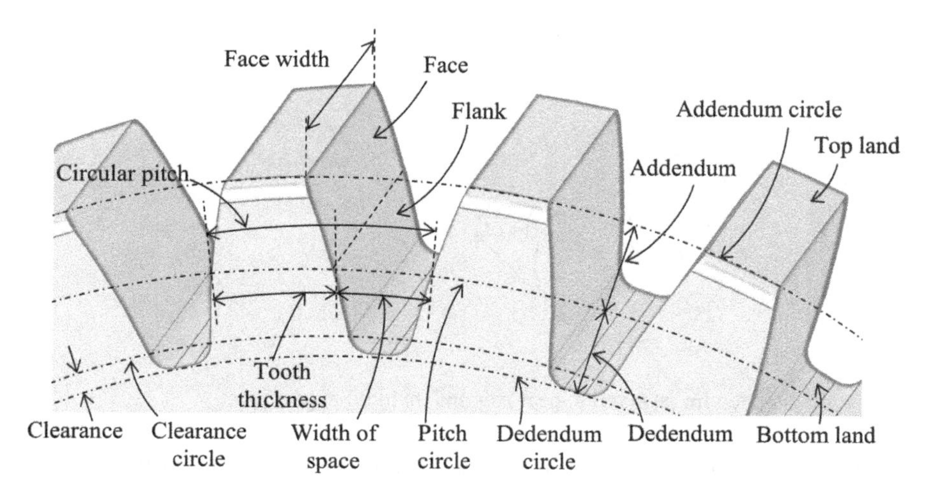

Figure 11.3 Gear terminology.

---

1 See, e.g. ANSI/AGMA 1012-G05:2005. *Gear Nomenclature, Definition of Terms with Symbols*. American National Standard Institute.

The ISO 54:1996 standard,[2] lists preferred values of modules in mm/*tooth*: 1, 1.25, 1.5, 2, 2.5, 3, 4, 5, 6, 8, 10, 12, 16, 20, 25, 32, 40, 50.

In the United States, we also sometimes use the *diametral pitch* P, which is the inverse of the module, i.e. the ratio of the number of teeth N and the pitch circle diameter d, typically expressed in number of teeth per inch. Finally, *backlash* is the amount by which the width of space exceeds the tooth thickness, measured along the pitch circle. Backlash not only accommodates manufacturing tolerances and imperfections but also reduces the positional accuracy of the meshing gears. It is particularly important when gears reverse direction because the gear tooth will traverse the backlash before engaging with the mating gear tooth, immediately after reversing direction.

Figure 11.4 shows a schematic representation of two meshing spur gears, with pitch circle radius r_1 and r_2, and rotating with angular velocity ω_1 and ω_2, respectively. Note that we refer to the smallest gear as the *pinion*. The pitch circles of the mating gears are tangent to each other. Figure 11.4a illustrates external spur gears, which cause the gears to rotate in opposite direction. In contrast, Figure 11.4b illustrates a combination of internal spur gears, which results in both gears rotating in the same direction.

The *line of centers* connects the center points of both gears, o_1 and o_2. The pitch circles of the meshing gears contact in point P. In P, the pitch circles of the meshing gears must roll without sliding. To satisfy that condition, the tangential velocity in P, V_P, must be identical for both gears. Thus,

$$V_P = |\omega_1 \times r_1| = |\omega_2 \times r_2|. \tag{11.3}$$

This reduces to

$$\left|\frac{\omega_1}{\omega_2}\right| = \frac{r_2}{r_1}. \tag{11.4}$$

Note that P is always located on the line of centers, too.

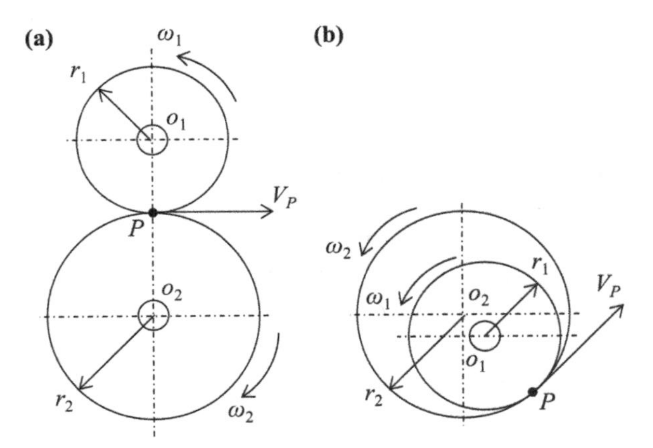

Figure 11.4 Gear rotation direction for (a) external gear pair and (b) internal gear pair.

2 ISO 54:1996. *Cylindrical Gears for General Engineering and for Heavy Engineering – Modules.* International Organization for Standardization.

11.2 Conjugate Gear Tooth Action

Gear design almost always requires a constant transmission or angular velocity ratio, i.e. $|\omega_1/\omega_2| = constant$, as this ensures "smooth" gear meshing. This condition is often referred to as *conjugate gear tooth action* or *conjugate action* for short. In theory, it is possible to select an arbitrary tooth profile and then calculate a mating tooth profile that results in conjugate gear tooth action, for instance by using the kinematic methods documented by Reuleaux.[3] However, one of the solutions to this problem is the *involute*, which is almost universally used today as gear tooth profile. Hence, we only cover involute gear tooth profiles in this textbook.

Involute gear tooth profiles have benefits in addition to satisfying the conjugate gear tooth requirement; for instance, we can move the center–center distance of the gears without changing the constant angular velocity ratio. Furthermore, when using a *rack-and-pinion* configuration, i.e. a small spur gear (pinion) that meshes with a spur gear with infinite pitch circle diameter (rack), the involute gear tooth shape of the rack is a straight line, which simplifies manufacturing processes.

Figure 11.5 shows two single gear tooth profiles in contact. Gear tooth 1 rotates with angular velocity ω_1 about rotation center o_1, whereas gear tooth 2 rotates with angular velocity ω_2 about rotation center o_2. The distances between the rotation centers and the contact point are l_1 and l_2, respectively. We also indicate the line of centers, which connects both rotation centers o_1 and o_2, and the common normal and common tangent in the contact point between both gear teeth. The intersection between the line of centers and the common normal is the pitch point P. When the angular velocity ratio is constant, the pitch point P remains fixed, i.e. all common normal lines for each instantaneous point of contact between both gear teeth pass through P.

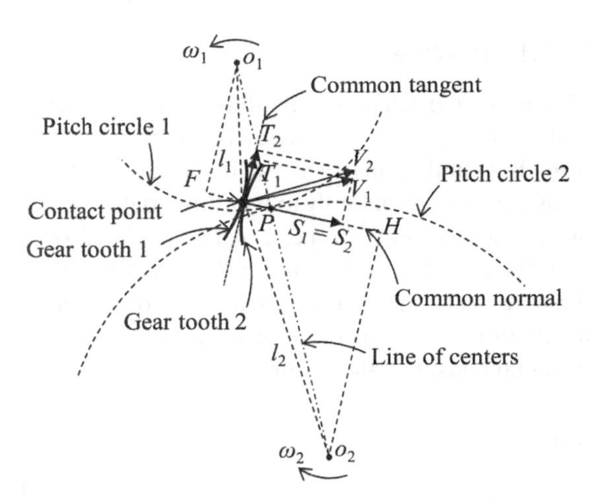

Figure 11.5 Two meshing gear teeth.

3 See, Reuleaux, F. (1963). *The Kinematics of Machinery: Outlines of a Theory of Machines*, Dover Publications.

Here, the velocity vectors of both contact points, V_1 and V_2, are different from each other because the instantaneous contact point between the gear teeth does not coincide with the pitch circle diameter, where $V_1 = V_2$. Note that $V_1 \perp l_1$ and $V_2 \perp l_2$. However, the velocity vector components along the common normal must be identical to avoid that the teeth separate or press against each other, i.e. $S_1 = S_2$. Hence, the velocity vector components along the common tangent must be different from each other and causes sliding between the gear teeth, i.e. $T_1 \neq T_2$. We emphasize that the vector sums $V_1 = T_1 + S_1$ and $V_2 = T_2 + S_2$, respectively.

We consider the requirement that $S_1 = S_2$ and write that

$$\begin{cases} S_1 = \omega_1 \times |Fo_1| \\ S_2 = \omega_2 \times |Ho_2| \end{cases} \tag{11.5}$$

and, thus,

$$\left|\frac{\omega_1}{\omega_2}\right| = \frac{|Ho_2|}{|Fo_1|} = \text{constant.} \tag{11.6}$$

Triangles ΔO_1FP and ΔO_2HP are congruent. Thus,

$$\left|\frac{\omega_1}{\omega_2}\right| = \frac{|Ho_1|}{|Fo_1|} = \frac{|Po_2|}{|Po_1|} = \text{constant.} \tag{11.7}$$

Here, $|Po_1|$ is the radius of the pitch circle of gear 1 and $|Po_2|$ is the radius of the pitch circle of gear 2. Thus, the angular velocity ratio of two gears is inverse proportional to their radii to the pitch point P. Furthermore, to transmit motion with constant angular velocity ratio the pitch point P must remain fixed.

11.3 Kinematics

11.3.1 Involute

Figure 11.6a–d schematically illustrates the development of an involute. Consider a flexible rope that we tightly wind around a cylinder. We trace the endpoint of the rope while we tautly unwind it from the cylinder; this trace is an involute (sometimes also referred to as an *evolvent*). Figure 11.6a–d shows a sequence of tautly unwinding a tightly wound rope *amn* from a cylinder, showing the trace of the endpoint of the rope: a, b, c, d, e. We also show the location where the rope loses contact with the cylinder: m, m', m'', m'''. We refer to the circle from which we develop the involute as the *base circle*. The radius of curvature of the involute changes continuously; it is zero in a and maximum in e. For instance, in c the radius of curvature of the involute is equal to $|m'c|$ because c instantaneously rotates around m'.

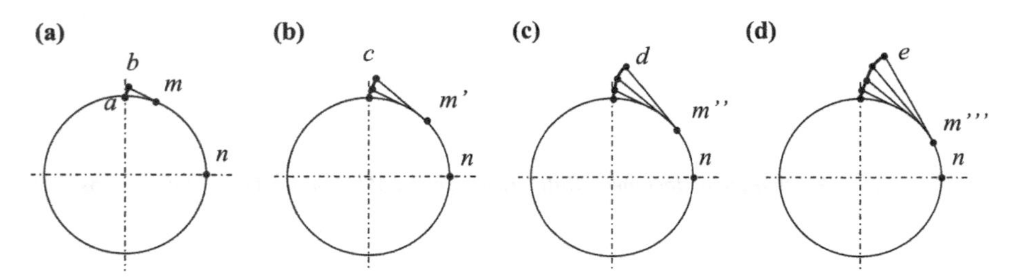

Figure 11.6 Schematic illustration of developing an involute by tracking the end point of tautly unwinding a rope from a cylinder.

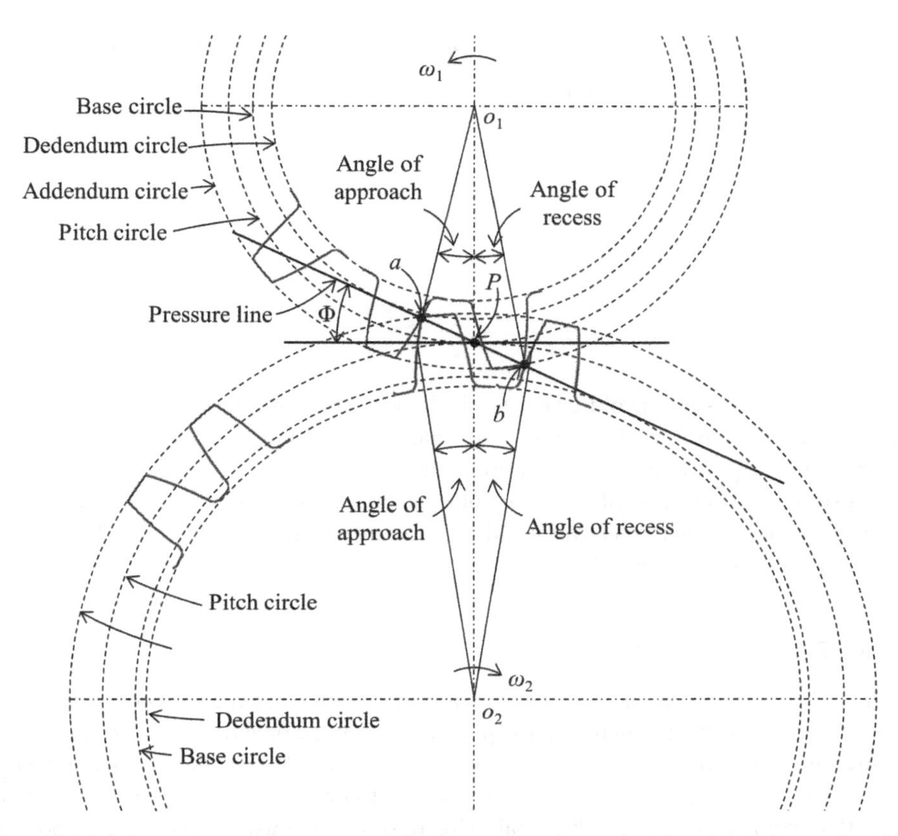

Figure 11.7 Gear tooth engagement and disengagement, between driver (pinion) gear 1 and driven gear 2.

Figure 11.7 illustrates how teeth of a meshing gear pair engage and disengage. The line $|ab|$ represents the *pressure line*, which is synonymous with the *generating line*, or the *line of action*. Point a marks the instance where the addendum circle of the driven gear crosses the pressure line, and the teeth first engage because the flank of the pinion tooth (driver gear) contacts the tip of the driven tooth. Correspondingly, point b marks the instance where the addendum circle of the pinion crosses the pressure line, and the teeth disengage because the tip of the pinion tooth loses contact with the flank of the driven tooth. The pressure line is tangent to the base circle of both gears, from which the involute gear profiles are developed. The contact point between the meshing gear teeth moves from a to b along the pressure line, and passes through the pitch point P, where the pitch circles of both gears are tangent to each other.

We refer to the angle between the pressure line and a line orthogonal to the line of centers $|o_1o_2|$ as the *pressure angle* Φ. Φ is typically either 20° or 25°. The pressure angle Φ also relates the pitch circle radius r_i to the base circle radius $r_{b,i}$ as

$$r_{b,i} = r_i \cos \Phi, \tag{11.8}$$

with $i = 1, 2$, referring to the gear number.

We also define the *angle of approach* ϕ_a and the *angle of recess* ϕ_r indicated in Figure 11.7. The sum of the angle of approach and the angle of recess is the *angle of action* $\phi_t = \phi_a + \phi_r$.

We observe that the involute gear tooth shape satisfies the conjugate gear tooth action requirement because at any contact point between the meshing teeth, the common normal runs through the fixed pitch point P. Furthermore, we observe that the contact point, as it moves between a and b

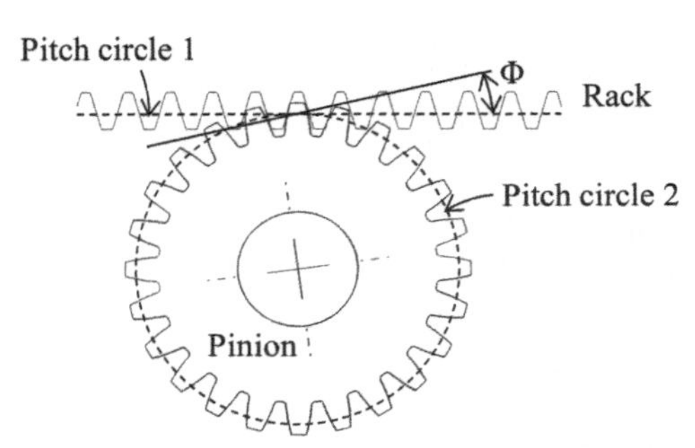

Figure 11.8 Rack and pinion configuration.

along the pressure line, traces the involute gear tooth shapes of both meshing gear teeth simultaneously. Note that only the portion of the gear tooth that makes contact with another gear tooth must satisfy the conjugate gear tooth action requirement.

Figure 11.8 illustrates the limit case of a rack and pinion configuration. A *rack* is a spur gear with a pitch circle diameter that approaches infinity and, thus, an infinite number of teeth. The involute tooth profile of a rack follows a straight line.

11.3.2 Contact Ratio

Referring to Figure 11.7, we observe that a pair of meshing gear teeth makes contact between a and b along the pressure line. This *zone of action* corresponds to the angle of action ϕ_t. If the arc corresponding to the angle of action is identical to the circular pitch p, i.e. $\phi_t = p$, then exactly one tooth pair is in contact at any given time. Thus, when the tip of one tooth makes contact with the flank of the meshing tooth in point a, another meshing pair of teeth simultaneously loses contact in point b. We define the contact ratio m_c as

$$m_c = \frac{\phi_t}{p}. \tag{11.9}$$

Thus, if $\phi_t = p$, then $m_c = 1$. Physically, this means that on average, one tooth pair is in contact at any time. However, typically, we design gear pairs such that the contact ratio is at least $m_c = 1.2$, because assembly and alignment inaccuracies could otherwise result in impact between the gear teeth, which increases noise and reduces the longevity of the gears. When the contact ratio exceeds one, two gear tooth pairs are simultaneously in contact for a brief period of time. When a gear tooth pair first makes contact in point a, the gear tooth pair that is already in contact has not yet reached point b.

Example problem 11.1

A drivetrain comprises an 18-tooth pinion and a 44-tooth gear. The module of the gears $m = 1.5$ mm, the addendum is m and the dedendum is 1.25 m, respectively. The pressure angle $\Phi = 20°$.

(a) Calculate the circular pitch p, the center distance between both gears, and their respective base circle radii, $r_{b,1}$ and $r_{b,2}$.
(b) Calculate the addendum circle diameter and the dedendum circle diameter.
(c) When mounting the gears, the center distance is incorrectly increased by 2 mm. Calculate the modified pressure angle Φ' and pitch circle diameters d_1' and d_2'.

Solution

(a) We first calculate the circular pitch p using Eq. (11.2), i.e.

$$p = \pi \times m = \pi \times 1.5 = 4.71 \text{ mm}.$$

We calculate the pitch circle diameters of gear and pinion using Eq. (11.1), i.e.

$$d_{pinion} = m \times N_p = 1.5 \times 18 = 27 \text{ mm, and}$$

$$d_{gear} = m \times N_g = 1.5 \times 44 = 66 \text{ mm}.$$

Then, we calculate the center distance between the pinion and gear as the sum of the pitch circle radii, i.e. $d_{pinion}/2 + d_{gear}/2$, because the pitch circles are tangential to each other. Thus, the center distance is $27/2 + 66/2 = 46.5$ mm.

We calculate the radii of the base circles using Eq. (11.8), i.e.

$$r_{b,pinion} = (d_{pinion}/2)\cos 20^\circ = 12.69 \text{ mm and,}$$

$$r_{b,gear} = (d_{gear}/2)\cos 20^\circ = 31 \text{ mm}.$$

(b) $d_{addendum} = d + 2 \times m$. Thus,

$$d_{addendum,pinion} = d_{pinion} + 2 \times m = 27 + 2 \times 1.5 = 30 \text{ mm}.$$

$$d_{addendum,gear} = d_{gear} + 2 \times m = 66 + 2 \times 1.5 = 69 \text{ mm}.$$

Similarly, $d_{dedendum} = d - 2 \times 1.25m$. Thus,

$$d_{dedendum,pinion} = d_{pinion} - 2 \times 1.25m = 27 - 2 \times 1.25 \times 1.5 = 23.25 \text{ mm}.$$

$$d_{dedendum,gear} = d_{gear} - 2 \times 1.25m = 66 - 2 \times 1.25 \times 1.5 = 62.25 \text{ mm}.$$

(c) The center distance is now modified to $d'_{pinion}/2 + d'_{gear}/2 = 48.5$ mm.

Additionally, the transmission ratio remains unchanged and, thus, according to Eq. (11.4), $|\omega_1/\omega_2| = d'_{gear}/d'_{pinion}$, and since $d = m \times N$, we also know that $d'_{gear}/d'_{pinion} = N_{gear}/N_{pinion}$.

Solving both equations simultaneously gives

$$d'_{pinion} = 28.16 \text{ mm},$$

$$d'_{gear} = 68.84 \text{ mm}.$$

The modified pressure angle $\Phi' = \arccos\left(\frac{r_{b,pinion}}{d'_{pinion}/2}\right) = \arccos\left(\frac{12.69}{14.08}\right) = 25.67^\circ$.

11.3.3 Gear Tooth System

Standards exist that prescribe the relationship between addendum, dedendum, tooth thickness, and pressure angle, to ensure interchangeability between gears of the same pressure angle and pitch, but different tooth number. For spur gears, we show (Table 11.1)[4,5]

11.3.4 Interference

We define *interference* as the contact between portions of meshing gear tooth profiles that are not conjugate. For instance, we construct the involute gear tooth profile from the base circle. Thus, the

---

4 Shigley, J.E. and Uicker, J.J. (1980). *Theory of Machines and Mechanisms*. McGraw-Hill.

5 ANSI/AGMA 1003-G93. (1993). *Tooth Proportions for Fine-Pitch Spur and Helical Gearing*. American National Standards Institute, and American Gear Manufacturers Association.

Table 11.1 Gear tooth system.

Pressure angle Φ (°)	Addendum *a* (mm)	Dedendum *b* (mm)
20	m	$1.25m$ or $1.35m$
22.5	m	$1.25m$ or $1.35m$
25	m	$1.25m$ or $1.35m$

Source: ANSI/AGMA 1003-G93. (1993). *Tooth Proportions for Fine-Pitch Spur and Helical Gearing.* American National Standards Institute, and American Gear Manufacturers Association.

portion of the gear tooth underneath the base circle is not an involute and does not satisfy the conjugate gear tooth action requirement. Interference occurs when contact exists within that portion of the gear tooth profile. As a result, the gear tip will attempt to cut into the flank of the nonconjugate portion of the gear tooth, which we refer to as *undercutting*. Undercutting can substantially reduce the strength of the gear teeth, thus limiting the power they can transmit.

Figure 11.9 illustrates the concept of interference. Figure 11.9a schematically shows two meshing involute gear teeth. We indicate the pressure line, which is tangent to the base circles of both gears, and the pressure angle Φ. The gear teeth engage in point a and disengage in point b. Furthermore, we identify the points t_1 and t_2, where the pressure line contacts the base circles. Because a and b fall within t_1 and t_2, all contact between the meshing gear teeth occurs on the involute profile and no undercutting occurs. In contrast, Figure 11.9b schematically shows two involute gear teeth meshing where the base circle has been displaced toward the pitch circle, i.e. the pressure angle

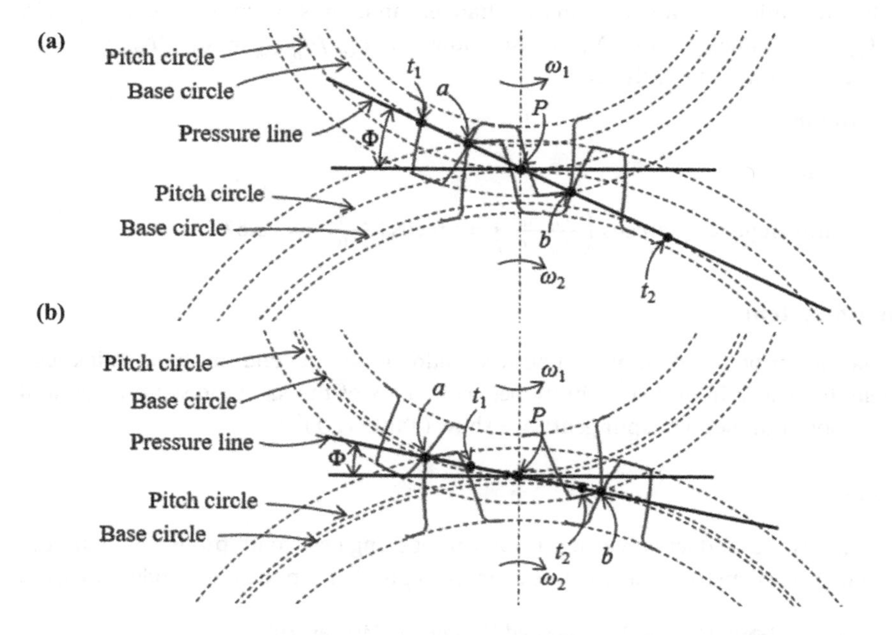

Figure 11.9 Schematic of interacting gear teeth illustrating (a) no interference when all contact occurs on the involute tooth profile and (b) interference when contact occurs outside the involute profile.

Φ is much smaller than in Figure 11.9a. Thus, the portions of the gear teeth below the base circle are not part of the involute. The gear teeth engage in point a and disengage in point b. However, because a and b fall outside the points t_1 and t_2, where the pressure line contacts the base circles, not all contact between the meshing gear teeth occurs on the involute portion of the tooth profiles and, thus, undercutting occurs.

We can avoid interference by (1) increasing the pressure angle Φ of the meshing gears, or increasing the number of teeth on the gears. Thus, a minimum number of gear teeth of the pinion N_p exists to avoid interference. The minimum number of teeth of the pinion N_p is given as

$$N_p = \frac{2}{(1+2i)\sin^2\Phi}\left(i + \sqrt{i^2 + (1+2i)\sin^2\Phi}\right). \tag{11.10}$$

Here, i is the gear ratio, i.e. the ratio of the number of teeth of the (driven) gear N_g and the pinion, $i = N_g/N_p$.

Example problem 11.2

Calculate the minimum number of teeth of the pinion, required to avoid undercutting, for a pinion/gear combination with a gear ratio of 3. The pressure angle of the gear combination is $\Phi = 20°$.

Solution

We start from Eq. (11.10), i.e.

$$N_p = \frac{2}{(1+2i)\sin^2\Phi}\left(i + \sqrt{i^2 + (1+2i)\sin^2\Phi}\right).$$

Using $i = 3$ and $\Phi = 20°$, we find

$$N_p = \frac{2}{(1+2\times 3)\sin^2(20°)}\left(3 + \sqrt{3^2 + (1+2\times 3)\sin^2(20°)}\right) = 14.98.$$

Thus, the minimum number of teeth of the pinion $N_p = 15$.

11.4 Gear Force Analysis

Gears transmit rotary motion and torque between two shafts. Therefore, forces exist between the meshing gear teeth. For instance, in a pinion and gear transmission, the pinion is the driving gear and, thus, the direction of its torque moment and its rotation is the same. In contrast, the driven gear resists motion because of the inertia of the shaft to which it is attached and, thus, the direction of its torque moment and its rotation are opposite. Much like the motor-machine torque transmission of Chapter 7.

Figure 11.10a schematically shows a pinion that rotates with angular velocity ω_1 and drives a gear with angular velocity ω_2. We also indicate their centers o_1 and o_2, and their pitch circle radii r_1 and r_2. Figure 11.10b shows a free-body diagram of both pinion and gear. Note that the forces between the gears act along the pressure line, defined by the pressure angle Φ.

The pinion is subject to a torque M_{t1} from the shaft to which it is connected, which rotates the pinion with ω_1 in the same direction as the torque moment. The pinion then drives the gear, i.e. it transmits torque and motion to the gear. As a result, it is subject to a force from gear 2 (driven gear) on gear 1 (pinion, driving gear), which we refer to as $F_{2,1}$. The shaft to which the pinion connects, in turn, exerts a reaction force on the pinion, which we refer to as $F_{1,1}$.

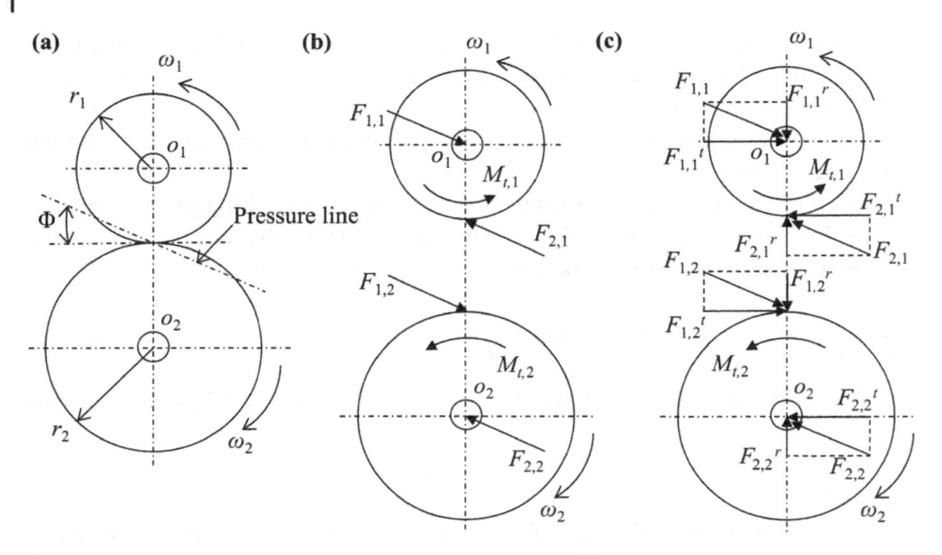

Figure 11.10 Schematic of forces acting on meshing spur gears.

The (driven) gear is subject to a torque M_{t2} from the shaft to which it is connected, which attempts to resist the motion of the gear because of inertia. Thus, the gear rotates with ω_2 in the opposite direction of the torque moment M_{t2}. Gear 2 is driven by gear 1 and, thus, gear 2 is subject to a force from gear 1 (driving gear), which we refer to as $F_{1,2}$. The shaft to which the gear connects, in turn, exerts a reaction force on the gear, which we refer to as $F_{2,2}$.

All force components act along the pressure line, which is oriented under an angle Φ with respect to the horizontal (or to a line orthogonal to the line of centers). However, we can project the force vectors in a component that is radial to the pinion and gear (which we indicate with superscript r), and a component that is tangential to the pinion and gear (which we indicate with superscript t). As such,

$$F^r_{1,1} = F_{1,1} \sin \Phi \tag{11.11}$$

$$F^t_{1,1} = F_{1,1} \cos \Phi \tag{11.12}$$

and similarly for $F^r_{2,1}$, $F^t_{2,1}$, $F^r_{1,2}$, $F^t_{1,2}$, $F^r_{2,2}$, and $F^t_{2,2}$.

The tangential force components are the only ones that are useful to transmit torque and motion, as the radial force components go through the center of the shafts and do not contribute to the torque moment. Thus, $F^t_{2,1} = F^t_{1,2}$ is the force that is transmitted between gear 1 (pinion) and gear 2 (driven gear). These forces relate to the torque moments as

$$M_{t1} = F^t_{2,1} \times r_1 \tag{11.13}$$

$$M_{t2} = F^t_{1,2} \times r_2 \tag{11.14}$$

with r_1 and r_2 the pitch circle radii of gear 1 and gear 2, respectively.

We calculate the power that the gears transmit as

$$P = M_{t1} \times \omega_1 = F^t_{2,1} \times r_1\omega_1, \tag{11.15}$$

$$P = M_{t2} \times \omega_2 = F^t_{1,2} \times r_2\omega_2. \tag{11.16}$$

Assuming an ideal transmission, in which the power input is fully transmitted to the output, we write that $M_{t1} \times \omega_1 = M_{t2} \times \omega_2$, or, alternatively,

$$\frac{M_{t1}}{M_{t2}} = \frac{\omega_2}{\omega_1}. \tag{11.17}$$

In reality, we must account for the efficiency of the gear transmission. For spur gears, the transmission efficiency η is approximately 98.0–99.5%.

Example problem 11.3

Consider a spur gear transmission that is 99% efficient. The pinion has $N_{pinion} = 20$ teeth and rotates at 100 RPM. The gear has $N_{gear} = 40$ teeth. The module of the gears $m = 1.25$ mm/*tooth*. A shaft connected to a motor, applies a torque moment of $M_{t,pinion} = 50$ Nm to the pinion.

(a) Calculate the torque moment $M_{t,gear}$ that is transmitted to the gear shaft.
(b) Calculate the power P that is transmitted to the gear shaft.
(c) Calculate the magnitude of the tangential force component between the meshing gears that transmits the torque moment.

Solution

(a) Combining Eqs. (11.1) and (11.4), we find that $\omega_{pinion}/\omega_{gear} = N_{gear}/N_{pinion}$.
We recognize the torque ratio of Eq. (11.17), i.e. $M_{t,pinion}/M_{t,gear} = \omega_{gear}/\omega_{pinion}$.
Thus, combining both equations, we find that $M_{t,pinion}/M_{t,gear} = N_{pinion}/N_{gear}$.
The torque moment that is transmitted to the gear shaft

$$M_{t,gear} = M_{t,pinion} \times N_{gear}/N_{pinion}.$$

Accounting for the transmission efficiency η, we modify the previous equation to $M_{t,gear} = \eta \times M_{t,pinion} \times N_{gear}/N_{pinion}$. Thus,

$$M_{t,gear} = 0.99 \times 50 \times 40/20 = 99\,\text{Nm}.$$

(b) We use Eq. (11.15) to calculate the input power

$$P = M_{t,pinion} \times \omega_{pinion} = 50 \times 2\pi \times 100/60 = 524\ \text{W}.$$

Accounting for the efficiency of the spur gear transmission, the power transmitted to the gear is $\eta \times P = 0.99 \times 524 = 518$ W.

We also find this answer using Eq. (11.16), i.e.

$$P = M_{t,gear} \times \omega_{gear} = M_{t2} \times N_{pinion}/N_{gear} \times \omega_{pinion}.$$ Hence,

$$P = 99 \times 20/40 \times 2\pi \times 100/60 = 518\ \text{W}.$$

(c) We know from Eq. (11.13) that $M_{t,pinion} = F^t_{gear,pinion} \times r_{pinion}$. Thus,

$$F^t_{gear,pinion} = M_{t,pinion}/r_{pinion}.$$

Equation (11.1) shows that $m = d_{pinion}/N_{pinion}$, or

$$F^t_{gear,pinion} = 2M_{t,pinion}/(mN_{pinion}) = 2 \times 50/(1.25 \cdot 10^{-3} \times 20) = 4000\ \text{N}.$$

11.5 Gear Manufacturing

Several sources exist that give a complete overview of gear manufacturing methods.[6] However, in this textbook, we only provide a brief summary. Gears can be manufactured by means of two categories of techniques: tooth forming or tooth machining. Each category comprises a large number of techniques (or combination thereof). The choice of manufacturing techniques further depends on the type of material of the gears.

11.5.1 Forming

In all tooth-forming operations, the gear teeth are formed all at once, by means of a mold or die, in which the tooth shape has been machined. Thus, the accuracy of tooth-forming manufacturing methods depends entirely on the accuracy and quality of the mold or die. In general, the precision is smaller than what we can obtain by means of machining. Most of the tooth-forming methods also have high tooling cost, making them suitable only for mass production volume. Processes include casting, molding, extrusion, cold drawing, forging, stamping, to name a few. Injection molding, for instance, is particularly suited for high-volume, low-precision, thermoplastic gears (e.g. nylon, polycarbonate), for applications with light-to-moderate load.

11.5.2 Machining

The majority of gears that transmit power in, e.g. machinery, are manufactured using machining processes from cast, forged, or hot-rolled gear blanks. These machining processes include milling, shaping, and hobbing. Sometimes, finishing operations are used for high-precision gears because imperfections cause additional forces that can be detrimental in high-speed and/or high-load applications. These finishing operations include burnishing and shaving for gears that have not been heat-treated, and grinding and lapping for hardened, heat-treated gear teeth.

Milling: Milling processes include disc- or end-mill cutters that fit in the gear tooth space. The disc- or end-mill cutter is fed through the gear blank to machine the tooth shape. Since the tooth space is different for each tooth size, one would need a different cutter for each different gear tooth size. However, a set of 8–10 standard sizes has shown to produce accurate gear teeth.

Gear shaping: A gear-shaped cutting tool reciprocates axially to cut a gear tooth until the pitch circles of the cutting tool and the gear blank are tangent to each other. Once the pitch circles are tangent, both the gear blank and the cutter rotate slightly to accommodate cutting the adjacent teeth. The gear is complete when the gear blank completes a full rotation. This is a shape-generating process because the gear-shaped cutting tool cuts itself in mesh with the stock material. One may use pinion- or rack-shaping. In the latter case, a "rack" of hardened steel with trapezium-shaped involute tooth shape cuts axially back-and-forth through the rotating gear blank, thus machining the involute shape on the gear blank. Because racks have straight gear teeth, they are easier to manufacture with high precision than pinions and, thus, can also be used to cut high-precision gears.

Hobbing: Hobbing involves using a "hob," which is a cutting tool shaped like a worm. Thus, the hob also rotates about an axis that is orthogonal to the axis of the gear blank. The hob teeth are trapezium-shaped similar to a rack. However, it must be rotated through the lead angle of the worm to cut straight spur gear teeth.

---

6 For an in-depth discussion of gear manufacturing processes see e.g., Groover, M.P. (2020). *Fundamentals of Modern Manufacturing: Materials, Processes, and Systems*, 7e. Wiley.

11.6 Key Takeaways

1. Two gears can operate together if their tooth profile satisfies the conjugate gear tooth action requirement, which requires a constant angular velocity ratio of both gears, and also defines that the common normal of any instantaneous contact point between the meshing gear teeth projects through the velocity pole. Modern gears almost all use the involute gear profile. As a result, two gears must have the same module to operate together.
2. When contact between gear teeth occurs that falls outside the tooth profile that satisfies the conjugate gear tooth action requirement, interference occurs, which may lead to undercutting. For that reason, a gear (or pinion) requires a minimum number of teeth to avoid undercutting.
3. Gear pairs or gear trains transmit power and rotary motion. Only the force components tangential to the gear contribute to transmitting torque and, thus, power.

11.7 Problems

11.1 A drivetrain comprises a 20-tooth pinion and a 40-tooth gear. The module of the gears $m = 2.5$ mm, the addendum is m and the dedendum is $1.25m$, respectively. The pressure angle $\Phi = 20°$.

(a) Calculate the circular pitch p, the center distance between both gears, and their respective base circle radii, $r_{b,1}$ and $r_{b,2}$.

(b) Calculate the addendum circle diameter, and the dedendum circle diameter.

(c) When mounting the gears, the center distance is incorrectly increased by 2 mm. Calculate the modified pressure angle Φ' and the pitch circle diameters d_1' and d_2'.

11.2 A drivetrain comprises a 25-tooth pinion and a 75-tooth gear. The module of the gears $m = 1.5$ mm, the addendum is m, and the dedendum is $1.25m$, respectively. The pressure angle $\Phi = 20°$.

(a) Calculate the circular pitch p, the center distance between both gears, and their respective base circle radii, $r_{b,1}$ and $r_{b,2}$.

(b) Calculate the addendum circle diameter, and the dedendum circle diameter.

(c) Calculate how much one must increase the center distance between the gears to obtain a modified pressure angle $\Phi' = 25°$. Also calculate the corresponding modified pitch circle diameters d_1' and d_2'.

11.3 Calculate the minimum number of teeth of the pinion required to avoid undercutting, for a pinion/gear combination with a gear ratio of 5. The pressure angle of the gear combination $\Phi = 25°$. Also, calculate how the minimum number of teeth changes if the pressure angle decreases to $\Phi = 20°$.

11.4 Consider a spur gear transmission that is 100% efficient. The pinion has $N_{pinion} = 24$ teeth and rotates at 300 RPM. The gear has $N_{gear} = 44$ teeth. The module of the gears $m = 1.50$ mm. A shaft connected to a motor applies a torque moment $M_{t,pinion} = 20$ Nm to the pinion.

(a) Calculate the torque moment that is transmitted to the gear shaft, $M_{t,gear}$.

(b) Calculate the power that is transmitted to the gear shaft, P.

(c) Calculate the magnitude of the tangential fore component between the meshing gears that transmits the torque moment.

11.5 Consider a spur gear transmission that is 98% efficient. The pinion has $N_{pinion} = 35$ teeth and rotates at 200 RPM. The gear has $N_{gear} = 65$ teeth. The module of the gears $m = 2.5$ mm. The pressure angle $\Phi = 20°$. A shaft connected to a motor applies a torque moment $M_{t,pinion} = 100$ Nm to the pinion.

(a) Calculate the circular pitch p, the center distance between both gears, and their respective base circle radii, $r_{b,1}$ and $r_{b,2}$.

(b) Calculate the torque moment that is transmitted to the gear shaft, $M_{t,gear}$.

(c) Calculate the power that is transmitted to the gear shaft, P.

(d) Calculate the magnitude of the tangential force component between the meshing gears that transmits the torque moment.

A

Area Moment of Inertia

A.1 Introduction

The area moment of inertia (second moment of inertia) quantifies the resistance of a shape against bending and deflection. It is not to be confused with the moment of inertia, which quantifies the rotational inertia of a rigid body.

Closed-form expressions for the area moment of inertia exist for simple geometries only. In many instances, we have to rely on numerical methods to calculate the area moment of inertia. In this appendix, we review the principal equations that define the area moment of inertia.

A.2 Terminology

We define the moment of inertia I_{oo} of a solid body with volume V and mass m that rotates about an axis o as

$$I_{oo} = \int_V r^2 \, dm, \tag{A.1}$$

where r is the orthogonal distance between the center of gravity of the solid body and the rotation axis o. The moment of inertia is expressed in units kgm^2. We calculate the mass m as

$$dm = \rho \, dV, \tag{A.2}$$

where ρ is the density of the solid body.

The area moment of inertia I_{xx} of a shape around an axis x is

$$I_{xx} = \int_A y^2 \, dA, \tag{A.3}$$

where y is the orthogonal distance between the infinitesimally small area element dA and the rotation axis x, as illustrated in Figure A.1. A is the area of the shape. The area moment of inertia is expressed in units m^4.

Equivalently, we express the area moment of inertia I_{yy} of a shape around an axis y as

$$I_{yy} = \int_A x^2 \, dA, \tag{A.4}$$

where x is the orthogonal distance between the infinitesimally small area element dA and the rotation axis y, as illustrated in Figure A.1.

Design of Mechanical Elements: A Concise Introduction to Mechanical Design Considerations and Calculations, First Edition. Bart Raeymaekers.

Companion website: www.wiley.com/go/raeymaekers/designofmechanicalelements

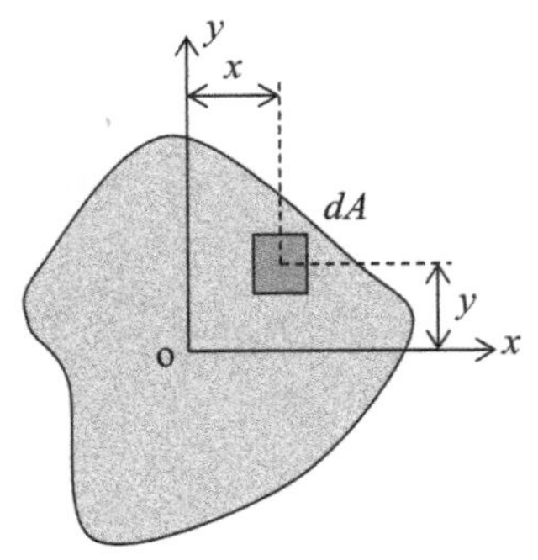

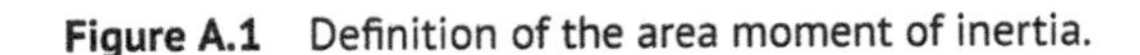
Figure A.1 Definition of the area moment of inertia.

We define the radii of gyration, i_x and i_y, as

$$I_{xx} = i_x^2 A, \tag{A.5}$$

$$I_{yy} = i_y^2 A. \tag{A.6}$$

The polar area moment of inertia I_o characterizes a shape's ability to resist torsion, and is defined around a point o as

$$I_o = \int_A r^2 \, dA, \tag{A.7}$$

where r is the Euclidean distance between the infinitesimally small area element dA and point o. Since $r = \sqrt{x^2 + y^2}$, we obtain

$$I_o = \int_A \left(x^2 + y^2\right) dA = I_{xx} + I_{yy}. \tag{A.8}$$

Thus, the polar area moment of inertia is the sum of the area moments of inertia about two orthogonal axes through point o located in the plane of the shape.

We define the product of inertia as

$$I_{xy} = \int_A xy \, dA, \tag{A.9}$$

where the subscripts denote the axes about which we calculate the product of inertia.

A.3 Parallel Axis Theorem

The area moments of inertia about the centroidal axes of a shape are of particular interest. We calculate the centroidal axes as

$$\bar{x}A = \int_A x \, dA, \tag{A.10}$$

$$\bar{y}A = \int_A y \, dA, \tag{A.11}$$

where $\bar{x}$ and $\bar{y}$ are the x and y coordinates of the centroidal axes. The *parallel axis theorem* defines the relationship between the area moment of inertia about the centroidal axes, and any axis parallel to the centroidal axes. If we denote the area moment of inertia about the centroidal axes as $\bar{I_{xx}}$ and $\bar{I_{yy}}$, then the area moment of inertia about an axis x parallel to the centroidal axis is

$$I_{xx} = \bar{I_{xx}} + d_y^2 A, \tag{A.12}$$

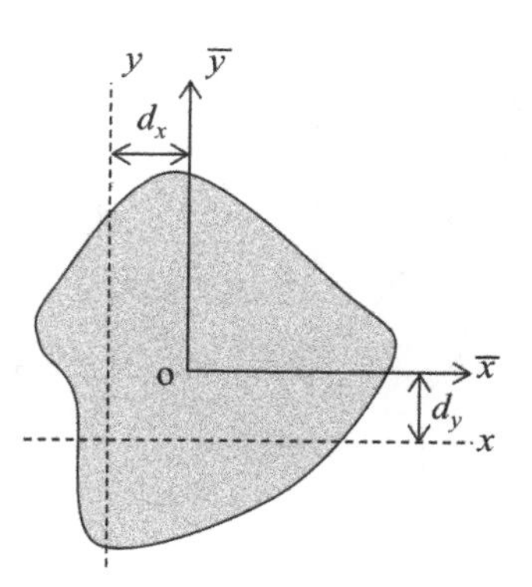

Figure A.2 Definition of the parallel axis theorem.

where d is the orthogonal distance between the centroidal and the parallel axis.

We verify Eq. (A.12) by calculating the area moment of inertia about an axis x, which is parallel to the centroidal axis $\bar{x}$, as shown in Figure A.2.

$$I_{xx} = \int_A y^2 \, dA, \tag{A.13}$$

$$I_{xx} = \int_A (d_y + \bar{y})^2 \, dA, \tag{A.14}$$

$$I_{xx} = d_y^2 \int_A dA + 2d_y \int_A \bar{y} \, dA + \int_A \bar{y}^2 \, dA, \tag{A.15}$$

$$I_{xx} = d_y^2 A + 0 + \bar{I}_{xx}, \tag{A.16}$$

$$I_{xx} = \bar{I}_{xx} + d_y^2 A. \tag{A.17}$$

A.4 Rotation About the Origin

We rotate two orthogonal axes about the origin, and the area moments and products of inertia are functions of the rotation angle. We define I_{xx}, I_{yy}, and I_{xy} as the original area moments and products of inertia. When we rotate to the xy-coordinate system about the origin over an angle α (see Figure A.3), then we denote the corresponding area moments and products of inertia as $I_{\xi\xi}$, $I_{\eta\eta}$, and $I_{\xi\eta}$, respectively.

$$I_{\xi\xi} = I_{xx} \cos^2 \alpha + I_{yy} \sin^2 \alpha - 2I_{xy} \sin \alpha \cos \alpha, \tag{A.18}$$

$$I_{\eta\eta} = I_{xx} \sin^2 \alpha + I_{yy} \cos^2 \alpha - 2I_{xy} \sin \alpha \cos \alpha, \tag{A.19}$$

$$I_{\xi\eta} = \frac{1}{2} \left(I_{xx} - I_{yy} \right) \sin 2\alpha + I_{xy} \cos 2\alpha. \tag{A.20}$$

We determine the rotation angle α_0 for which $I_{\xi\xi}$ and $I_{\eta\eta}$ reach extreme values, when $\delta I_{\xi\xi}/\delta\alpha = 0$ and $\delta I_{\eta\eta}/\delta\alpha = 0$. We calculate that

$$\tan 2\alpha_0 = \frac{2I_{xy}}{I_{yy} - I_{xx}}. \tag{A.21}$$

Whenever $I_{\xi\xi}$ is maximum, $I_{\eta\eta}$ is minimum and vice versa. Also, when $\alpha = \alpha_0$, $I_{\xi\eta} = 0$.

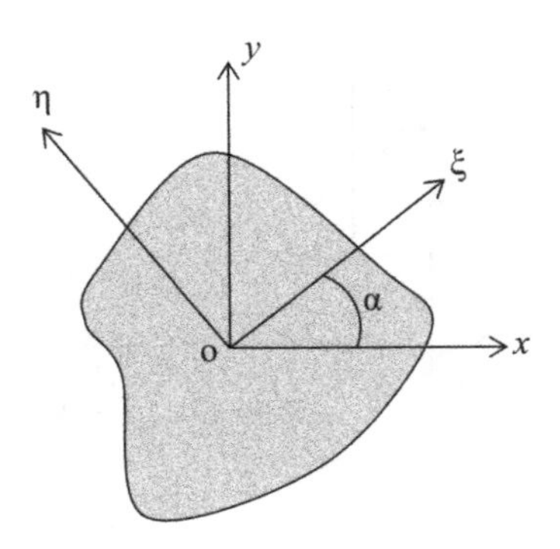

Figure A.3 Rotation of the area moment and product of inertia about the origin.

We define the orthogonal axes ξ and η about which the product of inertia vanishes and the moments assume maximum and minimum values (for a specific angle α_0) as the *principal axes of inertia*. If their origin coincides with the centroid of the shape, then we refer to them as *centroidal principal axes of inertia*. For a symmetric figure, the axis of symmetry coincides with a centroidal principal axis of inertia. We refer to the moments of inertia about the principal axes as the *principal moments of inertia*.

$$I_{max} = \frac{1}{2}\left(I_{xx} + I_{yy}\right) + \sqrt{\frac{1}{4}\left(I_x - I_y\right)^2 + I_{xy}^2}, \tag{A.22}$$

$$I_{min} = \frac{1}{2}\left(I_{xx} + I_{yy}\right) - \sqrt{\frac{1}{4}\left(I_x - I_y\right)^2 + I_{xy}^2}, \tag{A.23}$$

$$I_{max} + I_{min} = I_{xx} + I_{yy} = I_{\xi\xi} + I_{\eta\eta}. \tag{A.24}$$

Equation (A.24) shows that the sum of area moments of inertia of a shape about a system of orthogonal axes with a common origin is constant and independent of the orientation of the axes.

Example problem A.1
Rectangular shape (Figure A.4)

Solution
We first calculate the surface area and centroid of the rectangle:

Surface area $A = bh$.

$\bar{y} = \frac{1}{A}\int y \, dA$.

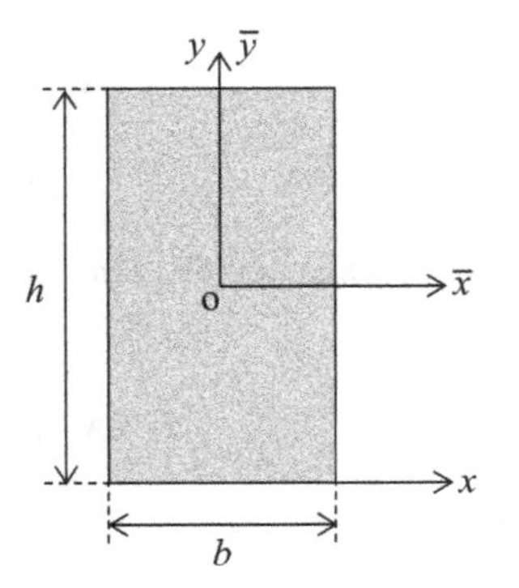

Figure A.4 Rectangle.

Since $dA = b\,dy$, we write that $\bar{y} = \frac{1}{A}\int by\,dy$.

$\bar{y} = \frac{b}{2A}y^2|_0^h = \frac{bh^2}{2bh} = \frac{h}{2}$.

Next, we calculate the area moment of inertia about the centroidal axes.

$\bar{I}_{xx} = \int_{-h/2}^{h/2} y^2\,dA,$

$\bar{I}_{xx} = 2b\int_0^{h/2} y^2\,dA = \frac{2b}{3}y^3|_0^{h/2},$

$\bar{I}_{xx} = \frac{bh^3}{12}$.

Finally, we calculate the area moment of inertia about axis x using the parallel axis theorem.

$I_{xx} = \frac{bh^3}{12} + \frac{bh}{2}\frac{h^2}{2}$.

$I_{xx} = \frac{bh^3}{3}$.

Example problem A.2
Triangular shape (Figure A.5)

Solution
We first calculate the surface area and centroid of the triangle:

Surface area $A = \int_0^b\int_0^y dx\,dy = \int_0^b y\,dx = \int_0^b\left(-\frac{h}{b}x + h\right)dx$.

$A = \frac{bh}{2}$.

$\bar{y}A = \int y\,dA = \int_0^b\int_0^y (y\,dx)(x\,dy) = \int_0^b \frac{y^2}{2}|_0^{-\frac{h}{b}x+h}dx$.

$\bar{y}A = \frac{1}{2}\int_0^b\left(-\frac{h}{b}x + h\right)^2 dx$.

$\bar{y}A = \frac{-b}{6h}\left(-\frac{h}{b}x + h\right)^3|_0^b$.

$\bar{y}A = \frac{1}{6}h^2b$.

$\bar{y} = \frac{1}{3}h$.

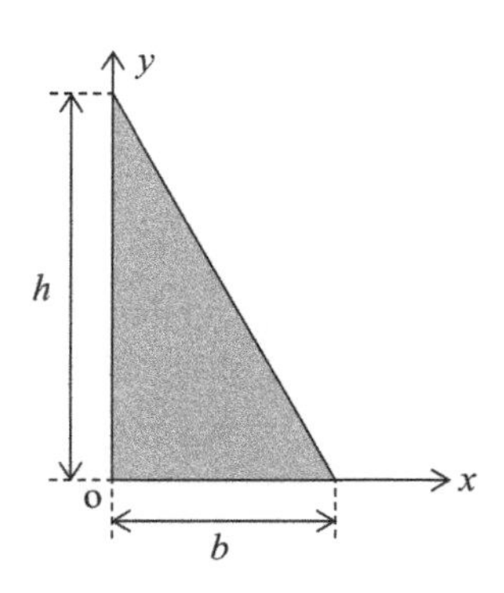

Figure A.5 Triangle.

Next, we calculate the area moment of inertia about the centroidal axes.

$\bar{I_{xx}} = \int_0^b \int_0^{-\frac{h}{b}x+h} y^2 \, dA.$

$\bar{I_{xx}} = \frac{1}{3}\int_0^b \left(-\frac{h}{b}x+h\right)^3 dx = -\frac{b}{12h}\left(-\frac{h}{b}x+h\right)^4 |_0^b.$

$\bar{I_{xx}} = \frac{1}{12}bh^3.$

Finally, we calculate the area moment of inertia about axis x using the parallel axis theorem.

$I_{xx} = \frac{bh^3}{12} + \frac{h^2}{9}A.$

$I_{xx} = \frac{bh^3}{36}.$

B

Internal Force Diagrams

B.1 Cantilever Beam with End Load

See Figure B.1

$$V_A = F, \tag{B.1}$$

$$M_A = FL, \tag{B.2}$$

$$M = F(x - L), \tag{B.3}$$

$$\delta = \frac{Fx^2}{6EI}(x - 3L), \tag{B.4}$$

$$\delta_{max} = -\frac{FL^3}{3EI}. \tag{B.5}$$

B.2 Cantilever Beam with Intermediate Load

See Figure B.2

$$V_A = F, \tag{B.6}$$

$$M_A = Fa, \tag{B.7}$$

$$M_{AB} = F(x - a), \tag{B.8}$$

$$M_{BC} = 0, \tag{B.9}$$

$$\delta_{AB} = \frac{Fx^2}{6EI}(x - 3a), \tag{B.10}$$

$$\delta_{BC} = -\frac{Fa^2}{6EI}(a - 3x), \tag{B.11}$$

$$\delta_{max} = \frac{Fa^2}{6EI}(a - 3L). \tag{B.12}$$

Design of Mechanical Elements: A Concise Introduction to Mechanical Design Considerations and Calculations, First Edition. Bart Raeymaekers.

Companion website: www.wiley.com/go/raeymaekers/designofmechanicalelements

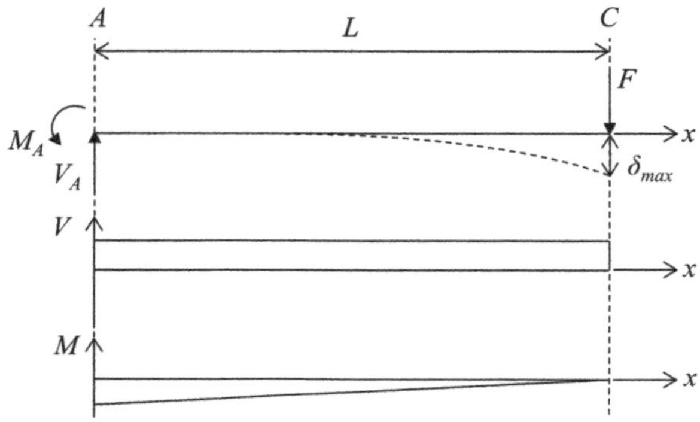

Figure B.1 Cantilever beam with end load.

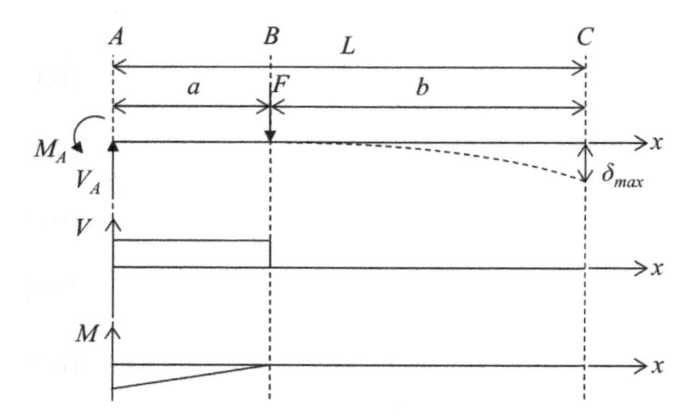

Figure B.2 Cantilever beam with intermediate load.

B.3 Simple Supported Beam with Center Load

See Figure B.3

$$V_A = V_B = F/2, \tag{B.13}$$

$$V_{AB} = V_A, \tag{B.14}$$

$$V_{BC} = -V_B, \tag{B.15}$$

$$M_{AB} = Fx/2, \tag{B.16}$$

$$M_{BC} = F(L-x)/2, \tag{B.17}$$

$$\delta_{AB} = \frac{Fx}{48EI}(4x^2 - 3L^2), \tag{B.18}$$

$$\delta_{max} = -\frac{FL^3}{48EI}. \tag{B.19}$$

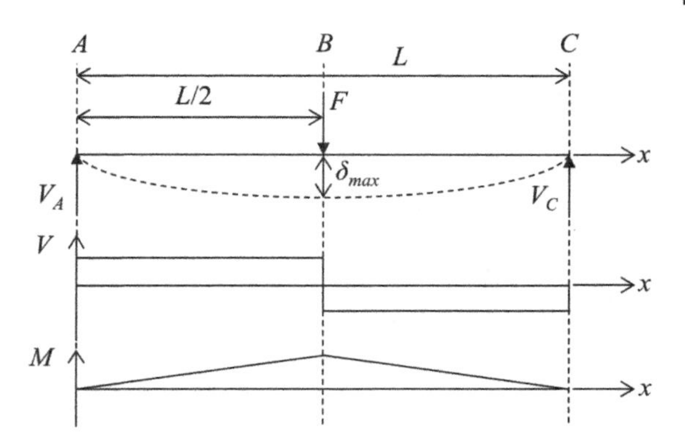

Figure B.3 Simple supported beam with center load.

B.4 Simple Supported Beam with Intermediate Load

See Figure B.4

$$V_A = Fb/L, \tag{B.20}$$

$$V_B = Fa/L, \tag{B.21}$$

$$V_{AB} = V_A, \tag{B.22}$$

$$V_{BC} = -V_B, \tag{B.23}$$

$$M_{AB} = Fbx/L, \tag{B.24}$$

$$M_{BC} = Fa(L - x)/L, \tag{B.25}$$

$$\delta_{AB} = \frac{Fbx}{6EIL}(x^2 + b^2 - L^2), \tag{B.26}$$

$$\delta_{AB} = \frac{Fa(L - x)}{6EIL}(x^2 + a^2 - 2Lx). \tag{B.27}$$

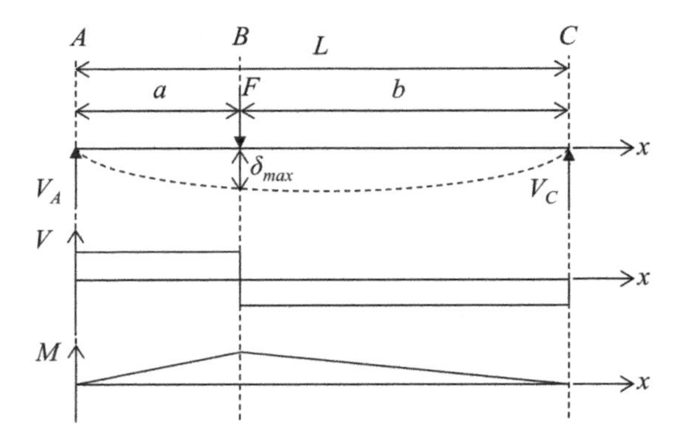

Figure B.4 Simple supported beam with intermediate load.

C

Elementary Stress Element

C.1 Introduction

Figure C.1 shows an elementary stress element. We use the following conventions:

1. Tensile stress is positive.
2. Compressive stress is negative.
3. Shear stress is positive if it make a stress element rotate in clockwise direction.

C.2 Principal Stresses

We calculate the principal stresses as

$$\sigma_{1,2} = \frac{\sigma_x + \sigma_y}{2} \pm \sqrt{\left(\frac{\sigma_x - \sigma_y}{2}\right)^2 + \tau_{xy}^2}. \quad \text{(C.1)}$$

We determine the orientation Φ_p of the principal stresses as

$$\tan 2\Phi_p = \frac{2\tau_{xy}}{\sigma_x - \sigma_y}, \quad \text{(C.2)}$$

with Φ_p positive in the same direction as the rotation from σ to τ.

C.3 Maximum Shear Stress

We calculate the maximum shear stress as

$$\tau_{1,2} = \pm\sqrt{\left(\frac{\sigma_x - \sigma_y}{2}\right)^2 + \tau_{xy}^2}. \quad \text{(C.3)}$$

We determine the orientation Φ of the maximum shear stress as

$$\tan 2\Phi = -\frac{\sigma_x - \sigma_y}{2\tau_{xy}}, \quad \text{(C.4)}$$

with Φ_p positive in the same direction as the rotation from σ to τ.

Design of Mechanical Elements: A Concise Introduction to Mechanical Design Considerations and Calculations, First Edition. Bart Raeymaekers.

Companion website: www.wiley.com/go/raeymaekers/designofmechanicalelements

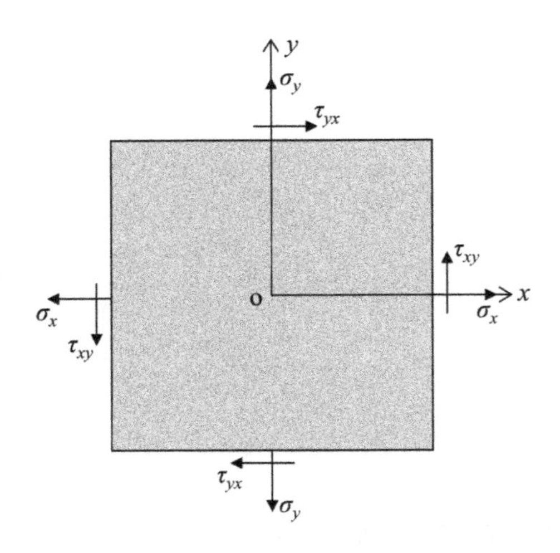

Figure C.1 Elementary stress element.

Index

Design of Mechanical Elements: A Concise Introduction to Mechanical Design Considerations and Calculations, First Edition. Bart Raeymaekers.

Companion website: www.wiley.com/go/raeymaekers/designofmechanicalelements